Integrated Frequency Synthesizers for Wireless Systems

The increasingly demanding performance requirements of communications systems, as well as problems posed by the continued scaling of silicon technology, present numerous challenges for the design of frequency synthesizers in modern transceivers.

This book contains everything you need to know for the efficient design of frequency synthesizers for today's communications applications. If you need to optimize performance and minimize design time, you will find this book invaluable.

Using an intuitive yet rigorous approach, the authors describe simple analytical methods for the design of phase-locked loop (PLL) frequency synthesizers using scaled silicon CMOS and bipolar technologies. The entire design process, from system-level specification to layout, is covered comprehensively. Practical design examples are included, and implementation issues are addressed.

A key problem-solving resource for practitioners in integrated-circuit design, the book will also be of interest to researchers and graduate students in electrical engineering.

Andrea Lacaita is Full Professor of Electrical Engineering at the Politecnico di Milano, Italy.

Salvatore Levantino is Assistant Professor of Electrical Engineering at the Politecnico di Milano.

Carlo Samori is Associate Professor of Electrical Engineering at the Politecnico di Milano.

Integrated Frequency Synthesizers for Wireless Systems

Andrea Lacaita, Salvatore Levantino, and
Carlo Samori
Politecnico di Milano, Italy

CAMBRIDGE UNIVERSITY PRESS
Cambridge, New York, Melbourne, Madrid, Cape Town, Singapore, São Paulo

Cambridge University Press
The Edinburgh Building, Cambridge CB2 8RU, UK

Published in the United States of America by Cambridge University Press, New York

www.cambridge.org
Information on this title: www.cambridge.org/9780521863155

First published 2007

Printed in the United Kingdom at the University Press, Cambridge

A catalogue record for this publication is available from the British Library

ISBN 978-0-521-86315-5 hardback

Contents

Preface

The phase-locked loop (PLL) concept is about 70 years old and a wealth of literature is already available on the subject. Someone may therefore ask why another book about PLLs.

The first reason is related to the specific application considered here, namely the silicon integration of frequency synthesizers. Classical texts do not deal in depth with issues related to the design of frequency synthesizers in modern transceivers. In particular, the design guidelines and the performance of some important building blocks and their impact on the whole system are sometimes barely mentioned. The attempt, here, has been instead to provide a broad description of the most typical circuit topologies of voltage-controlled oscillators, frequency dividers and phase and frequency detectors, and to discuss their performance in terms of power consumption, phase noise, spurs, and so forth. A chapter is also devoted to integrated passive components, such as varactors and inductors, since the ability to optimize their performance judiciously is becoming a key skill required of the RF designer.

The second reason is that the book attempts to provide an alternative approach to PLL theory and design. After years of research and study on the subject, the authors propose an analysis methodology that is both rigorous and intuitive. The ability to simplify the picture and to address schematically the impact of complex, often non-linear, effects is a fundamental skill of any good engineer. The PLL is a good training example for the designer. In this respect, the book provides many examples of models, starting from a schematic and simplified description of the circuit operation and then leading to estimates, which are compared with simulation results. These examples are intended not only to provide a deeper insight into complex and intriguing effects, but also to encourage students and young analogue designers to keep exercising the ability to figure out the consequences of technical choices before performing circuit simulations.

The book starts with three chapters addressing the PLL as a system. Chapter 1 points out the typical requirements of the frequency synthesizer in RF systems. Chapter 2 covers some PLL basics. It does not deal with the whole PLL theory, which is analyzed in depth in many classical books. The chapter highlights only the concepts needed for understanding the subsequent topics. Chapter 3 finally analyzes fractional-division PLLs, which are seldom discussed in other texts.

Chapters 4 to 9 are then devoted to discussing in detail the design issues related to the PLL building blocks. Chapters 4 to 7 deal with voltage-controlled oscillators and their practical implementations in bipolar and CMOS technologies, including resonator design and layout. Chapters 8 and 9 are focused on the design of programmable dividers and phase-comparison circuits, including issues related to non-linearities.

Acknowledgments

The research activity behind this book has been in progress for more than ten years. It was made possible thanks to the financial support provided by the Italian Ministry of Universities and Research and by industry. In this respect, many thanks go to Mario Paparo of STMicroelectronics (Catania, Italy), Maurizio Pagani of Ericsson Lab (Vimodrone, Italy), and to Mihai Banu of the (formerly) Silicon Circuit Research Department of Bell Laboratories (Murray Hill, NJ), for their strong support.

The authors are indebted to all their past graduate students, Mr Francesco Villa, Dr Alfio Zanchi, Dr Andrea Bonfanti, Dr Luca Romanò, Dr Stefano Pellerano, Dr Marco Milani and Dr Luigi Panseri, who shared the excitements of this research and contributed to most of the understanding reported in the following pages. The authors are also extremely grateful to their current Ph.D. students, Paolo Madoglio and Marco Zanuso, who worked out the examples and the simulations, providing critical revisions of the text. Clearly, only the authors are responsible for any errors that may still be present.

1 Local oscillator requirements

Personal wireless communications have represented, for the microelectronic industry, the market with the largest growth rate in the last ten years. The key for such a boom has been the standardization effort made by several organizations and the replacement of compound semiconductors with silicon technology in building radio front ends. This advancement was made possible by joint progress in communication theory, devices technology and system and circuit design. Silicon technology made it possible to attain lower fabrication costs, owing to the large production volumes and to the possibility of implementing complex digital functions together with radio-frequency (RF) signal manipulations, lowering the number of off-chip components.

Initially in the 1990s, cellular systems have been the driving application for this technology evolution. Further generations of cellular telephones have introduced the possibility of communicating not only by voice but also with text messages, images and videos. Later, a number of wireless technologies have emerged, not strictly belonging to the class of communication systems. Some examples are wireless local-area networks (WLAN), sensor networks, wireless USB applications and automotive radar.

Table 1.1 summarizes various high-level characteristics of the most common communication standards. Despite the variety of modulation formats and access methods, the basic structure of a typical transceiver has remained as shown in Figure 1.1. In both the receiving and the transmitting branch, frequency conversions are performed to move the signal from the RF band to the base band and vice versa. Up-conversions and down-conversions can be performed in one or more steps, and amplification and filtering can be distributed differently along the chains. Whatever architecture is adopted, the core of these operations is always the multiplication of the signal by sinusoids provided by the local oscillator (LO). This stage is therefore a key element of the overall transceiver.

What Table 1.1 does not point out is that the information is travelling in a 'hostile' time-varying channel, affected by noise and strong interferences, Doppler effects and multi-path fading. These effects impose severe requirements on the receiver and transmitter performance. Just to mention a popular example, the sensitivity (i.e., the minimum signal power to be detected at the antenna) in a GSM receiver is about -102 dBm. On the other hand, the largest blocker or interferer that the system must tolerate is 0 dBm. It follows that the GSM receiver has to be able to detect a weak signal even in the presence of an interferer with a power of about 10 orders of magnitude larger. Such a stringent requirement is quite uncommon in other fields of electrical engineering.

Table 1.1 *Characteristics of some communication standards*

Standard	RX band (MHz)	TX band (MHz)	Channel spacing (kHz)	Multiple access	Modulation
GSM	925–960	880–915	200	TDMA/FDMA	GMSK
DCS	1805–1880	1710–1785	200	TDMA/FDMA	GMSK
IS-95	869–894	824–849	1 250	CDMA/FDMA	QPSK/OQPSK
IS-54	869–894	824–849	30	TDMA/FDMA	π/4-DQPSK
IS-136	1930–1990	1850–1910	30	TDMA/FDMA	π/4-DQPSK
UMTS	2110–2170	1920–1980	5 000	W-CDMA	QPSK
Bluetooth	2400–2483.5		1 000	TDMA/FDMA	GFSK
802.11a	5180–5320, 5745–5805		20 000	TDMA/FDMA	OFDM
802.11b	2400–2483.5		20 000	CSMA	DSSS-CCK

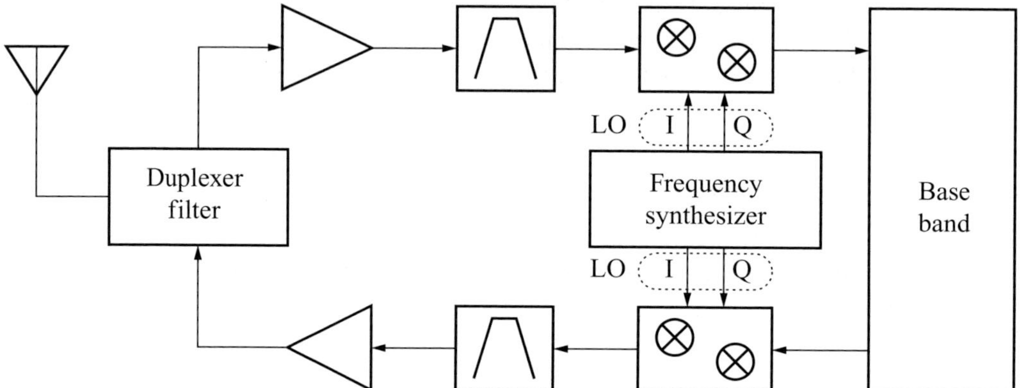

Figure 1.1 Generic structure of a transceiver

To achieve such extreme performance, the architecture of the transceiver has to be carefully selected, depending on the communication standard specifications, [1–3] and the design of its building blocks has to be carefully pursued. The local oscillator has to match tight levels of spectral purity so that the quality of the received signal is preserved, and it must be able to change its frequency so that various channels of the receiver band can be converted to the same frequency. For this reason, the local oscillator is, in practice, a frequency synthesizer, or a circuit that is able to synthesize harmonic reference waveforms in a certain frequency range. Several implementations of this stage exist; however, the phase-locked loop (PLL) is the most common.

1.1 AM and PM signals

The signals generated by the local oscillator are ideally sinusoidal or harmonic:

$$V_0(t) = A_0 \cdot \cos(\omega_0 t + \phi_0),$$

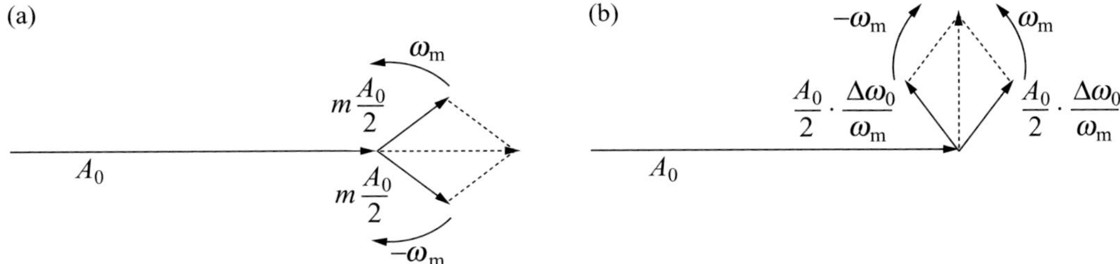

Figure 1.2 (a) Amplitude-modulated carrier and (b) phase-modulated carrier

where the amplitude A_0, the frequency ω_0 and the phase ϕ_0 are constant. The 'angular frequency' ω (measured in rad/s) will be referred to simply as 'frequency'. It will be clear from the symbol (either ω or f), and from the context, whether the term frequency refers to the angular frequency or to the actual frequency $\omega/2\pi$ (measured in Hz).

In a real synthesizer, the signal amplitude and frequency can suffer from modulation, owing to the presence of noise or disturbances. An amplitude-modulated (AM) signal may be written as:

$$V_0(t) = A_0 \cdot [1 + m \cdot \cos(\omega_m t)] \cdot \cos(\omega_0 t + \phi_0).$$

The spectral components of the AM signal are better identified, when the previous expression is written as:

$$V_0(t) = A_0 \cdot \cos(\omega_0 t + \phi_0) + \frac{mA_0}{2} \cdot \cos[(\omega_0 - \omega_m)t + \phi_0] + \frac{mA_0}{2} \cdot \cos[(\omega_0 + \omega_m)t + \phi_0].$$

The spectrum has two side tones at an offset $\pm\omega_m$ from the carrier at ω_0. Owing to the amplitude variations, the AM signal has a power larger than the original unmodulated harmonic. Using phasor notation and taking the carrier as a reference, the two tones will appear as in Figure 1.2(a). In most typical cases, $m \ll 1$ and $\omega_m \ll \omega_0$.

On the other hand, a frequency-modulated (FM) signal may be denoted as $\omega(t) = \omega_0 + \Delta\omega_0 \cdot \cos(\omega_m t)$. Since the phase is the integral of the frequency, the signal is also modulated in phase (PM). It is:

$$V_0(t) = A_0 \cos\left[\omega_0 t + \phi_0 + \frac{\Delta\omega_0}{\omega_m} \sin(\omega_m t)\right].$$

Since the amplitude is constant, this time the modulated signal has the same power as the original, unmodulated harmonic. The modulated phase is $\phi(t) = (\Delta\omega_0/\omega_m) \cdot \sin(\omega_m t)$. Under the assumption of a small modulation index $(\Delta\omega_0/\omega_m) \ll 1$ rad, it is $\cos[\phi(t)] \simeq 1$,

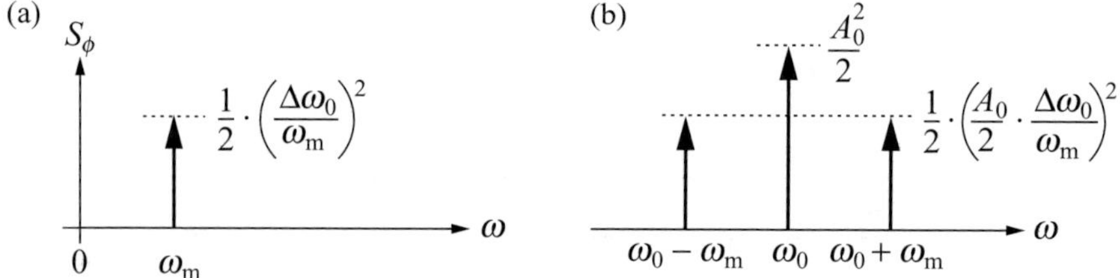

Figure 1.3 Power spectra (a) of the phase signal and (b) of the corresponding voltage signal

$\sin[\phi(t)] \simeq \phi(t)$, so the signal $V_0(t)$ can be approximated as:

$$\begin{aligned} V_0(t) &\cong A_0\cos(\omega_0 t+\phi_0) - A_0\sin(\omega_0 t+\phi_0)\cdot\frac{\Delta\omega_0}{\omega_m}\sin(\omega_m t) \\ &= A_0\cos(\omega_0 t+\phi_0) - \frac{A_0}{2}\cdot\frac{\Delta\omega_0}{\omega_m}\cos[(\omega_0-\omega_m)t] \\ &\quad + \frac{A_0}{2}\cdot\frac{\Delta\omega_0}{\omega_m}\cos[(\omega_0+\omega_m)t]. \end{aligned} \tag{1.1}$$

The approximation is usually referred to as narrow-band FM. Figure 1.2(b) shows the two side tones in the carrier frame. From the above assumptions, the carrier appears modulated not only in phase but also in amplitude. The resulting phasor has peak phase deviation equal to $\arctan(\Delta\omega_0/\omega_m)$ and peak amplitude equal to $A_0\cdot\sqrt{1+(\Delta\omega_0/\omega_m)^2}$, which approach $(\Delta\omega_0/\omega_m)$ and A_0, respectively, under the narrow-band FM approximation.

It is interesting to compare the power spectrum[1] S_ϕ of the phase signal $\phi(t)$ and the spectrum of the voltage signal $V_0(t)$. Since the phase signal is harmonic, its power spectrum is a δ-like function at ω_m, with area equal to the mean square value of the phase signal $(1/2)\,(\Delta\omega_0/\omega_m)^2$. It is schematically represented in Figure 1.3(a). Figure 1.3(b) shows the power spectrum of V_0 as derived from (1.1). The ratio between the power of each side tone and the power of the carrier is given by:

$$\frac{\text{Power of the single tone}}{\text{Carrier power}} = \frac{\left(\frac{A_0}{2}\cdot\frac{\Delta\omega_0}{\omega_m}\right)^2\cdot\frac{1}{2}}{\frac{A_0^2}{2}} = \frac{1}{4}\cdot\left(\frac{\Delta\omega_0}{\omega_m}\right)^2. \tag{1.2}$$

That is equal to half the power of S_ϕ. The tones due to a frequency modulation at ω_m are referred to as spurious tones or spurs. The above ratio is often called the spurious free dynamic range (SFDR). It is expressed in dB, and labelled as dBc, i.e., dB with respect to the carrier. For a given frequency deviation $\Delta\omega_0$, the S_ϕ amplitude is inversely proportional to the square of the offset ω_m. If the signal is both amplitude-modulated and frequency-modulated at ω_m, the voltage spectrum shows two side tones of different amplitudes.

The same arguments leading to (1.2) can be used to address the impact of every noise spectral component affecting the carrier frequency. The noise may be regarded as the superposition of tones $\Delta\omega_0\cdot\cos(\omega_m t)$. If the frequency noise is white, the peak frequency

[1] Here and in the following the power spectra are intended to be unilateral: they are defined only for positive frequencies.

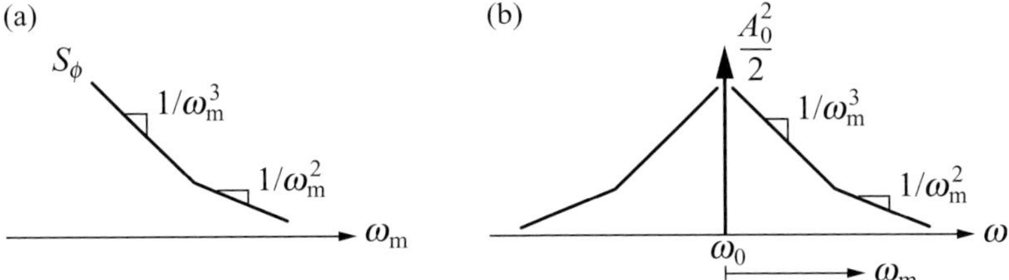

Figure 1.4 (a) Phase-noise spectrum and (b) corresponding voltage spectrum

deviation $\Delta\omega_0$ is constant. Since the phase is the integral of the frequency, the phase–power spectrum $S_\phi(\omega_m)$ shows a $1/\omega_m^2$ tail (−20 dB/decade slope in Figure 1.4(a)). If, instead, the frequency noise has a $1/f$ (flicker) component, the $\Delta\omega_0$ amplitude goes as $1/\omega_m$, and $S_\phi(\omega_m)$ shows a $1/\omega_m^3$ dependence (−30 dB/decade).

Under the small-angle approximation, the corresponding voltage signal features the same $S_\phi(\omega_m)$ shape (Figure 1.4(b)). The only difference is that it has two tails, for both positive and negative frequency offsets ω_m from the carrier. The voltage–power spectral density at $\omega_0 \pm \omega_m$ is $S_V(\omega_0 \pm \omega_m) \cong (S_\phi(\omega_m)/2) \cdot (A_0^2/2)$. Since the noise level of the sideband depends on the carrier power, the noise level is typically quantified as the noise power in a 1 Hz bandwidth at offset $+\omega_m$ or $-\omega_m$ from ω_0 divided by the carrier power. This figure is denoted as the single-sideband-to-carrier ratio (SSCR), or $\mathcal{L}$ (L script):

$$\mathcal{L}(\omega_m) = \frac{\text{Power in 1 Hz bandwidth}}{\text{Carrier power}} = \frac{S_V(\omega_0 \pm \omega_m)}{A_0^2/2} \cong \frac{S_\phi(\omega_m)}{2} \quad (\text{dBc/Hz}). \tag{1.3}$$

A factor of 1 Hz multiplies both S_V and S_ϕ and sets the correct physical dimensions. The term phase noise is often used indiscriminately for $\mathcal{L}$ and for S_ϕ, even though the two quantities are different (clearly, S_ϕ is 3 dB larger than $\mathcal{L}$). The phase noise is the most important characteristic of an oscillator used for RF applications.

It may be noticed that the power of the oscillator output voltage, which is obtained as the integral of the power spectral density[2] S_V in Figure 1.4(b), is infinite. This unphysical result comes from the small-angle approximation $\phi(t) \ll 1$ rad, which has been used to derive (1.1). At small offsets ω_m, S_ϕ goes to infinity, $\phi(t)$ does not satisfy the inequality $\phi(t) \ll 1$ rad any more and the voltage spectrum differs from S_ϕ. If the frequency noise is white, it can be shown that the voltage spectrum has a Lorentzian shape and its integral is equal to the power of the ideal carrier. [4]

For the approximation $\mathcal{L}(\omega_m) \cong S_\phi(\omega_m)/2$ to hold down to a certain frequency f_l it must be:

$$\int_{f_l}^{\infty} S_\phi(2\pi f_m) \cdot \mathrm{d}f_m \ll 1 \ (\text{rad})^2.$$

[2] Because the power spectral density of a signal is typically defined as the signal power in a *1 Hz* bandwidth, the frequency *f* and not the angular frequency has to be used in the integration of the spectral density.

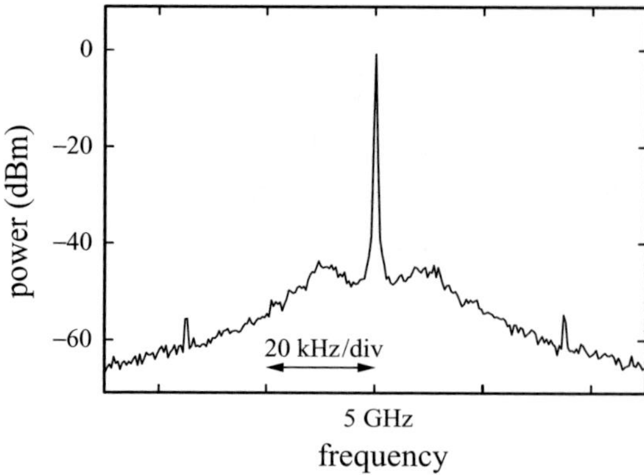

Figure 1.5 A typical PLL output spectrum

In RF oscillators for wireless systems, it is typically satisfied down to 100 Hz, which is a frequency limit low enough for most purposes. Equation (1.3) will therefore always be used in the following. Moreover, it should be taken into account that the oscillator is not a stand-alone circuit, but it is embedded in the PLL. In the next chapter, it will be shown that S_ϕ at the PLL output is high-pass filtered, for offsets smaller than the bandwidth of the PLL itself. The same holds for the S_V spectrum, thus removing the potential divergence close to the carrier. Figure 1.5 shows the typical S_V output spectrum of a PLL with a 10 kHz bandwidth. Note that the tails stop at about 10 kHz from the carrier and the spectrum does not show any divergence close to the carrier frequency. Two spurious tones at ± 35 kHz indicate a residual frequency (phase) modulation of the carrier.

1.2 Effect of phase noise and spurs

Both phase noise and spurs affect the spectral purity of the local oscillator. While the phase noise is characterized by a distributed spectrum, the spurs are instead well-defined undesired tones. Depending on the applications, care must be devoted to limit either the 'spot' value of the spectrum at a given frequency or the integral of the phase–power spectral density over a given spectral range.

Let us consider the simplified block diagram of a transceiver in Figure 1.1. In the receiving path, the signal at RF is down-converted to the base band or to an intermediate frequency (IF) by the mixer driven by the LO. Let us suppose that a strong interferer (blocker) at an offset ω_m is received together with the signal. This is a very realistic situation, taking place when the receiver also picks up the signal of a nearby transmitter. Assuming an ideal LO with a δ-like spectrum, the blocker will be down-converted at ω_m from the signal, and filtered out. When the LO phase noise is taken into account, the outcome changes drastically. The spectra of the two down-converted signals can overlap (Figure 1.6) and the desired signal can be corrupted by the tail of the interferer. This effect is called reciprocal mixing and degrades the signal-to-noise ratio (SNR). More quantitatively, the SNR may be

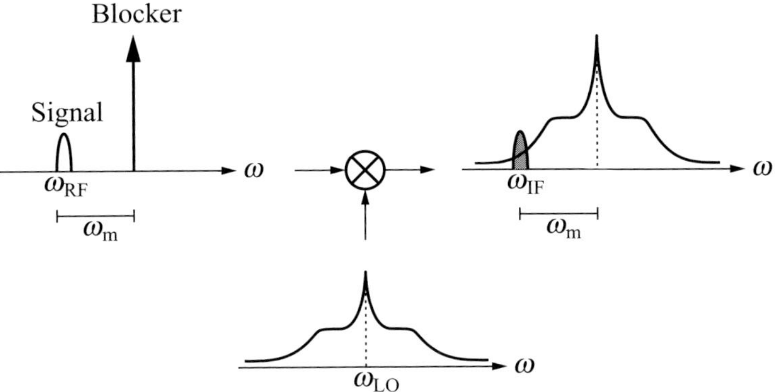

Figure 1.6 Reciprocal mixing

written as:

$$\mathrm{SNR} = \frac{P_S}{\mathcal{L}(\omega_m) \cdot B \cdot P_B}$$

where P_S and P_B are the powers of the desired signal and the blocker, respectively, and B is the signal bandwidth. The expression may be converted into dB, leading to:

$$\mathrm{SNR}|_{dB} = (P_S|_{dBm} - P_B|_{dBm}) - \mathcal{L}(\omega_m)|_{dBc} - 10 \cdot \log_{10} B. \tag{1.4}$$

Typically, a minimum value of SNR is required. If the ratio between the maximum blocker and the minimum signal power is large, the phase noise specification, $\mathcal{L}$, can be severe. That is the case of GSM, which is discussed in Example 1.1.

If an LO spur occurs at the same frequency offset between the signal and the blocker, the reciprocal mixing can be even more problematic. The blocker would be down-converted by the spur to the same IF of the signal. The signal-to-interference ratio can now be written by using the SFDR defined in (1.2):

$$\mathrm{SNR}|_{dB} = (P_S|_{dBm} - P_B|_{dBm}) - \mathrm{SFDR}|_{dB}.$$

Therefore, the occurrence of blockers defines the maximum level of the spot values of the LO phase noise spectrum at some well-defined frequencies.

Of course, even if blockers are not present, the LO phase noise corrupts the signal anyhow and leads to SNR degradation or detection loss. In this case, the SNR will be a function of the phase noise power, that is the integral of the LO spectrum. Let us consider, as an example, a generic M-QAM modulated carrier. It can be written as:

$$s(t) = \sum_k a_k \cdot p(t - kT) \cdot \cos(\omega_0 t) - \sum_k b_k \cdot p(t - kT) \cdot \sin(\omega_0 t),$$

where (a_k, b_k) are the symbols transmitted in the I/Q paths and $p(t)$ is the normalized Nyquist pulse. The signal $s(t)$ can be regarded as an amplitude-modulated and phase-modulated carrier at ω_0 and complex envelope $(a_k + \mathrm{j}b_k) \cdot p(t - kT)$. It is:

$$s(t) = \mathrm{Re}\left[\sum_k (a_k + \mathrm{j}b_k) \cdot p(t - kT) \cdot \mathrm{e}^{\mathrm{j}\omega_0 t}\right].$$

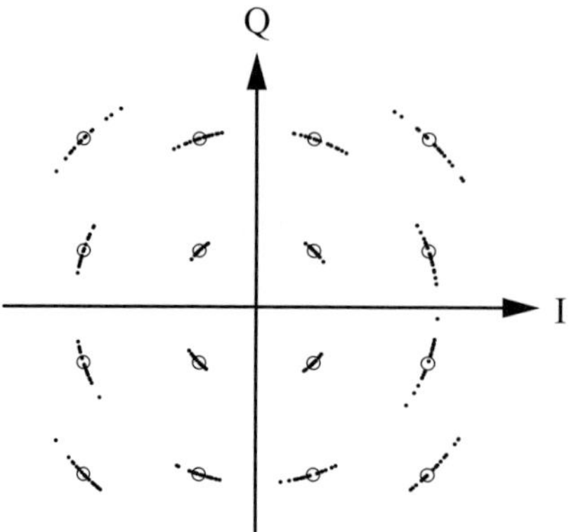

Figure 1.7 A 16-QAM constellation affected by phase noise

After down-conversion, coherent demodulation and sampling, the transmitted symbols (a_k, b_k), or constellation, are identified. Let us suppose that the LO in the transmitter is affected by phase noise $\phi_n(t)$. The transmitted signal $s(t)$ thus becomes:

$$s(t) = \sum_k a_k \cdot p(t - kT) \cdot \cos(\omega_0 t + \phi_n(t)) - \sum_k b_k \cdot p(t - kT) \cdot \sin(\omega_0 t + \phi_n(t)),$$

or, equivalently:

$$s(t) = \text{Re}\left[\sum_k (a_k + jb_k) \cdot e^{j\phi_n(t)} \cdot p(t - kT) \cdot e^{j\omega_0 t}\right].$$

Therefore, the demodulated signal will be $(a_k + jb_k) \cdot \exp[j\phi_n(t)]$, i.e., the received constellation is rotated by $\phi_n(t)$. Of course, this discussion holds even if the LO of the receiver is affected by phase noise. The phase-noise contributions of the receiver and the transmitter are added in power to get the overall phase noise. Figure 1.7 depicts, qualitatively, the effect of phase noise on a 16-QAM constellation. The detection loss is related to the r.m.s. value of $\phi_n(t)$. It is denoted as σ_ϕ and is given by:

$$\sigma_\phi = \sqrt{\int_{f_1}^{f_2} S_\phi(2\pi f_m) \cdot df_m}.$$

The phase-noise power spectral density is typically integrated between frequencies f_1 and f_2. The first lower limit is set by the bandwidth of a frequency-error correction algorithm, which is typically adopted in the digital base-band subsystem. The upper frequency f_2 is set approximately by the signal bandwidth. In practice, a phase noise at frequency offsets larger than the channel bandwidth has a negligible impact on detection loss. Even the spurious tones present in the phase spectrum contribute to the r.m.s. phase deviation σ_ϕ and should be accounted for in the design.

1.3 Frequency accuracy

The frequency generated by a synthesizer has to be extremely accurate. For instance, the mobile terminal in the GSM standard must transmit signals with frequency accuracy better than 0.1 parts per million (p.p.m.), which means an error of 100 Hz for a 1 GHz carrier frequency. This value is far beyond the performance of commercially available components. If the LO signal is locked to an off-chip temperature-controlled crystal oscillator (TCXO), the achievable frequency accuracy cannot be better than 20 ppm.

To reach the target performance, the base station broadcasts a tone for a short time (frequency control burst), which is derived from a more accurate frequency reference. The frequency error between the received tone and the mobile terminal LO is detected at the base band by a maximum-likelihood estimation algorithm. Frequency correction is then performed either by acting on the crystal oscillator, or by rotating the received constellation (that is by multiplying the base-band complex signal by $\exp[\mathrm{j}\Delta\omega \cdot t]$, $\Delta\omega$ being the frequency error). The former approach is adopted in GSM terminals, [5] while the latter can be found in some examples of WLAN clients.

A third option is to act on the input control of the frequency synthesizer, which generates the mobile terminal LO. This method requires a very-fine-tuning synthesizer, which can be achieved by the fractional-N PLL discussed in Chapter 3.

Example 1.1 Phase noise in GSM terminals

The GSM standard is the popular standard for cellular systems, which operates in the 900 MHz and 1800 MHz RF bands. The main characteristic of the GSM standard is its very tight blocking specification. The transceiver has to operate with blockers, which can be 76 dB more intense than the desired signal. Figure 1.8 shows the blocking signal level.

The GSM reference sensitivity has to be –102 dBm, but the receiver must meet the bit error rate (BER) for a useful signal 3 dB above the reference sensitivity in the presence of blockers, that is at −99 dBm. [6] Therefore, the LO phase noise specifications are set by the reciprocal mixing and not by the integral noise. The latter is also not an issue because the integration bandwidth is limited to a channel bandwidth B of only 200 kHz.

Equation (1.4) can be used to evaluate the required phase noise level $\mathcal{L}$ at a given offset. Taking $B \cong 200$ kHz and a minimum signal-to-noise ratio $\mathrm{SNR}|_{\mathrm{dB}} = 9$ dB, the resulting LO phase-noise requirements have been organized in Table 1.2. Assuming a $1/\omega_{\mathrm{m}}^2$ phase-noise

Table 1.2 *Local oscillator phase-noise requirements for GSM at some frequencies*

f_{m} (MHz)	Blocker power (dBm)	$\mathcal{L}(f_{\mathrm{m}})$ (dBc/Hz)
3	−23	−138
1.6	−33	−128
0.6	−43	−118

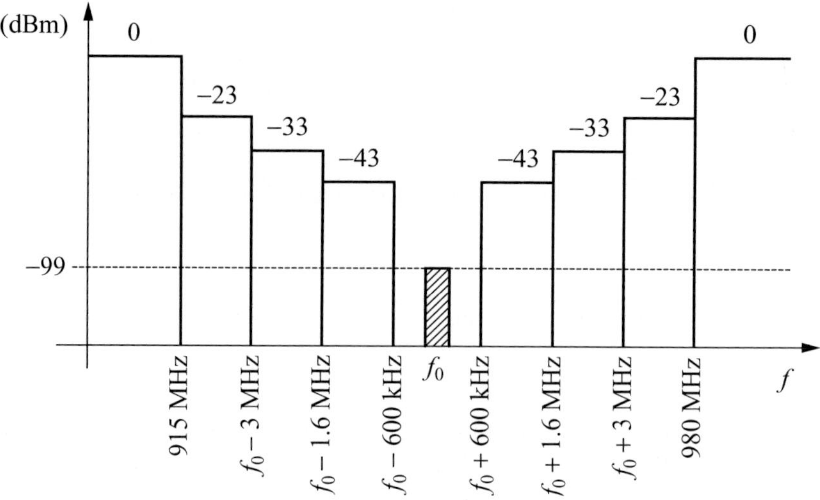

Figure 1.8 Blocking signal level for GSM

shape, the most stringent specification is −138 dBc/Hz at 3 MHz. If this spot value is guaranteed, the other specifications are also met.

In reality, a more careful design must take into account the gain compression or desensitization caused by the blocker. [1–3] Moreover, the SNR at the sampler is not only set by the LO noise, but also by noise contributions from other stages (filters, LNA, mixer). For these reasons, the LO noise requirements must be tighter than the values shown in Table 1.2. A typical realistic requirement is about −139.5 dBc/Hz at 3 MHz, [6] which is only 1.5 dB more stringent. This correction may appear to be a minor change, but this is not the case. A reduction of any single dB causes a considerable increase in the power dissipation of the overall synthesizer. As a rule of thumb, to lower the LO noise by 3 dB, its power dissipation has to be doubled.

Example 1.2 Phase noise in UMTS terminals

The Universal Mobile Telecommunication System (UMTS) is the third-generation cellular standard, and allows for a higher data bit rate. It operates in the 1.9–2.1 GHz band and it is a frequency division duplex (FDD) system, continuously transmitting and receiving. The strongest interferer is the leakage of the transmitted signal into the receiver, which causes reciprocal mixing with the LO noise. [7] However, as the minimum distance between the transmission and receiving bands is 135 MHz, the stringent phase noise performance is at that offset. Taking into account the 3.84 MHz channel bandwidth, the sensitivity of −99 dBm and reasonable noise figure and linearity requirements, the tolerable LO phase noise at 135 MHz is −150 dBc/Hz. [7] This is equivalent to −117 dBc/Hz at 3 MHz, which is much more relaxed than the GSM requirement.

Example 1.3 Phase noise in 802.11a/g clients

In these communication standards, the modulation schemes are much more complex than in cellular phone standards. The channels have large bandwidths, therefore the LO phase noise

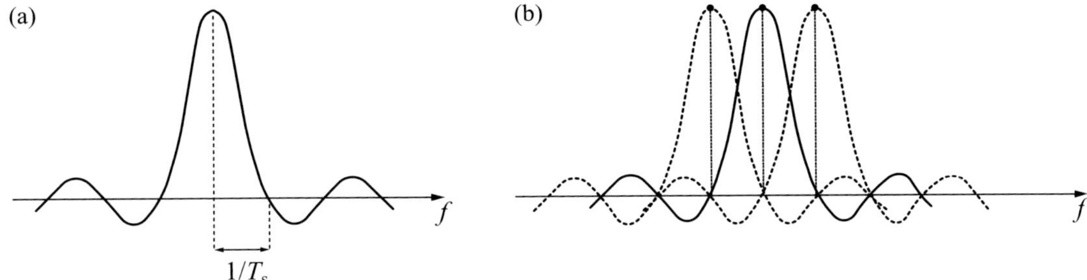

Figure 1.9 OFDM modulation: (a) spectrum of one modulated subcarrier and (b) superposition of several subcarriers

specifications are set by the integral noise and their calculations often require numerical simulations. [8–10] Wireless LAN (WLAN) standards and other high-data-rate systems operating in multi-path fading channels use Orthogonal Frequency Division Multiplexing (OFDM). A discussion of the impact of phase noise in such modulations is outside the scope of this book. However, some qualitative insight can easily be derived in the case of the IEEE 802.11a standard.

In this standard, the OFDM signal is composed of $N = 52$ equally spaced QAM-modulated subcarriers over a 20 MHz bandwidth. The spectrum of a QAM-modulated signal has a $(\sin x)/x$ shape, with zeros at multiples of $1/T_S$, where T_S is the symbol duration (Figure 1.9(a)). The spacing among the subcarriers is selected to be exactly equal to $1/T_S$ (Figure 1.9(b)). In this way, although the spectra of the subcarriers overlap, each subcarrier is precisely aligned to the zeros of the other spectra. Thus, if the overall spectrum is sampled at $1/T_S$, the subcarrier signals do not interfere with each other.

The N complex numbers from a QAM constellation are the magnitudes and the phases of the N subcarriers. It is obvious that these N numbers form the frequency-domain representation of the OFDM symbol. The time-domain symbol is obtained by performing an inverse fast Fourier transform (IFFT) of the N samples.

The receiver samples the signal at a rate NT_S, obtaining N complex time-domain samples. These samples are transformed into the original N complex numbers by an FFT algorithm. The FFT algorithm samples, implicitly, the signal spectrum at $1/T_S$, avoiding inter-carrier interference.

Essentially, phase noise affects the OFDM signal quality in two ways:

- Slow LO phase noise (at frequency offsets lower than $1/T_S$) adds the same phase error to each subcarrier.
- Fast LO phase noise produces inter-carrier interference. The subcarriers are no longer orthogonal, because of the LO noise, and they interfere with each other. Since many subcarriers are modulated by data in a stochastic way, the received constellation appears to be affected by Gaussian noise.

In Figure 1.10, these two contributions are visible. The slow-noise component has rotated the constellation and it can be easily cancelled by introducing pilots within the OFDM symbol. The fast-noise component has caused random noise around each point and it cannot be compensated. A typical specification for the 5/2.5 GHz LO synthesizer for an 802.11a/g

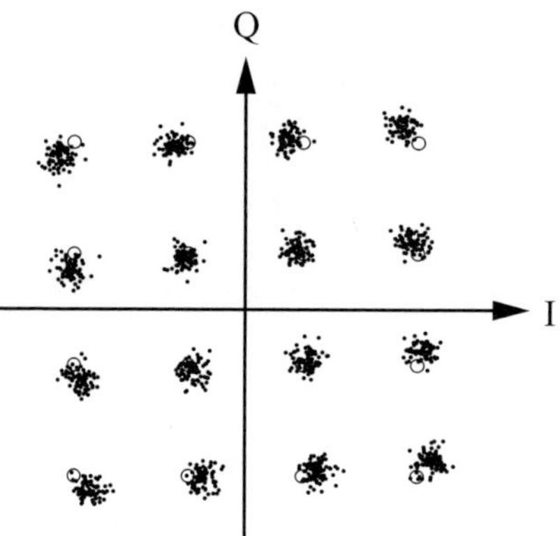

Figure 1.10 Effect of phase noise for an OFDM-modulated signal, with 16-QAM modulation per carrier

transceiver is that the phase noise integrated from 10 kHz to 10 MHz has to be less than 1° r.m.s. or −35 dBc.

1.4 Switching speed

A very limited amount of time is typically allowed to make the frequency synthesizer switch from one frequency to another one. In a phase-locked loop, fast switching implies a need for a large bandwidth, which may degrade phase noise and spur performance.

These switching requirements are quite different depending on the communication standard. For instance, the GSM works in time-division multiple-access (TDMA) within each frequency channel and time-division duplex (TDD). Both the transmission and receiving channels have eight time slots of 577 μs. A mobile terminal receives at a certain frequency during a slot and transmits at another frequency after two slots or 1.1 ms. However, the most critical switching time for the LO takes place between the transmission and the system-monitoring slots. The time allowed for this is about 865 μs. Since some time should also be left to other stages to settle after the LO, the maximum frequency-switching time for the LO must be reduced to the duration of a single slot or 577 μs. [11] Speed requirements are even more severe in GSM base stations, where the switching time must be less than 10 μs. [12]

In wireless LAN transceivers, if the transmitter and the receiver share the same LO but have different architectures, the frequency switching between TX and RX may take place only during the preamble and the time left to the synthesizer is as little as a few μs. [13]

1.5 References

[1] B. Razavi, *RF Microelectronics*, New York, NY: Prentice-Hall, 1997.

[2] T. H. Lee, *The Design of CMOS Radio-Frequency Integrated Circuits*, Cambridge, UK: Cambridge University Press, 1998, 2nd edn, 2004.

[3] J. B. Groe and L. E. Larson, *CDMA Mobile Radio Design*, Norwood, MA: Artech House, Inc., 2000.

[4] M. Lax, Classical noise. V. Noise in self-sustained oscillators, *Phys. Rev.*, **160**, Aug. 1967, 290–307.

[5] J. Lin, A low-phase-noise 0.004-ppm/step DCXO with guaranteed monotonicity in the 90-nm CMOS process, *IEEE J. Solid-St. Circ.*, **40**, Dec. 2005, 2726–34.

[6] E. Ngompe, Computing the LO phase noise requirements in a GSM receiver, *Appl. Microwave & Wireless*, **11**, Jul. 1999, 54–8.

[7] F. Gatta, D. Manstretta, P. Rossi and F. Svelto, A fully-integrated 0.18 – μm CMOS direct-conversion receiver front-end with on-chip LO for UMTS, *IEEE J. Solid-St. Circ.*, **39**, Jan. 2004, 15–23.

[8] C. Muschallik, Influence of RF oscillators on an OFDM signal, *IEEE T. Consum. Electr.*, **41**, Aug. 1995, 592–603.

[9] A. Garcia Armada, Understanding the effect of phase noise in orthogonal frequency division multiplexing (OFDM), *IEEE T. Broadcast.*, **41**, Jun. 2001, 153–9.

[10] L. Piazzo and P. Mandarini, Analysis of phase noise effects in OFDM modems, *IEEE T. Commun.*, **50**, Oct. 2002, 1696–1705.

[11] W. S. T. Yan and H. C. Luong, A 2-V 900-MHz monolithic CMOS dual-loop frequency synthesizer for GSM receivers, *IEEE J. Solid-St. Circ.*, **36**, Feb. 2001, 204–16.

[12] M. Keaveney, P. Walsh, M. Tuthill, C. Lyden and B. Hunt, A 10 μsec fast switching PLL synthesizer for a GSM/EDGE base station, *Digest of Technical Papers of the 51st IEEE International Solid-State Circuits Conference*, Feb. 2004, 192–3.

[13] S. L. J. Gierkink, D. Li, R. C. Frye and V. Boccuzzi, A 3.5-GHz PLL for fast low-IF/zero-IF LO switching in an 802.11 transceiver, *IEEE J. Solid-St. Circ.*, **49**, Sep. 2005, 1909–21.

2 Phase-locked loops

2.1 Basics

2.1.1 Introduction

The earliest research on phase-locked loops dates back to 1932, when British researchers developed the homodyne receiver as an alternative to Edwin Armstrong's superheterodyne receiver. The homodyne or synchrodyne system has fewer elements than the superheterodyne system: just a local oscillator, a mixer and an audio amplifier. The local oscillator is tuned to the desired input frequency and its output is multiplied by the input signal, thus producing a replica of the original audio modulation signal.

However, the performance of such a simple receiver was affected by the slight drift of the oscillator frequency. Following the seminal work of a French scientist, [1] an automatic correction signal was introduced, to keep the oscillator signal at the same phase and frequency as the desired signal. Even if the solution fixed the limitations of the homodyne receiver, the application of PLLs remained limited, mainly because of the cost of the circuit.

In TV receivers, the set-up was adapted to synchronize horizontal and vertical sweep oscillators to the transmitted sync pulses, but only in the 1960s did the appearance of monolithic circuits with complete phase-locked loop systems on a chip spur its widespread use.

Since then, a number of books and countless papers have been devoted to PLLs. References [2–15] are some of the most popular books on the topic, [16–19] are collections of classical papers, while [20] and [21] are special issues dedicated to PLLs. Such a wealth of publications is justified by both the importance of PLLs in many applications and the conceptual challenges posed by the circuit analysis and design. Even in its simpler version, a PLL is a non-linear system; it is a blend of digital and analogue techniques in one package, which makes difficult a rigorous mathematical description of its steady and dynamic behaviour.

In several cases, however, linear approximations are enough to address the system performance quantitatively. We will follow a mixed approach: whenever possible, the PLL building blocks will be described in terms of transfer functions (i.e., as linear time-invariant systems). The non-linear or linear time-variant operation of the single building block will be taken into account to determine its own performance and parameters. As an example, the voltage-controlled oscillator (VCO) will be studied in Chapters 4, 5 and 7 as a time-variant

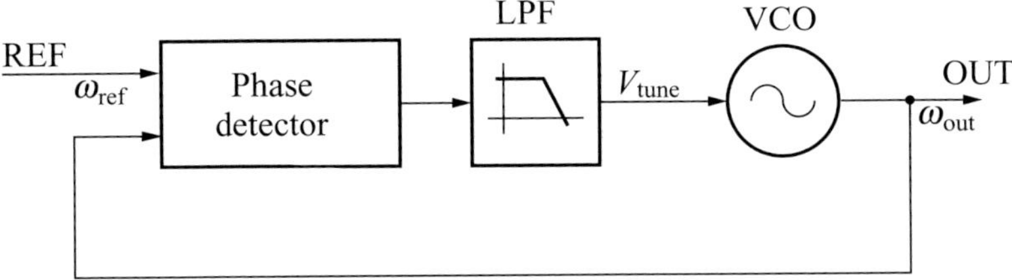

Figure 2.1 The phase-locked loop

system and so its noise will be derived. However, such a noise will be transferred to the PLL output by using the linear transfer function of the whole PLL. This approach is convenient and valuable since it provides quantitative guidelines, while still relying on conceptual tools and analysis procedures belonging to the usual background of any circuit designer.

This chapter is largely devoted to a discussion of the basic circuit operation. The linear model will be introduced and will be used as a reference framework even if its limits will be properly highlighted.

2.1.2 Simple loop

The role of a frequency synthesizer is to generate any of a range of frequencies from a single fixed time base or oscillator. For this purpose, the adoption of a VCO may be considered sufficient. If the VCO is implemented using an LC-tuned topology, the VCO frequency may be set to the desired value ω_{out} by acting on a voltage-controlled capacitor (i.e., a varactor) placed in the LC resonant cell. The varactor capacitance can be controlled by changing a tuning voltage V_{tune}. The corresponding output VCO waveform may be written as:

$$V_{out}(t) = A_0 \cos(\omega_{out} t + \phi_{out}), \text{ with } \omega_{out} = \omega_{FR} + K_{VCO} V_{tune}. \tag{2.1}$$

In practical circuits, the value of V_{tune} is referred to the midpoint of its dynamic range. At such a bias, which is denoted as $V_{tune} = 0$, the VCO frequency is equal to the free-running frequency ω_{FR}. By changing V_{tune} the output VCO frequency spans the so-called *tuning range* and the term K_{VCO} in (2.1), measured in rad/(sV), is usually referred to as the *VCO gain* or *tuning constant*, even if, in practice, it is not constant at all, featuring a dependence on V_{tune}.

However, the adoption of a VCO as frequency synthesizer is not enough. In many cases, the VCO frequency has to match a desired value within parts-per-million accuracy, which is out of the question owing to the typical tolerance of the passive components. Two replicas of the same VCO circuit would need different values of V_{tune} to generate the same frequency. Moreover, the frequency stability of a VCO is poor. Its oscillation frequency suffers from short-term drift (essentially the phase noise introduced in Chapter 1) as well as from long-term drift, for instance because of thermal and aging effects. These limitations are solved by *locking* the VCO signal to a very stable reference source by a feedback loop. The reference can be a crystal oscillator, and may sometimes even be temperature-controlled.

Figure 2.1 shows the basic scheme of the phase-locked loop (PLL). The phase detector (PD) compares the two input signals, the input reference (REF) and the output loop signal

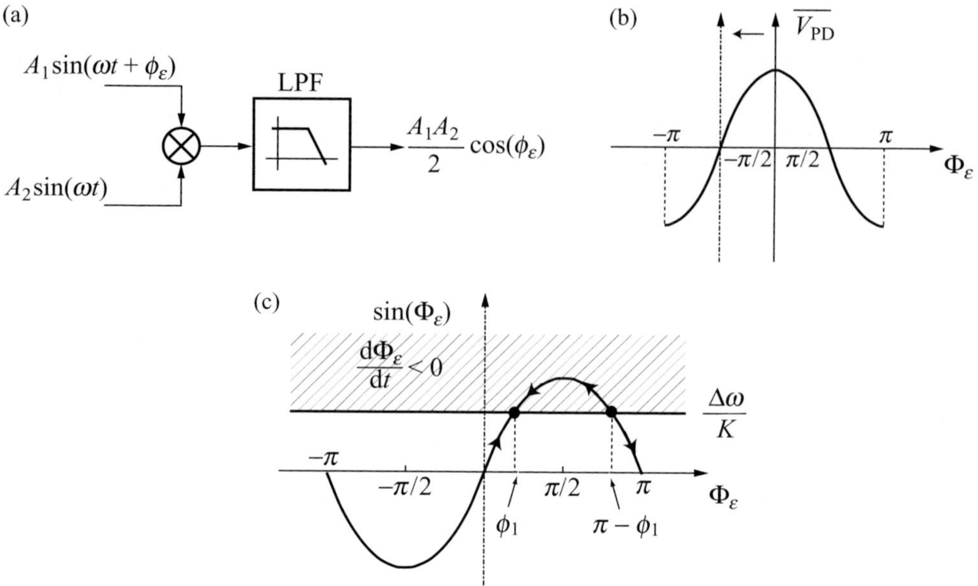

Figure 2.2 Mixer phase detector (PD): (a) schematic, (b) static characteristic, (c) graphic representation of the steady-state solution of (2.4)

(OUT), providing a measure of their phase difference. The phase difference is the error signal in the loop. Ideally, the feedback should force the two signals to maintain a constant phase difference. Since the feedback loop preserves the phase coherence between the two signals, they both run at the same frequency. The output frequency ω_{out} is therefore locked to the reference value, as desired.

Let us consider a phase detector implemented by a mixer. [15] Figure 2.2(a) shows the output of the detector. It can be written as:

$$A_1 \sin(\omega t + \phi_\varepsilon) \cdot A_2 \sin(\omega t) = -\frac{A_1 A_2}{2} \cos(2\omega t + \phi_\varepsilon) + \frac{A_1 A_2}{2} \cos(\phi_\varepsilon).$$

If the following low-pass filter cuts off the high-frequency component, the output depends on the phase difference ϕ_ε. Figure 2.2(b) shows the mixer input–output characteristic. The curve gives the dependence of the d.c. output value from the mixer as a function of the phase error. Note that the error signal is zero only if the two input signals have the same frequency and are in quadrature ($\phi_\varepsilon = \pi/2$ or $-\pi/2$). It will, therefore, be more convenient in the following to already account for such a $\pi/2$ shift, thus writing:

$$V_{\text{ref}}(t) = A_{\text{ref}} \sin\left[\Phi_{\text{ref}}(t)\right] = A_{\text{ref}} \sin(\omega_{\text{ref}} t + \phi_{\text{ref}}),$$
$$V_{\text{out}}(t) = A_{\text{out}} \cos\left[\Phi_{\text{out}}(t)\right] = A_{\text{out}} \cos(\omega_{\text{out}} t + \phi_{\text{out}}).$$

Moreover, note that the mixer output depends on the signal amplitudes. Assuming unity amplitudes, the filter output, which corresponds to the tuning voltage, can be written as:

$$V_{\text{tune}}(t) = K_{\text{PD}} \cdot \sin(\Phi_{\text{ref}} - \Phi_{\text{out}}) = K_{\text{PD}} \cdot \sin(\Phi_\varepsilon), \tag{2.2}$$

where K_{PD} is referred to as the mixer conversion gain and $\Phi_\varepsilon = \Phi_{ref} - \Phi_{out}$ is the overall phase error. In the more general case, K_{PD} should be replaced by $A_{ref}A_{out}K_{PD}$. Accounting for the input delay between REF and OUT is equivalent to shifting the y axis of the input–output characteristic of the mixer by $-\pi/2$ (the dash–dotted axis in Figure 2.2(b)).

The time derivative of $\Phi_{out}(t)$ gives the instantaneous frequency of the output signal, which can be written in terms of the free-running VCO frequency ω_{FR} and the instantaneous V_{tune} value:

$$\frac{\mathrm{d}\Phi_{out}}{\mathrm{dt}} = \omega_{FR} + K_{VCO}V_{tune}(t). \tag{2.3}$$

Using a similar expression for Φ_{ref}, the time derivative of the overall phase error Φ_ε can be written as:

$$\frac{\mathrm{d}\Phi_\varepsilon}{\mathrm{d}t} = \frac{\mathrm{d}\Phi_{ref}}{\mathrm{d}t} - \frac{\mathrm{d}\Phi_{out}}{\mathrm{d}t} = (\omega_{ref} - \omega_{FR}) - K_{VCO}V_{tune}(t).$$

Denoting the constant $(\omega_{ref} - \omega_{FR})$ as $\Delta\omega$ and using (2.2), the differential equation describing the loop becomes:

$$\frac{\mathrm{d}\Phi_\varepsilon}{\mathrm{d}t} = \Delta\omega - K \cdot \sin[\Phi_\varepsilon(t)], \tag{2.4}$$

where the product $K = K_{VCO}K_{PD}$ is an angular frequency measured in rad/s. Equation (2.4) highlights the fact that the PLL is a non-linear system, with non-linearity arising from the mixer.

Some further considerations can be derived from (2.4) without solving it. If the loop locks and a steady-state condition is reached, $(\mathrm{d}\Phi_\varepsilon/\mathrm{d}t) = 0$, and the phase error is given by:

$$\sin(\Phi_\varepsilon) = (\Delta\omega/K).$$

This condition is represented graphically in Figure 2.2(c). The loop can lock only if $|(\Delta\omega/K)| < 1$. At the steady state (for $t \to \infty$), $\omega_{out} = \omega_{ref}$ and the *excess* phase error ϕ_ε is $\arcsin(\Delta\omega/K)$, which has two solutions ϕ_1 and $(\pi - \phi_1)$ in the range $[-\pi, \pi]$.

Only one of these solutions is the steady-state phase error and it depends on the sign of K. If $K > 0$ and $\sin[\Phi_\varepsilon(t)] > \Delta\omega/K$, then:

$$\frac{\mathrm{d}\Phi_\varepsilon}{\mathrm{d}t} = \Delta\omega - K \cdot \sin[\Phi_\varepsilon(t)] < 0.$$

Therefore, in that region of the plane in Figure 2.2(c), the phase error decreases. The only stable steady-state condition occurs at phase ϕ_1. If instead $K < 0$, then the stable steady-state solution has phase error $(\pi - \phi_1)$. In the following, K will be assumed to be positive.

The limit frequency range (also called hold-in range) $\Delta\omega = \pm K$ sets the PLL *lock frequency range*. If $|(\Delta\omega/K)| > 1$, the loop never locks. Nevertheless, the PLL does not run at ω_{FR}. The average value of the output frequency ω_{out} gets closer to ω_{ref} than the original free-running frequency. The conclusion can be grasped intuitively. Without any loss of generality, let us assume that $\Delta\omega > 0$ (a similar argument holds for $\Delta\omega < 0$). The VCO is running at a frequency slower than the reference and the loop raises the tuning voltage value V_{tune} to speed the VCO up. During the loop operation, the sine function in (2.4) sweeps from positive to negative values. Any time the sine value is negative, a positive term is added to $\Delta\omega$ in (2.4), thus increasing the time derivative of the phase error Φ_ε. When the sine is positive, the time dependence of Φ_ε slows down. It follows that $\sin(\Phi_\varepsilon)$ stays positive longer than it stays negative. Based on (2.2), this corresponds to keeping V_{tune} positive for a longer time. It follows that even if the loop does not lock and the output frequency ω_{out} does not settle, its average value moves from the initial free-running value and gets closer to ω_{ref}.

Let us now go back to the case $|(\Delta\omega/K)| < 1$, when the locking condition is satisfied. Starting from the initial condition, the loop undergoes a transient response, which should be studied by solving (2.4). The task may be accomplished numerically. It is, however, more useful for our purposes to linearize (2.4) by taking $\sin(\Phi_\varepsilon) \cong \Phi_\varepsilon$, thus getting the simple solution:

$$\Phi_\varepsilon(t) = \mathrm{e}^{-K\cdot t}\left[\Phi_{\varepsilon 0} - \frac{\Delta\omega}{K}\right] + \frac{\Delta\omega}{K}.$$

As already noticed above, Φ_ε starts from the initial $\Phi_{\varepsilon 0}$ value and will eventually approach $(\Delta\omega/K)$. The transient time constant is set by K.

Such a linear approximation is useful to describe the PLL transients in the so-called *tracking regime*, where small frequency variations are induced in an already locked PLL. The approximation assumes that the PD works in its linear region and that the high-frequency component at the PD output is well cut off by the loop filter.

The system response to small perturbations may also be studied using the Laplace transform in the s domain. Let us consider, for instance, a PLL originally locked, with the VCO running at $\omega_{\text{out}} = \omega_{\text{FR}} + K_{\text{VCO}} V_{\text{tune}}^0$. Owing to a small step increment of the reference frequency, a phase error starts to build up, the PD output drives the VCO tuning node and the instantaneous output frequency departs from ω_{out} by $K \cdot [\phi_{\text{ref}} - \phi_{\text{out}}]$. Based on (2.3) its value may be written as:

$$\frac{\mathrm{d}\Phi_{\text{out}}}{\mathrm{d}t} = \omega_{\text{out}} + \frac{\mathrm{d}\phi_{\text{out}}}{\mathrm{d}t} = \omega_{\text{FR}} + K_{\text{VCO}} V_{\text{tune}}^0 + K \cdot [\phi_{\text{ref}} - \phi_{\text{out}}]\,.$$

Since $\omega_{\text{out}} = \omega_{\text{FR}} + K_{\text{VCO}} V_{\text{tune}}^0$, it turns out that

$$\frac{\mathrm{d}\phi_{\text{out}}}{\mathrm{d}t} = K \cdot [\phi_{\text{ref}} - \phi_{\text{out}}]\,,$$

which leads in the s domain to:

$$\frac{\phi_{\text{out}}}{\phi_{\text{ref}}}(s) = \frac{K}{K + s}.$$

It turns out that the phase variations of the output and reference signals are linked by a transfer function with a single pole. The PLL is therefore referred to as a first-order PLL.

2.1.3 Linear models of the loop-building blocks

To investigate the linear response of more complex PLLs, it is convenient to derive the transfer functions of all building blocks in the s domain and then to proceed using the standard methodology adopted in the analysis of linear electrical circuits.

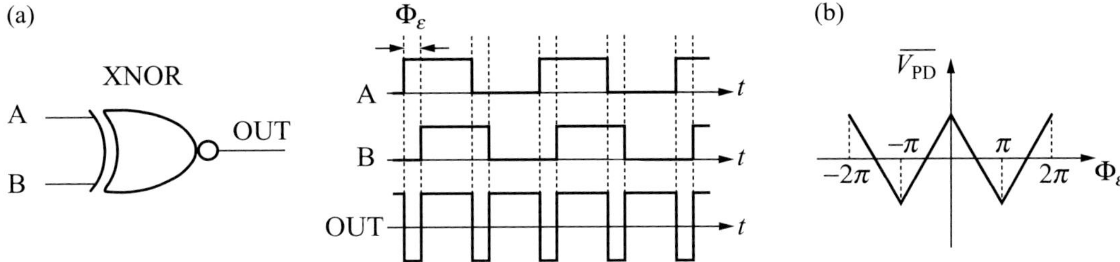

Figure 2.3 XNOR PD: (a) schematic, (b) static characteristic

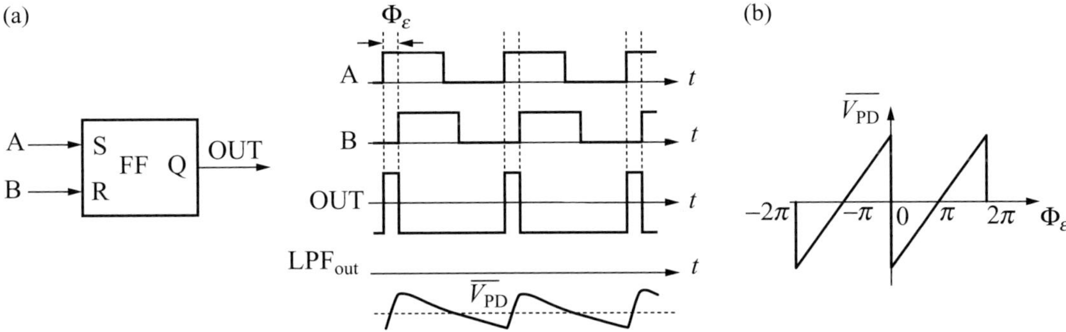

Figure 2.4 SR flip-flop PD: (a) schematic (the low-pass filter output is also given), (b) static characteristic

The phase detector acts as an error amplifier in the loop with an output signal V_{PD} proportional to the phase difference between the two inputs. The input–output characteristic of a mixer was already discussed above (Figure 2.2(b)). If the loop is locked, the OUT and the REF signals run at the same frequency, they are in quadrature except for a residual steady-state phase error given by $\arcsin(\Delta\omega/K_{PD}K_{VCO})$. The phase detector gain K_{PD}, measured in V/rad, can be taken as the slope around the points at $\pm\pi/2$. For a mixer, K_{PD} depends on the amplitudes of the input signals.

Digital versions of a mixer are implemented using exclusive-OR (XOR) or exclusive–NOR (XNOR) gates. Figure 2.3(a) shows the time dependence of the input and output signals at the XNOR terminals. Since the gate is a digital stage, the output is only affected by the zero crossing of the input signals. This is why the inputs have been represented as logic square waves. Note that, unlike the mixer, the output signal is indeed a square wave and its amplitude does not depend on the amplitude of the input signals. These stages overcome this limitation of the mixer stage. The PD characteristic (Figure 2.3(b)) is similar to that of the mixer (Figure 2.2(b)). It has zero-crossing points corresponding to $\pm\pi/2$ phase shifts. However, the dependence is linear within the phase range $[0-\pi]$. The PD gain K_{PD} has a magnitude equal to V_{FS}/π, where V_{FS} is the full-scale output voltage swing. An exclusive-OR (XOR) circuit would work similarly. A PD can be also made using a set–reset flip-flop (SR FF) (Figure 2.4). In this case, the characteristic shows an input linear range that is twice as wide, up to a phase error of 2π.

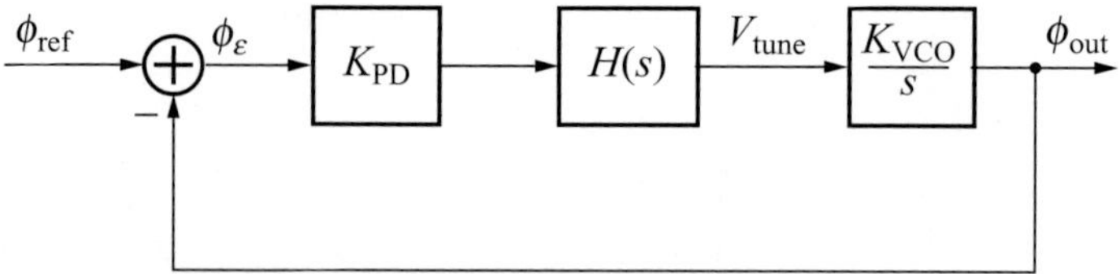

Figure 2.5 Linear model of the PLL in Figure 2.1

As far as the representation in the s domain is concerned, all these PDs may be represented to the first order by their gain K_{PD} without any dependence on s. In practice, they are considered to be memoryless stages.

Let us now consider the VCO. According to (2.1) the excess phase of the VCO output signal can be written as

$$\phi_{out}(t) = K_{VCO} \int_{-\infty}^{t} V_{tune}(\eta) \cdot d\eta,$$

which in the s domain becomes:

$$\frac{\phi_{out}}{V_{tune}}(s) = \frac{K_{VCO}}{s}.$$

Therefore, the VCO acts as an integrator in the loop. Even if the oscillator frequency immediately responds to a V_{tune} change, it takes some time for the phase signal to build up.

2.1.4 PLL linear model

Figure 2.5 shows the complete PLL block diagram in the s domain, where the low-pass filter is represented by its transfer function $H(s)$. The open-loop gain is given by $G_{loop}(s) = -K_{PD} \cdot H(s) \cdot (K_{VCO}/s)$. In the simplest PLL scheme, the low-pass filter has just a single pole $H(s) = 1/(1 + s\tau)$; hence, the input–output transfer function has second-order low-pass dependence:

$$\frac{\phi_{out}}{\phi_{ref}}(s) = \frac{-G_{loop}(s)}{1 - G_{loop}(s)} = \frac{1}{s^2/\omega_P^2 + 2\xi s/\omega_P + 1}. \tag{2.5}$$

This means that the output is able to track slow input variations of the phase signal. High-frequency input signals are filtered instead. The system has two poles with magnitude $\omega_P = \sqrt{K/\tau}$ and a damping factor $\xi = 1/(2\sqrt{K\tau})$.

Very often, the input signal consists of a variation of the reference frequency. In the Laplace transform domain the signal is represented by $\omega_{ref}(s)$, while $\omega_{out}(s)$ denotes the Laplace transform of the output frequency variation. Since $(\omega_{out}/\omega_{ref}) = (s\phi_{out})/(s\phi_{ref}) = \phi_{out}/\phi_{ref}$, it turns out that the transfer function in (2.5) also links the frequency variations. If the reference frequency has a step-like change with amplitude $\Delta\omega_{ref}$, it is $\omega_{ref}(s) = \Delta\omega_{ref}/s$. The output transient $\omega_{out}(t)$ follows the step response of a second-order system, eventually reaching the final value $\omega_{out}(t \to \infty) = \Delta\omega_{ref}$. In principle, the transient $\omega_{out}(t)$ can be

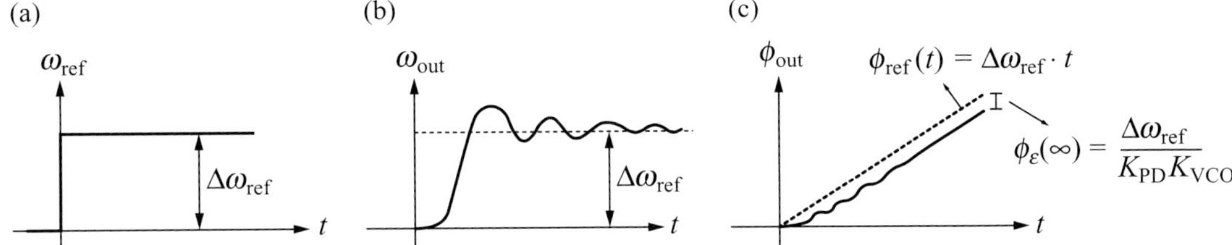

Figure 2.6 (a) Input frequency step, (b) corresponding output frequency variation, (c) excess phase signals

measured by monitoring the transient of the signal $V_{\rm tune}(t)$. Figures 2.6(a) and (b) show the ideal time dependence of the signals $\omega_{\rm ref}(t)$ and $\omega_{\rm out}(t)$.

As far as the phase error $\phi_\varepsilon = \phi_{\rm ref} - \phi_{\rm out}$ is concerned, the transfer function is given by:

$$\frac{\phi_\varepsilon}{\phi_{\rm ref}}(s) = \frac{1}{1 - G_{\rm loop}(s)} = \frac{2\xi}{\omega_{\rm P}} \cdot \frac{s\left(\dfrac{s}{2\xi\omega_{\rm P}} + 1\right)}{s^2/\omega_{\rm P}^2 + 2\xi s/\omega_{\rm P} + 1}.$$

A frequency step corresponds to an input phase signal with a linear ramp dependence, $\phi_{\rm ref}(s) = \Delta\omega_{\rm ref}/s^2$, and $\phi_{\rm out}$, after the initial transient, will eventually approach a ramp dependence with the same $\phi_{\rm ref}$ slope. For $t \to \infty$, the difference between $\phi_{\rm out}$ and $\phi_{\rm ref}$ can be derived from the above expression, using the limit theorem of the Laplace transform: $\lim_{t\to\infty} \phi_\varepsilon(t) = \lim_{s\to 0} s\phi_\varepsilon(s)$. It turns out that $\phi_\varepsilon(t \to \infty) = \Delta\omega_{\rm ref}/K$. Figure 2.6(c) shows the phase signals.

To avoid large overshoots in the time domain, the damping factor is usually set to $\xi = 1/\sqrt{2}$. This corresponds to setting $K = 1/(2\tau)$ and to having a closed-loop bandwidth, given by $\omega_{\rm P} = \sqrt{2} \cdot K$. Note that the condition on the damping factor has set all the loop parameters. It is evident that such a system has too few degrees of freedom to accommodate further requirements independently. To filter out input disturbances, the closed-loop bandwidth should be narrow, thus leading to a low value for the K product and, in turn, to a narrow *lock frequency range* of the PLL. On the other hand, if K is set larger than $1/(2\tau)$, the damping factor may be not enough, leading to a large overshoot and a long settling time. Figure 2.7(a) shows a typical Bode diagram of the open-loop gain magnitude, the input-to-output phase transfer function and the root locus of a PLL with a single-pole filter.

An additional degree of freedom can be gained by adding a zero in the loop-filter transfer function. The transfer function $H(s)$ of the filter in Figure 2.7(b) is given by:

$$H(s) = \frac{1 + s\tau_2}{1 + s\tau_1}.$$

Figure 2.7(b) shows the magnitude of the open-loop gain, $|G_{\rm loop}|$, the magnitude of the phase transfer function, $|\phi_{\rm out}/\phi_{\rm ref}|$, and the corresponding root locus. The transfer function has no overshoot if the open-loop gain is high enough to push the closed-loop poles to be

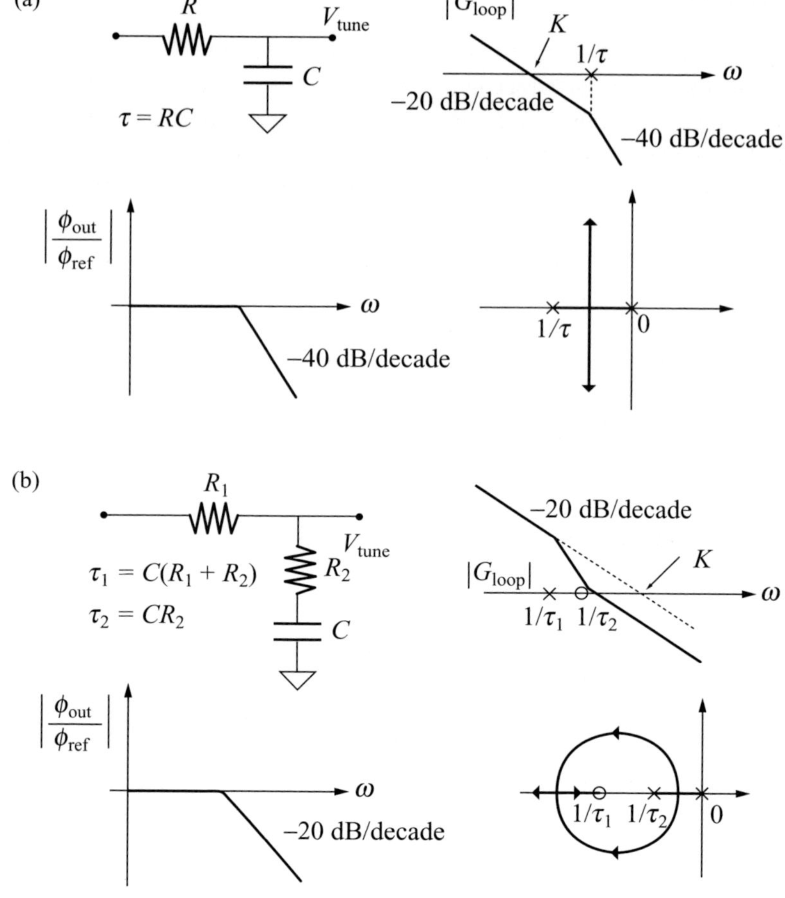

Figure 2.7 Loop filter schematic, open-loop gain magnitude, input–output transfer function and root locus of a second-order PLL with: (a) single-pole filter, (b) zero-pole filter

real. The input–output transfer function is now:

$$\frac{\phi_{\text{out}}}{\phi_{\text{ref}}}(s) = \frac{s\,(2\xi - \omega_{\text{P}}/K)\,/\omega_{\text{P}} + 1}{s^2/\omega_{\text{P}}^2 + 2\xi s/\omega_{\text{P}} + 1}$$

where the pole magnitude and the damping factor are given by:

$$\omega_{\text{P}} = \sqrt{\frac{K}{\tau_1}}, \quad \xi = \frac{\omega_{\text{P}}}{2}\left(\tau_2 + \frac{1}{K}\right).$$

In this loop the lock range, the closed-loop bandwidth and the pole-damping factor (parameters K, ω_{P} and ξ) can be set independently.

2.1.5 Type-I and type-II PLLs

The steady-state phase error which occurs when the PLL deviates from the VCO free-running frequency causes inaccuracy in the phase lock between OUT and REF and it makes the *lock frequency range* limited by the output voltage headroom of the PD. These drawbacks can be avoided by placing a further pole at d.c. in the loop, thus forcing a zero phase error at the steady state. For this purpose, the low-pass loop filter may be replaced by an active integrator implemented with an operational amplifier. In integrated circuits, however, the same result is obtained using a charge-pump (CP) circuit and a load capacitor. Doing so, even if the VCO runs far from its free-running frequency, the integrator stores the proper tuning voltage and the PD average output approaches zero at steady state. As a result, the *lock frequency range* is only limited by the VCO tuning range. A PLL featuring an extra integrator in the loop is called a *type-II PLL*, while the PLLs discussed in the previous section are called *type-I PLLs*.

2.2 PLL for frequency synthesis

2.2.1 Integer-*N* divider PLL

The previous section has been devoted to PLL principles. However, so far, the system does not appear to be useful at all for the synthesis of a desired frequency. After all, the output signal is just the replica of an already available reference signal at ω_{ref}, which could then be directly used in place of ω_{out}. The system becomes interesting once a frequency divider is placed in the feedback path (Figure 2.8). Now, the divider output runs at the PLL output frequency divided by N, i.e., $\omega_{div} = \omega_{out}/N$. When the PLL is locked, $\omega_{div} = \omega_{ref}$, so the output is the reference frequency multiplied by N:

$$\omega_{out} = N \cdot \omega_{ref}.$$

The loop is therefore able to synthesize an output frequency, which is a multiple of a highly accurate reference at a lower frequency. Moreover, by changing the integer value of N, the output frequency can be changed with a minimum frequency step equal to ω_{ref}. The

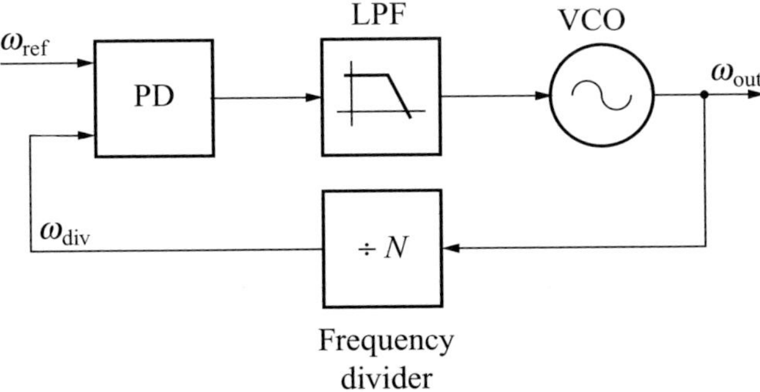

Figure 2.8 PLL with integer-N frequency divider

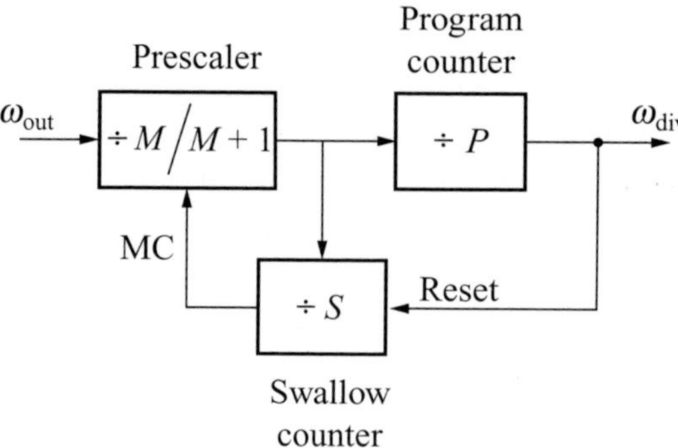

Figure 2.9 Pulse swallow divider

arrangement is called *integer-N PLL.* In wireless applications, the output signal is in the GHz range while the reference may be derived from a stable quartz oscillator, usually at several MHz. The magnitude of N is of the order of 10^2–10^3 or even larger.

Figure 2.9 shows the basic structure of the frequency divider whose design issues are discussed in detail in Chapter 8 (Section 8.3.2). It is essentially made up of counters. The program counter and the swallow counter change their outputs after they have accumulated P and S input pulses, respectively. The prescaler, instead, divides either by M or by $(M + 1)$, depending on the output of the swallow counter, i.e., the logic signal indicated as modulus control (MC).

When a cycle begins, the prescaler divides its input by $(M + 1)$. After S periods of the prescaler output, the MC changes its status; in the meantime, the program counter has accumulated S input pulses. For the remaining $(P - S)$ pulses of the program counter input, the prescaler divides by M. After that, the program counter overflows; it produces an output pulse, which resets the swallow counter, and the entire cycle restarts.

The prescaler output period lasts:

$$N = (M + 1)S + M(P - S) = P \cdot M + S$$

input periods. N is the division factor. The product $P \cdot M$ is fixed and is as large as necessary. The factor S is, instead, a small integer. As $S = 0, 1, 2, \ldots$, the output frequency changes. This scheme is often referred to as a *pulse swallow divider*. Because of its simple structure, the set-up has been widely adopted.

After the introduction of the divider, the PLL transient response must be briefly reconsidered. For example, to change N by one is equivalent to driving the loop with an input frequency step of amplitude ω_{ref}. Moreover, the divider-by-N affects the open-loop gain. If ω_{out} is divided by N, ϕ_{out} is also divided by the same number. The model in Figure 2.5 must, therefore, be modified by placing a block with a gain of $1/N$ in the feedback path. The open-loop gain is correspondingly divided by N. In the case of a single-pole filter (the circuit in Figure 2.7(a)), the input–output transfer function becomes:

$$\frac{\phi_{out}}{\phi_{ref}}(s) = \frac{N}{s^2/\omega_P^2 + 2\xi s/\omega_P + 1}.$$

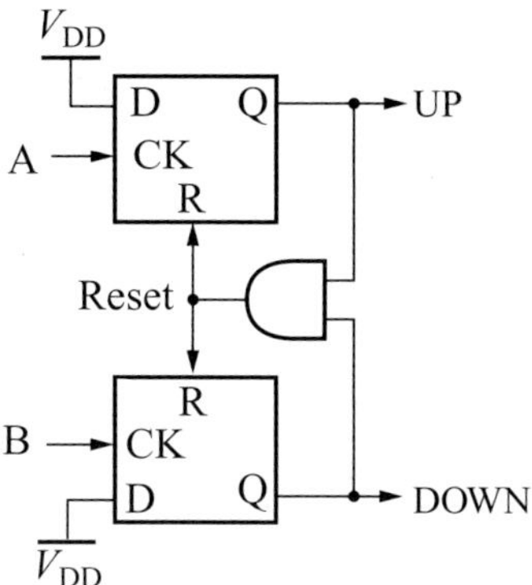

Figure 2.10 A phase/frequency detector (PFD)

The closed-loop bandwidth and the pole-damping factor are now $\omega_P = \sqrt{K/(N \cdot \tau)}$ and $\xi = \sqrt{N}/(2\sqrt{K \cdot \tau})$, respectively.

Note that the value of the low-frequency PLL gain is now equal to N. Consequently, any input phase disturbance, which is slow enough to be within the PLL bandwidth, is transferred to the output amplified by N independently of the choice of the loop filter.

2.2.2 Phase frequency detector and charge pump

In the design of PLLs for frequency synthesis, two additional issues usually lead to substantial modifications of the building blocks considered so far.

The first one is related to phase detection. The PDs in Figures 2.2, 2.3 and 2.4 are not able to provide an average output proportional to the difference between the input frequencies. Their output signal has low-frequency components, which follow the build-up of the phase error. The larger the VCO detuning, the larger the frequency of these components (e.g., see (2.2)). In some cases the PD output signal may, therefore, fall outside of the loop filter bandwidth, and be severely cut off, and the PLL, lacking a strong driving signal, may retard or even fail to lock. The adoption of a phase/frequency detector (PFD), which provides a d.c. signal proportional to the difference of the input frequencies, substantially improves the lock acquisition performance.

The second issue is related to the presence of spurious tones in the PLL output spectrum. When the PLL is locked, the conventional PDs generate a high-frequency component at either $2\omega_{\text{ref}}$ (for the mixer and the XOR) or ω_{ref} (for the SR flip-flop), which causes the *ripple* schematically shown in Figure 2.4(a). The adoption of a type-II loop does not solve this issue, since it only makes the PD *average* voltage zero. This ripple causes a frequency modulation (FM) of the VCO output, which appears in the PLL output spectrum as spurious tones at $\omega_{\text{out}} \pm k\omega_{\text{ref}}$, with $k = 1$ or 2 depending on the PD implementation. The adoption of a tri-state PFD in a type-II loop ideally solves this issue. At the steady state, the PFD provides a zero output signal with no high-frequency component.

Figure 2.10 shows the most popular implementation of a PFD, [22] though alternative solutions are discussed in Chapter 9. This circuit is characterized by two edge-triggered

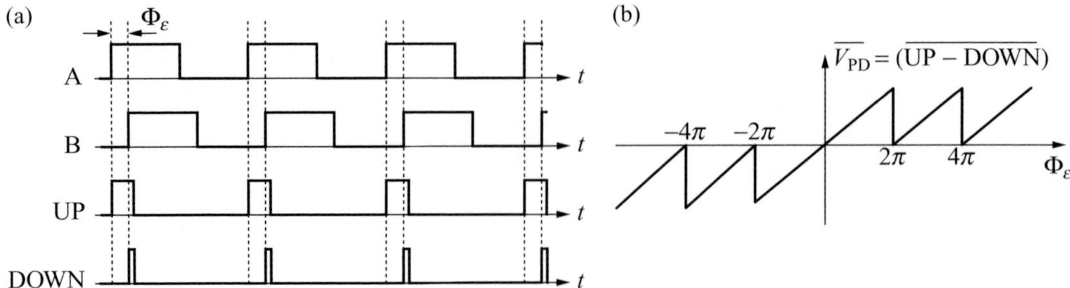

Figure 2.11 PFD: (a) time diagrams of inputs and outputs for phase-delayed inputs and (b) static characteristic

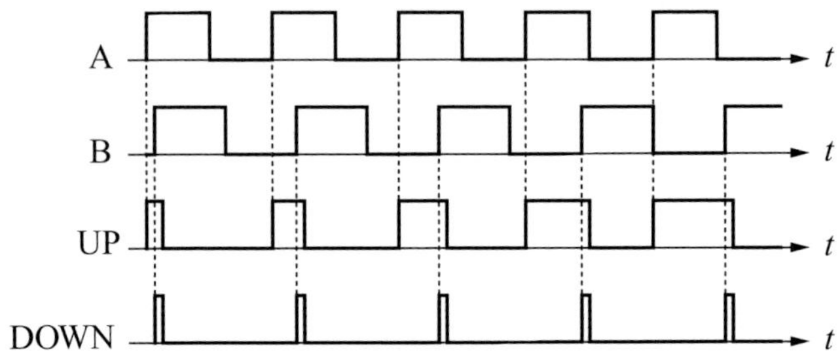

Figure 2.12 PFD: time diagrams of inputs and outputs for inputs at different frequencies

D flip-flops, with 1 on both the D inputs. The input signals A and B are the clock inputs. As an example, when the signal A goes high, the output UP goes high too. Then it remains unchanged for any further transition of A. Instead, when B goes high, DOWN also goes high and the AND gate resets both UP and DOWN outputs to the low state. Figure 2.11(a) shows the circuit operation when the signals A and B have the same frequency but different phases. For a short time, depending on the gate delay of the circuits, both of the outputs are high.

The output signal of the PFD will be the average value, $\overline{V_{PD}}$, of the difference between UP and DOWN. To the purpose, the two signals may be fed to a differential amplifier driving a low-pass filter. Figure 2.11(b) shows the ideal characteristic of this arrangement. The key property of the circuit is most obvious when the two signals have different frequencies, as in Figure 2.12. The frequency of the A signal is larger than the frequency of the B signal, and positive pulses always appear at the output UP, while only very short pulses appear at DOWN, because of gate delay. As desired, the d.c. value of the signal (UP – DOWN) is not zero; therefore, the circuit also detects the difference in frequency. Equivalently, when the input frequencies are different, the phase error increases. Hence, the static characteristic in Figure 2.11(b) is swept in time and the output always has the same sign. So its time average is not zero. The only limit to the lock range is represented by the VCO tuning range.

Let us now consider how to introduce an additional pole at d.c. in the loop. Figure 2.13(a) shows the tri-state charge pump (CP) placed after the PFD. The UP and DOWN signals

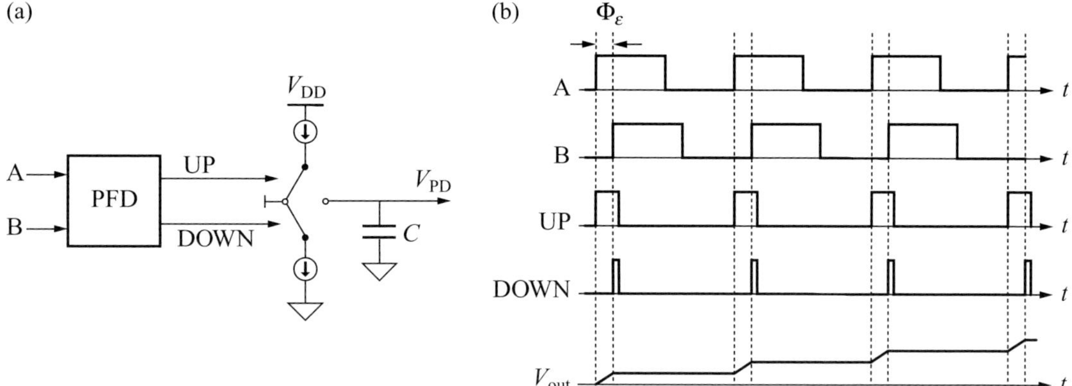

Figure 2.13 PFD driving a tri-state charge pump: (a) schematic, (b) corresponding input and output time diagrams for phase-delayed inputs

drive the corresponding switches, and, therefore, the current generators inject a current I_P in the capacitor. In the figure the UP signal positively charges the capacitor, while the DOWN signal discharges it. When either the UP or DOWN signal is zero, the corresponding current is diverted to ground, or to a fixed voltage source. Figure 2.13(b) shows the PFD outputs when the input signals have the same period but different phase. The phase error is integrated leading to a rising output voltage. When the loop is closed, instead, the phase error must be zero; in this situation, no current reaches the capacitance. Figure 2.13(b) also suggests that, to a first-order continuous-time approximation, the cascade of PFD and CP may be approximated by an integrator of the phase error, with a transfer function K_{PD}/s.

The PFD/CP cascade should be used in the PLL in Figure 2.8 to replace the PD and the low-pass filter. However, the resulting loop is not stable. The PLL open-loop gain is given by $G_{loop} = -(K_{PD}/s)(K_{VCO}/s)(1/N)$, with two poles placed at d.c. Drawing the root locus is effortless. The poles start from $s = 0$ and move along the imaginary axis. The system has imaginary poles at $s = \pm j(K_{PD}K_{VCO}/N)^{1/2}$. The stability may be recovered by replacing the capacitance in Figure 2.13(a) with another impedance, $Z(s)$.

2.2.3 Linear continuous-time analysis of the charge-pump PLL

Before studying the stability of the new loop, let us point out the condition under which the circuit in Figure 2.13(a) can be regarded as a continuous-time integrator. After all, the phase comparison is a discrete-time process, with a $T_{ref} = 2\pi/\omega_{ref}$ period. It is intuitive, however, that a continuous-time description may be recovered if the loop response is slow with respect to T_{ref}. In this case, the loop reacts to the average current 'pumped' by the PFD/CP stage into the loop filter impedance, $Z(s)$. This approach was followed by Gardner in a classical paper. [23] He showed that the continuous-time approximation works well if the closed-loop bandwidth is narrower than $1/10$ of ω_{ref}. For a wider bandwidth, the analysis must be performed, instead, using a discrete-time linear model, to correctly design the loop and avoid potential instabilities. In practical designs, the closed-loop bandwidth is always chosen within the Gardner limit. Often even $1/20$ of ω_{ref} is taken.

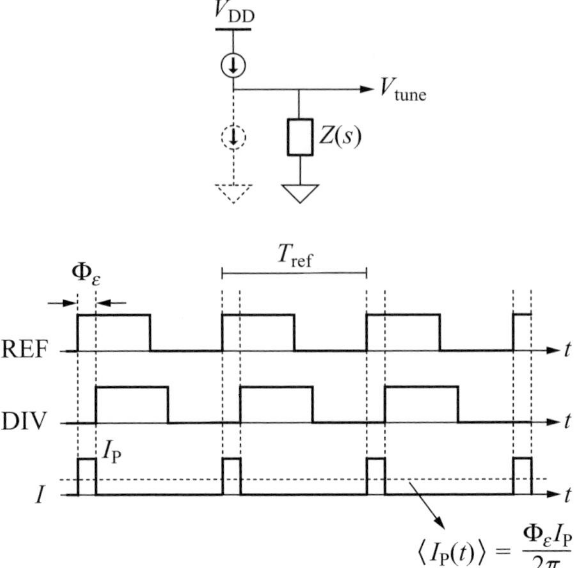

Figure 2.14 Continuous-time approximation for charge-pump transfer function

Let us now assume that we will work within the Gardner limit. A constant phase error ϕ_ε makes the PFD/CP stage inject an average current $\phi_\varepsilon \cdot (I_P/2\pi)$ into the loop filter (Figure 2.14). The PFD/CP cascade may, therefore, be modelled as a stage with an $(I_P/2\pi)$ gain. The loop performance now depends on the impedance $Z(s)$. Figure 2.15(a) shows the case corresponding to $Z(s) = 1/sC$. The presence of an additional pole in $s = 0$, from the VCO, calls for a modification of $Z(s)$. In Figure 2.15(b) a stabilizing zero is added by placing a resistance R in series to C, thus giving $Z(s) = (1 + sRC)/sC$.

This solution is not completely satisfactory, yet. At high frequencies, the impedance $Z(s)$ approaches a finite value set by R. Two limitations may be met:

- Any current pulse I_P injected into the loop filter causes a step of the output voltage V_{tune}. The step has amplitude $I_P R$, which may push the current generators, or the VCO tuning node, out of their linear dynamic ranges.
- When the PLL is locked, ideally no current should reach the filter. Unavoidable circuit mismatches, however, cause the PFD/CP to generate fast current pulses at ω_{ref}. These signals are responsible for spurious side tones in the output spectrum, whose amplitude is proportional to $|Z(j\omega_{ref})|$.

These effects may seriously affect both settling transient and spectral purity and they can be reduced by placing a further capacitor in parallel to the RC branch. Figure 2.15(c) shows such a design option. The role of the capacitor is to further reduce the filter impedance $Z(s)$ at high frequency. $Z(s)$ has now two poles and one zero:

$$Z(s) = \frac{1}{s(C_1 + C_2)} \cdot \frac{1 + sC_1 R}{1 + sC_1 C_2 R/(C_1 + C_2)}.$$

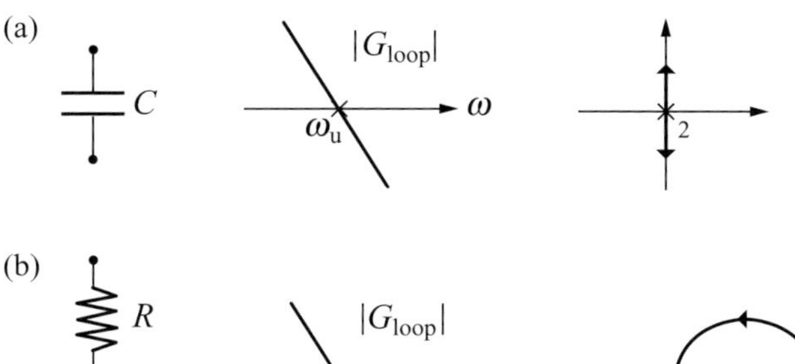

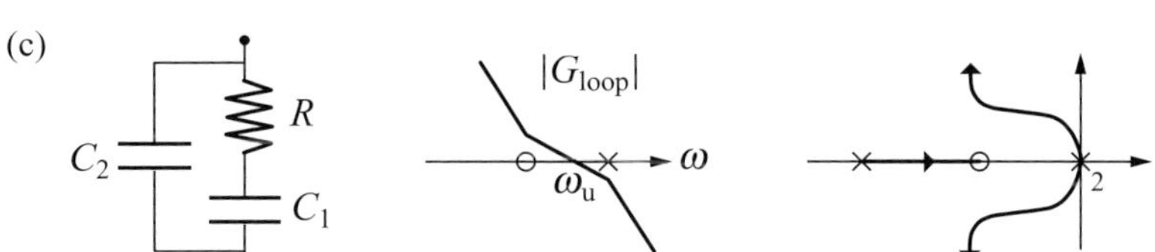

Figure 2.15 Open-loop gain and root locus of the CP PLL for three different impedances, $Z(s)$

Since the task of C_2 is to attenuate the signals out of the PLL bandwidth (e.g., at ω_{ref}), this additional capacitance is typically chosen to be $C_2 \ll C_1$. At low frequencies, $Z(s)$ is $1/s(C_1 + C_2)$. At intermediate frequencies, $Z(s)$ is practically equal to R. At high frequencies, it reduces to $1/sC_2$.

The open-loop gain is given by:

$$G_{\mathrm{loop}}(s) = -\frac{I_{\mathrm{P}}}{2\pi} \cdot Z(s) \cdot \frac{K_{\mathrm{VCO}}}{s} \cdot \frac{1}{N}. \tag{2.6}$$

Its magnitude is shown in Figure 2.15(c). The unity gain frequency ω_{u}, highlighted in the figure, can be used as a coarse estimate of the closed-loop bandwidth. For frequencies well above ω_{u}, the loop is practically open. On the other hand, at ω_{u}, $Z(s)$ is well approximated by R, so it is:

$$\begin{aligned} |G_{\mathrm{loop}}(j\omega_{\mathrm{u}})| &\cong \frac{I_{\mathrm{P}} R}{2\pi \cdot N} \cdot \frac{K_{\mathrm{VCO}}}{\omega_{\mathrm{u}}}, \; |G_{\mathrm{loop}}(j\omega_{\mathrm{u}})| = 1 \\ \Rightarrow \omega_{\mathrm{u}} &\cong \frac{I_{\mathrm{P}} \cdot R \cdot K_{\mathrm{VCO}}}{2\pi \cdot N}. \end{aligned} \tag{2.7}$$

This estimate works properly, at least for standard design parameters, both for the unity gain frequency and for the –3 dB closed-loop bandwidth.

The system has now three poles. The actual shape of the root locus depends on the system parameters. Figure 2.15(c) shows one of the possible loci. The highest phase margin is achieved when the unity gain frequency equals the geometric mean of the zero and the pole frequencies. Following [23], the number of symbols can be reduced by defining

$$b = 1 + \frac{C_1}{C_2}, \quad \tau_{\mathrm{Z}} = RC_1, \quad \beta = \frac{K_{\mathrm{VCO}} I_{\mathrm{P}} (b-1)}{2\pi \cdot N \cdot C_1}, \tag{2.8}$$

so that the open-loop gain can now be written as:

$$G_{\text{loop}}(s) = -\beta \cdot \frac{s + 1/\tau_Z}{s^2(s + b/\tau_Z)}. \tag{2.9}$$

This expression may not be as intuitive as (2.6), but it is useful for the loop design. The parameter b is the ratio between the frequencies of the *third* pole and the zero of the open-loop gain, G_{loop}. The approximate expression for the unity gain frequency given in (2.7) can be now written as $\omega_u = \beta\tau_Z/(b-1)$.

The input-output transfer function becomes:

$$\frac{\phi_{\text{out}}}{\phi_{\text{ref}}} = \frac{N \cdot \beta \cdot (s + 1/\tau_Z)}{s^3 + s^2 b/\tau_Z + s\beta + \beta/\tau_Z}. \tag{2.10}$$

Based on (2.10), let us now check the steady-state phase error after a frequency step. The phase error is given by:

$$\phi_\varepsilon(s) = \phi_{\text{ref}}(s) - \frac{\phi_{\text{out}}(s)}{N} = \phi_{\text{ref}}(s) \cdot \frac{s^2(s + b/\tau_Z)}{s^3 + s^2 b/\tau_Z + s\beta + \beta/\tau_Z}.$$

An input frequency step with amplitude $\Delta\omega_{\text{ref}}$ is equivalent to a phase ramp whose Laplace transform is $\phi_{\text{ref}} = \Delta\omega_{\text{ref}}/s^2$. Therefore, the steady-state phase error is $\lim_{t\to\infty} \phi_\varepsilon(t) = \lim_{s\to 0} s\phi_\varepsilon(s) = 0$, which is one of the principal benefits of a type-II PLL.

Example 2.1 Settling time calculation from linear continuous-time model

Let us consider a PLL with a 1 GHz output frequency and a 10 MHz output step. Based on the previous equations, let us estimate the settling time within 0.1 parts per million (0.1 ppm) of the output frequency as a function of the parameters b, β and τ_Z of (2.9).

In a second-order type-I PLL, with pole magnitude ω_P and damping factor ξ, such a settling time will be given by:

$$t_S \cong \frac{1}{\xi\omega_P} \cdot \ln\left(\frac{\Delta\omega_{\text{out}}}{\omega_\varepsilon\sqrt{1-\xi^2}}\right).$$

The time t_S depends on the loop parameters and on the ratio between the output step, $\Delta\omega_{\text{out}} = 2\pi \cdot 10\,\text{Mrad/s}$, and the settling accuracy, $\omega_\varepsilon = 2\pi \cdot 100\,\text{rad/s}$.

For the third-order loop, there is no simple formula to rely on, and numerical simulations of (2.10) must be performed. The numerical procedure is, however, speeded up if the root locus dependence on circuit parameters is fully exploited.

It should be first noted that by changing only the zero time constant $\tau_Z = RC_1$ by a factor λ, the root locus in Figure 2.15(c) scales, but preserves the same shape. In fact, as τ_Z scales by a factor λ, the ratio between pole and zero remains constant and equal to b. Since β does not depend on τ_Z, the open-loop gain becomes:

$$G_{\text{loop}}(s) = -\beta \cdot \frac{s + 1/\lambda\tau_Z}{s^2(s + b/\lambda\tau_Z)} = -\beta\lambda^2 \cdot \frac{s\lambda + 1/\tau_Z}{(s\lambda)^2\,(s\lambda + b/\tau_Z)}.$$

It turns out that scaling τ_Z is equivalent to scaling the axes (from s to $s\lambda$). If s_1 is a solution of the equation $1 - G_{\text{loop}}(s) = 0$ for $\beta = \beta_1$, then s_1/λ is a solution of the same equation

Table 2.1 *PLL settling time of the third-order type-II PLL*

b	$\beta\tau_Z^2$	t_s/τ_Z
6	16	5.8
8	22.7	4.5
12	37.5	5.2

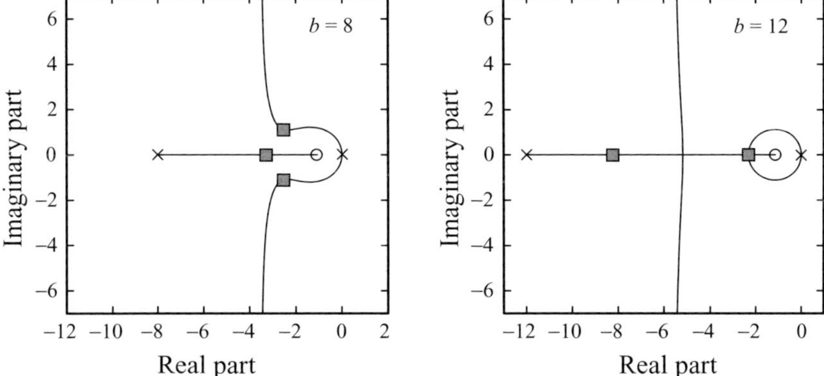

Figure 2.16 Root loci for the third-order CP PLL for different values of the parameter b; squares represent the closed-loop poles for the optimal settling time

for $\beta = \beta_1/\lambda^2$. The root locus shape depends instead on b. If b increases too much, the third open-loop pole moves away from the zero, and the plot 'closes' on the real axis for $b = 9$. Figure 2.16 shows the root loci for $b = 8$ and $b = 12$.

Based on the above observations, the optimum pole placement, i.e., the search for the β value giving the fastest settling, can be performed by numerical simulations of (2.10) working with normalized parameters ($N = 1$, $\tau_Z = 1$). The search starts by taking a value for b and then computing the settling time $t_{S,\min}$ to a given accuracy as a function of β. Once a value for β is obtained, τ_Z can then be scaled from $\tau_Z = 1$ to the real value, computing the λ factor. The whole process may be repeated for different b values until the required performance is matched.

Table 2.1 shows some numerical results for a 0.1 ppm accuracy. The best case was obtained for $b = 8$. Results for two other b values are shown, together with the corresponding optimum β values. The minimum 0.1 ppm settling time is about five times τ_Z. Note that for $b = 8$ and $\beta \approx 23$ the three poles lie almost at the same distance from the axis origin, as in the first root locus in Figure 2.16. It is easy to check that for a given value of b (e.g., $b = 8$) if β is incremented beyond the value in Table 2.1, the complex poles move to higher frequency and the real pole approaches zero. Owing to the zero-pole doublet, [24] the zero-pole cancellation is not perfect and the settling time increases slightly. Placing the poles along the 45° line in the complex plane, the resulting settling time would increase to about $8\tau_Z$. On the other hand, if β is made too small, the complex poles move close to the origin,

Table 2.2 *PLL 3 dB bandwidth and unity gain bandwidth*

b	$\beta\tau_Z^2$	$\omega_{(-3\text{dB})}\tau_Z$	$\omega_u\tau_Z$	$\omega_{u(\text{approx})}\tau_Z$
6	16	4.4	2.6	3.2
8	22.7	4.7	2.6	3.2
12	37.5	5	3.1	3.4

limiting the speed of the step response. In a practical design, it is not easy to make the value of β accurate and repeatable.

The PLL bandwidth may be estimated using the approximation $\omega_u \approx \omega_{u(\text{approx})} = \beta\tau_z/(b-1)$. Table 2.2 reports the -3 dB bandwidth and the unity gain frequency of the open loop gain, ω_u, as derived by simulations and by the approximated formula. It may be concluded that the latter is accurate enough to be used during the design flow.

2.3 Discrete-time and non-linearity effects

2.3.1 Limit of the continuous-time approximation

The reference frequency has not been considered so far. Of course, it has an impact on the value of the division factor. Once the desired β is calculated, the product $K_{VCO}I_P$ must scale proportionally to N, see (2.8). However, except for this practical issue, the reference frequency seems to be out of the game. Moreover, the above procedure may suggest that the settling time can be reduced simply by increasing the frequencies of both the zero and the third pole, taking the factor b as a constant.

These conclusions are erroneous, since they do not take into account the discrete-time nature of the PLL. As the zero frequency rises, the PLL bandwidth gets wider, moving closer to the Gardner limit $\omega_{ref}/10$, and the continuous-time approach breaks down. The Gardner condition $\omega_{(-3\text{dB})} < \omega_{ref}/10$ sets, in practice, a bound to the loop bandwidth and to the settling time.

For example, Table 2.1 and Table 2.2 show that the optimum loop with $b = 8$ is characterized by $\tau_Z = 4.7/\omega_{(-3\text{dB})} = t_s/4.5$. Since $\omega_{ref} = 2\pi/T_{ref}$, the Gardner condition $\omega_{(-3\text{dB})} < \omega_{ref}/10$ gives $t_s > 34T_{ref}$. Sometimes a safer condition $\omega_{(-3\text{dB})} < \omega_{ref}/20$ is adopted, thus doubling up the number of reference cycles required for settling.

As the closed-loop bandwidth of the PLL exceeds the Gardner limit the circuit can still be considered a linear system, but it must be analysed in the discrete-time domain, using the zeta transform. After Gardner's work, this approach can be found for instance in [25] and in some books, such as [11].

The analysis is not straightforward. For the present purposes, it is sufficient to estimate the settling time from the continuous-time model and to compare it with the result of the behavioural simulations while the closed-loop bandwidth increases. The comparison is carried out in Figure 2.17 for $b = 8$. The solid line represents the results from the linear continuous-time model, which are very close to the results of behavioural simulations

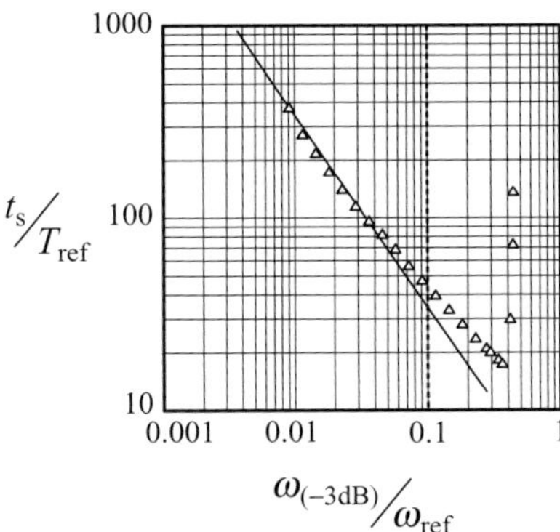

Figure 2.17 Settling time to 0.1 ppm versus PLL bandwidth

(triangles) when the bandwidth is less than $\omega_{ref}/20$ ($\omega_{(-3dB)}/\omega_{ref} < 0.05$). A similar dependence is also followed by loops with different b values. At wider bandwidths, the settling time increases, because of the large voltage spikes on the tuning node caused by the small filter capacitances. When the bandwidth is larger than $\omega_{ref}/5$, the system is very close to instability, even if the root locus of (2.10) does not give any warning. It can be shown, instead, that the analysis performed in the z domain correctly predicts the incipient instability and makes it possible to point out the proper design choices, even if the PLL works beyond, to some extent, the Gardner limit. [11, 26]

2.3.2 Limit of the linear approximation

At the beginning of this chapter, it has already been pointed out that non-linear effects come into play as the linear dynamics of the PD, or of the PFD, are exceeded. These conditions are usually met when large input frequency steps take place or when the PLL is switched on. Typical consequences are settling times that are longer than the value calculated from the linear model, or even the failure of the locking process.

This issue is usually investigated with circuit simulations; yet good approximations can be helpful to the designer to gain some preliminary insight. To this purpose, let us define first the working ranges of a PLL. Several working ranges may be identified. Here, we will focus on two of them: the hold-in range and the pull-in range, following the nomenclature also used in [15].

The hold-in range, already introduced in Section 2.1.2, is the maximum frequency range that a PLL can cover. For a type-I PLL, this range can be limited by the output range of the PD. As an example, the largest voltage output for the PD in Figure 2.4 is $K_{PD}\pi$. The magnitude of the maximum offset from the free-running frequency is $K_{VCO}K_{PD}\pi$; thus the hold-in range goes from ($\omega_{FR} - K_{VCO}K_{PD}\pi$) to ($\omega_{FR} + K_{VCO}K_{PD}\pi$).

Clearly, the VCO must be able to reach these bounds. If not, the hold-in range coincides with the tuning range of the VCO. The latter is the typical situation for a type-II PLL with PFD.

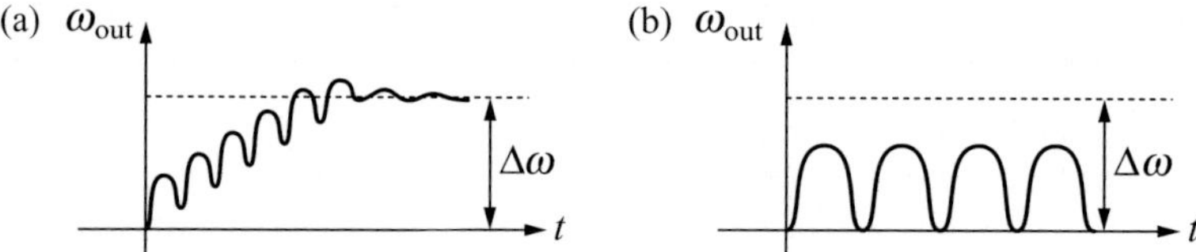

Figure 2.18 Output frequency transients in presence of non-linear behaviour

The hold-in range can be measured by slowly varying the reference frequency or, equivalently, changing the division factor by a small amount (by one for instance) and waiting every time for the PLL to lock.

The pull-in range is defined by referring to the response to a large frequency step $\Delta\omega$. Let us assume that we have a type-I PLL with PD, already locked, and that we apply a frequency step $\Delta\omega$. Even if the new reference frequency falls within the hold-in range, it may happen that the PLL fails to lock. In fact, just after the input step, the PD output is a signal running at $\Delta\omega$. If this frequency is much larger than the loop filter bandwidth, the output voltage is cut off by the filter and may be too small for the PLL to lock. The pull-in range is, therefore, defined as the maximum frequency step for which the PLL recovers the lock.

For type-I PLLs with PD, an approximation of the pull-in range can be derived by taking into account the filter attenuation $H(\mathrm{j}\Delta\omega)$. The VCO peak frequency variation due to V_{tune} must be larger than the frequency step $\Delta\omega$:

$$K_{\mathrm{PD}}\,|H\,(\mathrm{j}\Delta\omega)|\,K_{\mathrm{VCO}} > |\Delta\omega|\,.$$

If the division factor N is larger than one, the r.h.s. term is multiplied by N. The preceding inequality, which can be solved either numerically or graphically, provides a pull-in range narrower than the hold-in range, since $H(\Delta\omega) < 1$.

This is, however, only a rough approximation. What happens in reality is a non-linear locking process, already discussed in Section 2.1.2. Let us assume that the reference frequency is increased by $\Delta\omega$. The tuning voltage V_{tune} just after the step, is a periodic waveform. When V_{tune} is positive (negative), the VCO frequency gets closer to the reference frequency, thus reducing the frequency deviation $\Delta\omega$ and slowing down (accelerating) $V_{\mathrm{tune}}\,(t)$. Therefore, the resulting $V_{\mathrm{tune}}\,(t)$ stays positive longer than it stays negative or, in other words, a d.c. component pulls the system towards the lock state. Figure 2.18(a) shows the impact on a typical transient of the output frequency. In this case, $\Delta\omega$ is within the pull-in range, and the lock state is achieved. In Figure 2.18(b), the system does not lock. As observed in Section 2.1.2, even if the lock condition is not reached, the average output frequency is increased.

Let us now consider the CP PLL with PFD. Since the stage is able to deliver a d.c. output proportional to the frequency difference, the phenomenon just described above does not take place. Nevertheless, owing to the limited dynamic range of the PFD, another non-linear transient may occur. In Figure 2.19, the REF and DIV signals are initially locked. Then DIV abruptly increases its frequency. The phase error begins to build up and the area of the UP pulses increases with time, while the DOWN signal is practically zero and is not shown. If the loop bandwidth is too narrow, i.e., the PLL is very slow in recovering the frequency difference, the phase error first reaches and then exceeds 2π. The area of the UP

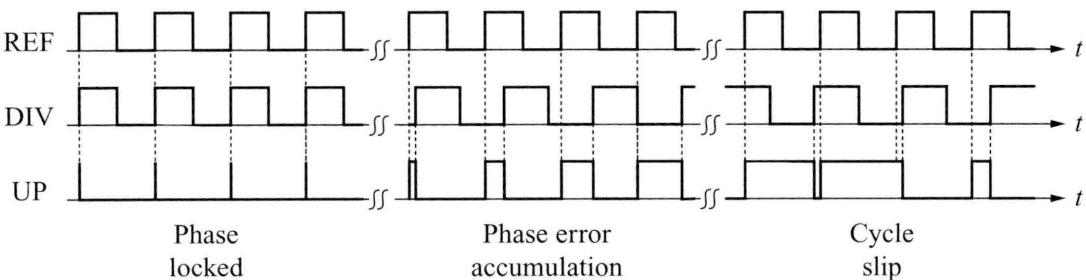

Figure 2.19 Time diagram of PFD signals when a cycle slip occurs: the PLL is initially locked, then the phase error accumulates, and finally it exceeds 2π

impulses drops and so does the time derivative of the tuning voltage. The effect may also be described by referring to Figure 2.11(b). As the phase error increases, the tuning voltage rises until the phase error reaches a multiple of 2π and then suddenly falls. This effect, known as cycle slip, is seldom discussed in the literature. It causes a peculiar ripple of the tuning voltage transients and slows down the PLL settling time.

An accurate estimate of the settling time needs the support of system-level simulations. However, it is useful to derive at least a limiting condition for the cycle slips to occur. For a frequency step $\Delta\omega_{\mathrm{ref}} = \Delta\omega_{\mathrm{out}}/N$, the first part of the transient may be approximated by a linear ramp. Then, the frequency error as a function of time is given by:

$$\omega_{\mathrm{err}}(t) = \frac{\Delta\omega_{\mathrm{out}}}{N} \cdot \left(1 - \frac{\mathrm{t}}{\tau}\right).$$

To a first-order approximation, the term $1/\tau$ can be taken to equal the $-3\,\mathrm{dB}$ bandwidth obtained from the linear model. The phase error sampled from the PFD is the integral of $\omega_{\mathrm{err}}(t)$, that is:

$$\phi_{\varepsilon}(t) = \frac{\Delta\omega_{\mathrm{out}}}{N} \cdot t \cdot \left(1 - \frac{\mathrm{t}}{2\tau}\right).$$

The waveform has a maximum for $t = \tau$. which must not exceed 2π:

$$\phi_{\varepsilon}^{\max} = \frac{\Delta\omega_{\mathrm{out}}}{N} \cdot \frac{\tau}{2} < 2\pi \Rightarrow \frac{\Delta\omega_{\mathrm{out}}}{N} < \frac{4\pi}{\tau}. \tag{2.11}$$

It turns out that cycle slips do not occur provided that the magnitude of the input reference step $\Delta\omega_{\mathrm{out}}/N$ is lower than about 4π times the –3 dB bandwidth $\omega_{(-3\mathrm{dB})}$. Given the maximum value of $\Delta\omega_{\mathrm{out}}$, (2.11) forces a condition on the loop bandwidth or vice versa. Of course, (2.11) is just a guideline for first-order estimations. A more careful assessment should rely on numerical simulations.

Example 2.2 Impact of non-linearity and discrete-time behaviour

Let us consider the dependence of the PLL settling time as a function of the reference frequency. Based on the linear continuous-time approximation, the settling time should only depend on the loop bandwidth. It follows that, according to the linear model and assuming that the other loop parameters do not vary, the settling time remains constant even

Table 2.3 *Settling time in presence of cycle slips and in discrete-time behaviour*

ω_{ref} (rad/s)	$\Delta\omega_{ref}$ (rad/s)	t_s (s)
628	6.28	4.5
6280	62.8	9.5
62.8	0.628	5.8

when the reference frequency is increased. However, the discussion in Section 2.3.1 has highlighted that the discrete-time behaviour of the PLL may slow down the settling time, as the closed-loop bandwidth gets wider. On the other hand, if the PLL bandwidth is too narrow, the cycle slips discussed in this section may degrade the settling time as well.

Table 2.3 reports some system-level simulation results for the third-order PLL in Example 2.1. The parameters $b = 8$, $\tau_Z = 1$s, $\beta = 22.7$ rad^2/s^2 ensure the shortest settling time. In all three cases, the division factor N is set equal to one, a 1% step variation is applied to the reference frequency and the settling time for an accuracy of 0.1 ppm has been computed. Since the closed-loop bandwidth is always the same ($\omega_{(-3\text{dB})} \approx 4.7$ rad/s), the linear model would predict the same settling time, $t_s = 4.5$ s, as Table 2.1. This result is reached, instead, only for $\omega_{ref} = 628$ rad/s. For $\omega_{ref} = 6280$ rad/s, the condition (2.11) is not fulfilled and the PLL has cycle slips. Taking the –3 dB bandwidth as $(1/\tau)$, the inequality becomes $\Delta\omega_{ref}/\omega_{(-3\text{dB})} < 4\pi$, or $13.3 < 4\pi$, which is false. The Gardner limit is instead satisfied, being $\omega_{(-3\text{dB})}$ (i.e., 4.7 rad/s), much lower than $\omega_{ref}/10$ (i.e., ~ 628 rad/s).

For $\omega_{ref} = 62.8$ rad/s (third row of Table 2.3), the situation is the opposite. Cycle slips are avoided, but the bandwidth is too wide compared with the reference frequency: $\omega_{ref}/10$ (i.e., 6.28 rad/s) is very close to $\omega_{(-3\text{dB})}$. In this case, the settling is longer than expected from the continuous-time model.

Note that even if the simulations have been performed with $N = 1$ the results also apply to PLL with different division factors but having the same bandwidth relative to the reference frequency, as far as the Gardner limit is concerned, and the same input-referred frequency step, as far as the non-linearities are concerned.

Figure 2.20 shows the output frequency transients simulated for the three cases in Table 2.3. The curves are normalized to their asymptotic values. Figure 2.20(a) shows the first two cases in Table 2.3. Note the ripple due to the cycle slips in the lower curve. As the transient goes on, the difference between the reference and the output frequencies is reduced, increasing the duration of the cycle slip.

Figure 2.20(b) shows the transient in the third case. The wide bandwidth implies large spikes on the tuning node. Even if the system is still stable, the settling time is increased.

Example 2.3 Design of a third-order PLL

Let us comment on the design of a PLL with the following specifications:

- Minimum output step = 1 MHz.
- Output central frequency $f_{out} = 2.4$ GHz.

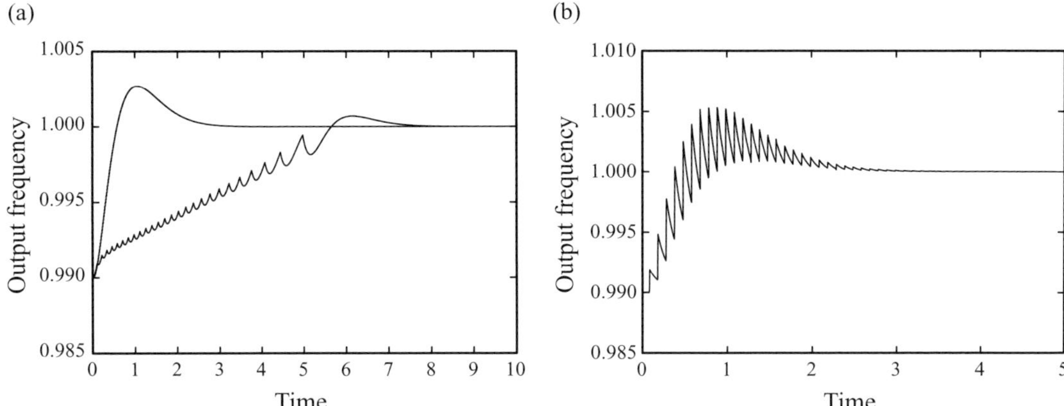

Figure 2.20 Output frequency transients for different ratios between bandwidth and reference frequency: (a) highly non-linear settling time compared with a 'well-designed' loop and (b) the case of a very large bandwidth PLL

- Tuning range $\Delta f_{\text{out}} = 64\,\text{MHz}$.
- Settling time for 0.1 ppm $t_S = 0.2\,\text{ms}$.
- Power supply of the circuit $V_{DD} = 3\,\text{V}$.

The minimum output step sets the reference frequency $f_{\text{ref}} = 1\,\text{MHz}$ and the division factor $N = (f_{\text{out}}/f_{\text{ref}}) = 2400$. More precisely, this is the division factor corresponding to the central output frequency. Since the tuning range is 64 MHz, the maximum division factor is 2432, while the minimum is 2368.[1]

To get the fastest settling, we should have $b = (1 + C_1/C_2) = 8$ and $\beta\tau_Z^2 = 23$ (see Table 2.1). Based on the results in Tables 2.1 and 2.2, the -3 dB bandwidth can be written in terms of the settling time: $\omega_{(-3\text{dB})} \approx (4.7 \cdot 4.5)/t_s \approx 21/t_s$. Therefore, the required settling time of 0.2 ms is reached with $\omega_{(-3\text{dB})} \approx 100\,\text{krad/s}$.

The next step is to set the values of β and τ_Z. From Table 2.1, it follows that $t_S/\tau_Z \approx 4.5$, hence $\tau_Z = 44\,\mu\text{s}$. Then, the value of β is derived from $\beta\tau_Z^2 \approx 23$. From the β expression in (2.8), the following condition applies:

$$\frac{K_{\text{VCO}} I_P}{C_2} \approx 1.79 \cdot 10^{14}\,(\text{rad/s})^2\,. \tag{2.12}$$

Since 3 V is the voltage range available to tune the VCO, $K_{\text{VCO}}/2\pi$ must be at least $\Delta f_{\text{out}}/V_{DD} \approx 22\,\text{MHz/V}$. Let us take a value of 30 MHz/V.

After choosing the charge pump current $I_P = 0.8\,\text{mA}$, the capacitance C_2 is obtained from (2.12): $C_2 = 0.85\,\text{nF}$. The other two components of the filter are the other capacitance $C_1 = C_2(b-1) = 6\,\text{nF}$ and the resistance $R = \tau_Z/C_1 = 7.4\,\text{k}\Omega$.

As a final step, the validity of the linear continuous-time model has to be checked. Since $\omega_{(-3\text{dB})}/\omega_{\text{ref}} \approx 1/62.8$ the Gardner limit is satisfied. Moreover, no cycle slips are expected, since the maximum input frequency step (i.e., 64 MHz/2400 = 26.7 kHz) fulfils the condition

[1] These values of the division factor are typically too high for practical use, as will be shown.

Table 2.4 *Design of Example 2.3*

f_{ref}	1 MHz
N	2368–2432
$K_{\text{VCO}}/2\pi$	30 MHz/V
I_{P}	0.8 mA
C_2	0.85 nF
C_1	6 nF
R	7.4 kΩ

for the linear approximation to hold: $\Delta\omega_{\text{ref}}/\omega_{(-3\text{dB})} = 1.7 < 4\pi$. The PLL parameters are summarized in Table 2.4.

2.4 Spectral purity: spurs and phase noise

The output spectrum of a frequency synthesizer is always corrupted by noise with various frequency dependences and by tones, called spurs, placed at specific offsets from the synthesized signal. Noise and spurs have different origins, but both are small signals affecting the phase of the output signal. They can, therefore, be studied in the framework of the linear model for phase transfer.

2.4.1 Phase noise spectra transfer functions

In the evaluation of the output phase noise, we assume that we know the noise spectra of the main building blocks (Figure 2.21(a)). The spectra $S_{\phi_{\text{ref}}}$, $S_{\phi_{\text{div}}}$ and $S_{\phi\text{VCO}}$ are usually calculated as *phase* noise spectra, measured in rad^2/Hz, while $S_{I_{\text{P}}}$ and $S_{V_{\text{tune}}}$ are *current* and *voltage* noise spectra, measured in A^2/Hz and V^2/Hz respectively.

In most cases, all the spectra in the figure are white within the frequency range of interest, with the important exception of $S_{\phi\text{VCO}}$. The oscillator phase noise $S_{\phi\text{VCO}}$, in fact, features at least two components, shaped as $1/f^3$ and $1/f^2$, the former clearly becoming dominant as f approaches d.c.

Some of the PLL building blocks are non-linear (above all the oscillator). These features must be taken into account in the evaluation of the noise spectra. After that, however, these spectra can be transferred to the output using the linear continuous-time model.

It is convenient to refer some of the noise sources to the PLL input and others to the output, as in Figure 2.21(b). It turns out that:

$$\begin{aligned} S^{\text{OL}}_{\phi\text{in}}(f) &= S_{\phi_{\text{ref}}}(f) + S_{\phi_{\text{div}}}(f) + S_{I_{\text{P}}}(f)\left(\frac{2\pi}{I_{\text{P}}}\right)^2, \\ S^{\text{OL}}_{\phi\text{out}}(f) &= S_{\phi\text{VCO}}(f) + S_{V_{\text{tune}}}(f)\left(\frac{K_{\text{VCO}}}{2\pi f}\right)^2. \end{aligned} \tag{2.13}$$

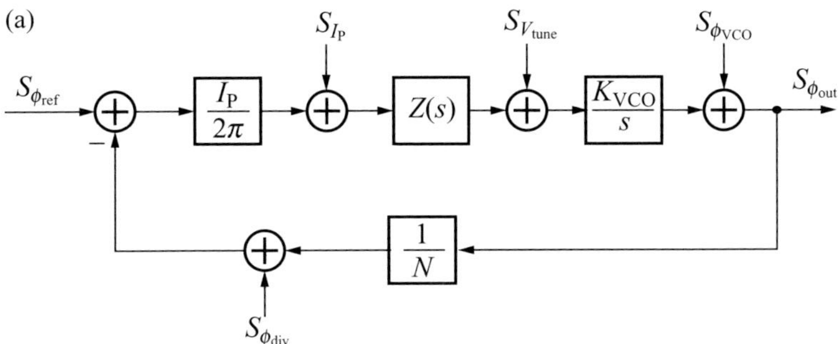

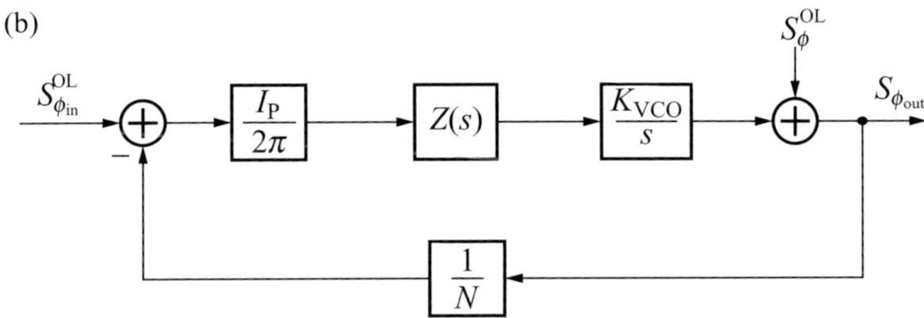

Figure 2.21 Linear model of a CP PLL: (a) with noise of building blocks, (b) with equivalent input-referred noise and output-referred noise

The superscript OL points out that these phase noise generators, measured in rad^2/Hz, are calculated when the loop is open. For the third-order type-II PLL, the overall output phase noise is given by:

$$S_{\phi_{out}} = S_{\phi_{in}}^{OL}(f) \cdot |\mathrm{LP}\,(\mathrm{j}2\pi f)|^2 + S_{\phi_{out}}^{OL}(f)\,|\mathrm{HP}\,(\mathrm{j}2\pi f)|^2\,, \tag{2.14}$$

where

$$\begin{aligned} \mathrm{LP}(s) &= \frac{-N \cdot G_{\mathrm{loop}}(s)}{1 - G_{\mathrm{loop}}(s)} = \frac{N\beta(s + 1/\tau_Z)}{s^3 + s^2 b/\tau_Z + s\beta + \beta/\tau_Z}, \\ \mathrm{HP}(s) &= \frac{1}{1 - G_{\mathrm{loop}}(s)} = \frac{s^2(s + b/\tau_Z)}{s^3 + s^2 b/\tau_Z + s\beta + \beta/\tau_Z}. \end{aligned} \tag{2.15}$$

The first equation has already been given in (2.10), and it is reported here for completeness. Note that the input-referred noise is low-pass shaped and its d.c. value is amplified by N^2. The output-referred noise, which is typically dominated by the VCO noise, is high-pass shaped and, beyond the loop bandwidth, it reaches the output unfiltered. It turns out that the low-frequency noise and slow drifts of the VCO frequency, falling within the loop bandwidth, are corrected by the feedback. Higher-frequency noise is, instead, too fast to be tracked by the loop.

In many practical cases, the PLL bandwidth is set not by requirements on the settling time but by the noise performance. Figure 2.22 shows typical input and output noise spectra. The input noise is white, while the output noise has two components (dashed lines) shaped as $1/\omega^3$ and $1/\omega^2$. The input noise in Figure 2.22(a) is amplified by N^2 within the bandwidth,

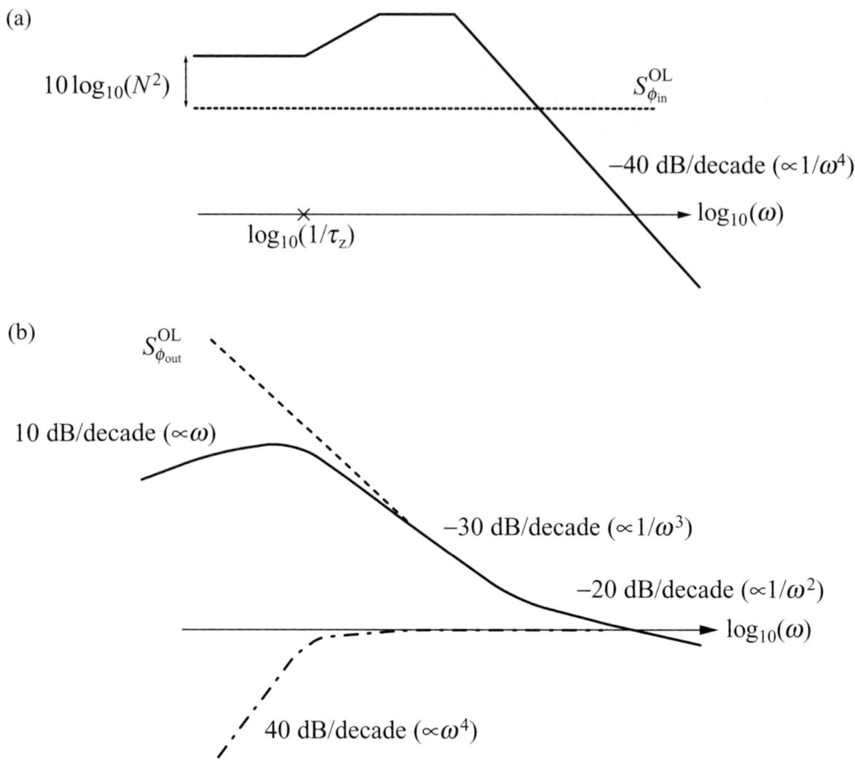

Figure 2.22 Sketch of noise transfer functions in a PLL: for (a) input-referred and (b) output-referred noises. Dashed lines represent the open loop noises, solid lines the corresponding output spectra. In (b) the magnitude of the HP transfer function is also represented with a dash-dotted line (frequency dependence of *power* spectra in parenthesis)

and then drops as $-40\,\text{dB/decade}$ ($1/\omega^4$ in power). If the frequency of the zero is too low with respect to the frequency of the poles, the in-band peaking can significantly increase the integral of the phase noise, which is an important measurement in some applications. In Figure 2.22(b), the output noise features a corner frequency (i.e., the crossing point between the spectra at $1/\omega^3$ and $1/\omega^2$) higher than the PLL bandwidth. This is a common situation, unfortunately, when using submicrometre CMOS technologies.

Let us now superimpose the two spectra and set the PLL bandwidth to the value where the VCO noise is about equal to the input noise multiplied by N^2. In this case, the spectrum should appear as in Figure 2.23: within the PLL bandwidth the dominant noise term is the input-referred noise amplified by N^2, while outside the PLL bandwidth, the output-referred noise (usually from the VCO) becomes dominant. Such a choice for the PLL bandwidth usually provides the lowest noise, in terms of both spot and integral noise.

The adoption of a wider loop bandwidth (Figure 2.24(a)) makes the noise from the input overcome the VCO noise out of the PLL bandwidth. The $1/\omega^4$ slope in the output spectrum is a clear signature of this condition. For a narrower bandwidth (Figure 2.24(b)), the output spectrum can feature a peak, due to the VCO noise. In the latter case, the output-referred noise, though filtered, is dominant within a portion of the PLL bandwidth.

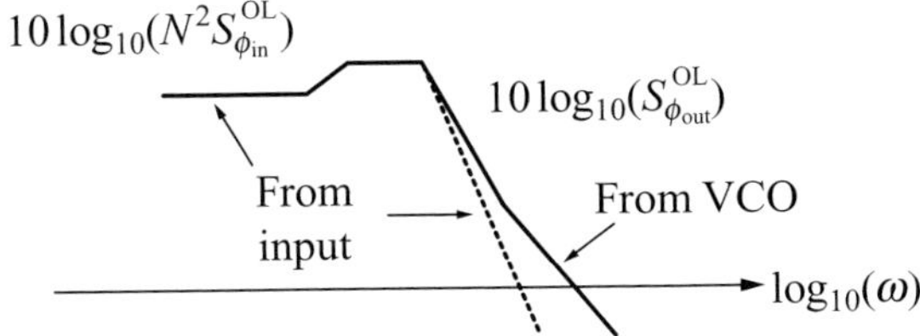

Figure 2.23 'Optimal' bandwidth for noise performance

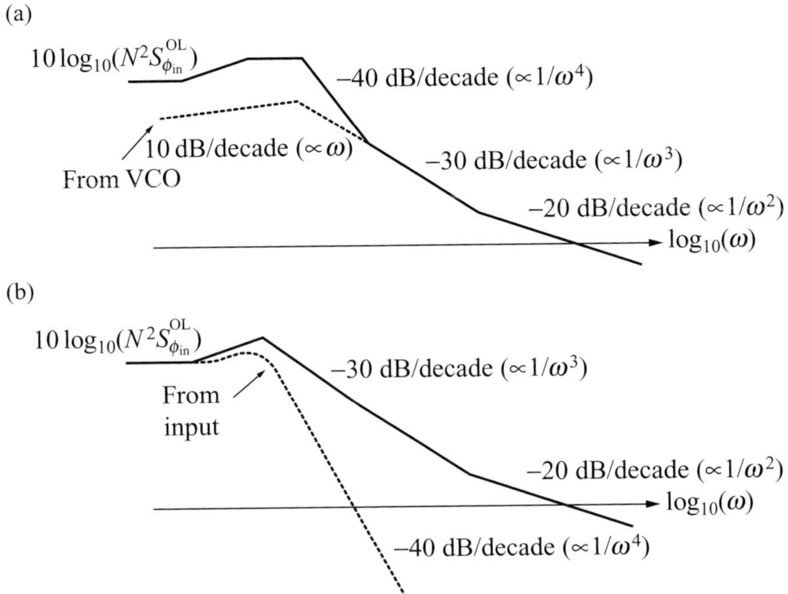

Figure 2.24 Sketch of output spectrum in two cases: (a) the bandwidth is 'wide' and the input amplified noise is dominant; (b) the bandwidth is 'narrow', and the output-referred noise is more detrimental (dashed lines represent the noise contribution of less importance)

It should, however, be pointed out that the optimization of the noise performance cannot be decoupled from the decision made on the loop bandwidth. The parameters setting the PLL bandwidth (e.g., K_{VCO}, I_P and so forth), in fact, also define the noise of the building blocks. The noise optimization must be performed, in practice, through an iterative procedure.

2.4.2 Reference spurs

In most cases, there are different mechanisms leading to spur generation. For instance, fractional-N synthesizers may be affected by the 'fractional' spurs, as discussed in Chapter 3. Let us consider here the origin of the reference spurs, i.e., the spurious tones appearing at frequency offsets equal to integer multiples of $\pm\omega_{ref}$ from the ω_{out} carrier.

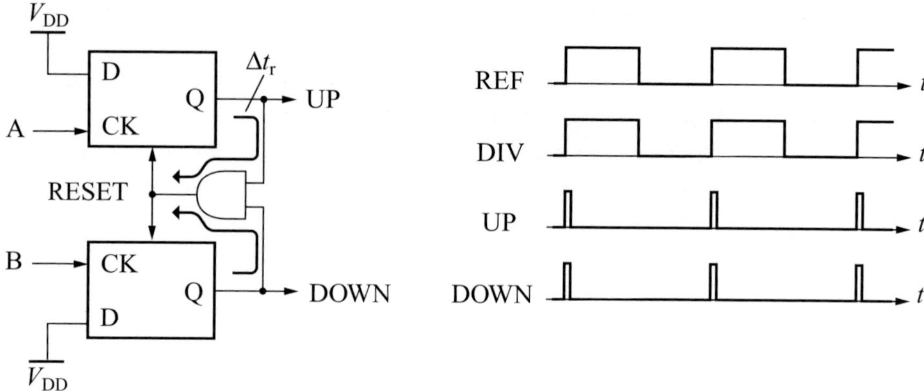

Figure 2.25 Effect of the non-zero reset time of PFD on UP and DOWN signal

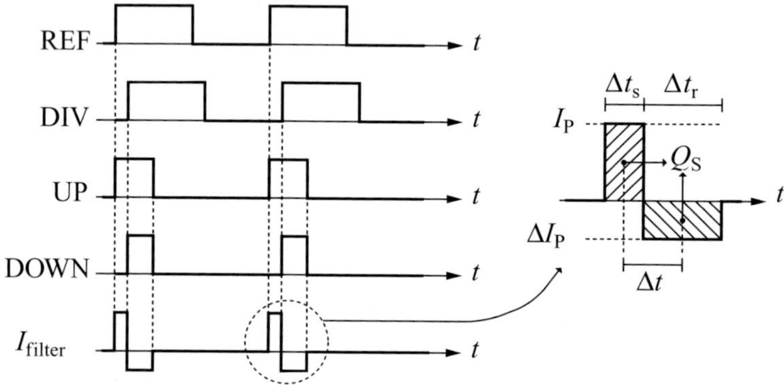

Figure 2.26 Steady-state condition in the PLL, in presence of spur charge injection

In a locked CP PLL, the edges of the two input signals in the PFD would ideally be synchronous. However, at the input rising edges, two short pulses, produced as a result of the finite propagation delay through the AND gate and the flip-flop reset (Figure 2.25), drive the UP and DOWN lines. The two CP switches in Figure 2.13(a) close at the same time, for a short interval Δt_r. If the two switches were perfectly matched as well as the current generators, that would have no impact on the following filter: the switches would be crossed by the same current and no current would be forced into the loop filter.

In the real world, mismatches cause a net (spurious) charge to be injected into the loop filter, instead. Mismatches may result from a ΔI_P difference between the two CP current generators or from the gate-to-channel capacitances of the two switches. The spurious injection occurs at a rate T_{ref}. However, since at the steady state, no net charge should be injected into the loop filter, the feedback loop reacts by shifting the edges of the REF and DIV signals (Figure 2.26). The Δt_s shift balances the charge injection created by mismatches. In Figure 2.26, the positive current pulse is a result of the phase shift, $2\pi \cdot \Delta t_s / T_{ref}$, between the signals. The negative pulse is, instead, created by mismatches.

For the charge injection to be balanced, the pulses must have the same area $Q_S = I_P \Delta t_s = \Delta I_P \Delta t_r$.

This effect ultimately causes a periodic disturbance at ω_{ref}. It modulates the tuning node, giving rise to spurious tones in the output spectrum at integer multiples of $\pm\omega_{ref}$.

Another contribution to reference spurs comes from the so-called 'varactor leakage'. It occurs, for instance, any time a reverse biased pn junction is adopted as voltage-controlled capacitor in the VCO. The leakage current of the diode discharges the filter capacitors, hence the tuning voltage between two phase comparisons changes. Even in this case, the feedback loop produces a short time shift Δt_s to restore the charge lost during a time T_{ref}. Once more, a periodic disturbance at ω_{ref} modulates the frequency of the output signal. The adoption of MOS varactors greatly reduces the leakage current, making negligible such a source of reference spurs. The different varactor typologies are discussed in Chapter 6.

Last but not least, reference spurs may be created by disturbances at ω_{ref} coupled to the tuning voltage node or to the power supply and ground lines of the VCO. A careful design, in which the tuning voltage line is well shielded and the VCO has low sensitivity to power supply, is the basic prerequisite for moderating the effect.

2.4.3 Reference spur magnitude: CP current mismatch

It is instructive to compute the approximate spur level created as result of the mechanisms discussed above and to highlight their dependence on the loop parameters. In Chapter 1, the spurious free dynamic range (SFDR) has been defined as the ratio between the power of a single-side tone at ω_{ref} and the carrier power. For frequency modulation with frequency deviation $K_{VCO} V_{tune}^{(\omega_{ref})}$ it can be written as:

$$\mathrm{SFDR} = \left(\frac{K_{VCO} \cdot V_{tune}^{(\omega_{ref})}}{2 \cdot \omega_{ref}} \right)^2, \tag{2.16}$$

where $V_{tune}^{(\omega_{ref})}$ denotes the first harmonic of $V_{tune}(t)$. The feedback loop has a negligible effect on the spur transfer, since the loop is practically open at ω_{ref}.

Let us consider first the spurs arising from the mismatch described in Figure 2.26. The aim of the following calculations will be to estimate the contribution to $V_{tune}^{(\omega_{ref})}$ and to use it in (2.16). For this purpose, note first that a train of rectangular current pulses with amplitude I_P and duty cycle $D = \Delta t_s / T_{ref}$ has a fundamental harmonic with amplitude $(2I_P/\pi)\sin(\pi D)$. In the limit $D \ll 1$, the pulses become δ-like and the harmonic amplitude is given by

$$(2I_P/\pi)\sin(\pi D) \approx 2I_P \cdot \Delta t_s / T_{ref} = 2Q_S / T_{ref}.$$

The expression is convenient, because it only depends on the pulse area.

The current waveform in Figure 2.26 is characterized by two pulse trains. They have the same area and they are shifted by $\Delta t = (\Delta t_s + \Delta t_r)/2$, i.e., the distance between their centroids. Assuming ideal δ-like pulses, the first harmonic of the current may be approximately given by:

$$I_P^{(\omega_{ref})} = (2Q_S/T_{ref}) \cdot [1 - e^{(-j2\pi\Delta t/T_{ref})}]$$
$$\approx (2Q_S/T_{ref}) \cdot (j2\pi\Delta t/T_{ref}).$$

The value of $V_{tune}^{(\omega_{ref})}$ in (2.16) is then obtained by multiplying $I_P^{(\omega_{ref})}$ for the filter impedance. The resulting SFDR is:

$$\text{SFDR} = \left(\frac{K_{VCO} \cdot |I(\omega_{ref})| \cdot |Z(j\omega_{ref})|}{2\omega_{ref}}\right)^2.$$

At ω_{ref}, the filter impedance $Z(j\omega)$ is practically $1/(j\omega_{ref}C_2)$ (see Figure 2.15(c)), thus reducing to:

$$\text{SFDR} \approx \left(\frac{K_{VCO}Q_S \cdot \Delta t}{2\pi C_2}\right)^2. \tag{2.17}$$

Let us now make some remarks:

- Note that C_2 considerably reduces the spur level. Without such a capacitance, the impedance $Z(j\omega_{ref})$ would be practically equal to R, that is, much higher in magnitude.
- Large K_{VCO} values are detrimental. This is not surprising, since a large tuning constant makes the oscillation frequency very sensitive to any disturbance. On the other hand, a large synthesizer tuning range demands large K_{VCO} values. Since K_{VCO} enters the expression for the bandwidth in (2.7), a lower K_{VCO} also implies a higher CP current or resistance R.
- The frequency ω_{ref} does not seem to play any role; this is not completely true. If ω_{ref} is reduced, it seems reasonable to scale down proportionally the frequencies of the loop singularities and to increase the values of the filter components, the C_2 value included. With this dependence in mind, it can be stated that the spur level becomes proportional to ω_{ref}.
- Any reduction of Q_S and Δt lowers the spur level. A well-designed CP can reduce the injected charge Q_S. Instead, the nominal value of the pump current I_P has little effect on Δt, if Q_S is only a function of the current mismatch. In this case, denoting ΔI_P as the current mismatch, it is $\Delta t_s = Q_S/I_P = (\Delta I_P/I_P)\Delta t_r$. In practice, since the relative current error $(\Delta I_P/I_P)$ is substantially less than 0.1, the delay is $\Delta t \approx \Delta t_r/2$ and the spur level is independent of I_P.
- Regarding Δt, it must be noted that the value of Δt_r is usually increased by design, to avoid a PFD problem usually known as 'dead zone', which is discussed in Chapter 9.

In general, because of the presence of K_{VCO} and C_2 in (2.17), a trade-off exists between bandwidth and spur level. To relax this trade-off partially, sometimes an additional filter, a simple first-order low-pass RC filter, is placed between the loop filter and the VCO tuning node. This additional pole further filters the tuning voltage harmonic at ω_{ref}, without reducing the bandwidth too much. The pole frequency of this additional filter is chosen to be much higher than the unity-gain frequency of G_{loop}, so that it does not affect the system stability, but substantially lower than ω_{ref}, to filter the reference spurs effectively. The resulting loop is of fourth order, and the settling time and the noise transfer functions should be calculated again.

2.4.4 Reference spur magnitude: leakage current

The spur generation mechanism due to the varactor leakage current is very similar to the effect of CP current mismatch. Let us refer again to Figure 2.26 and imagine, in place of ΔI_P, a small d.c. leakage current, i_l, flowing over the entire period T_{ref}. The condition for zero charge injection within the period T_{ref} implies that $Q_S = i_l \cdot T_{ref}$. Therefore, the first harmonic of the current pulse train is about $I_P^{(\omega_{ref})} = 2Q_S/T_{ref} = 2 \cdot i_l$ and the SFDR is given by:

$$\text{SFDR} = \left(\frac{K_{VCO} \cdot 2i_l \cdot |Z(j\omega_{ref})|}{2\omega_{ref}} \right)^2$$

$$\Rightarrow \text{SFDR} = \left(\frac{K_{VCO} \cdot i_l}{\omega_{ref}^2 \cdot C_2} \right)^2. \tag{2.18}$$

The main difference between (2.18) and (2.17) is the explicit dependence on the reference frequency. Remembering the practical link between C_2 and ω_{ref}, it can be concluded that the spur due to the varactor leakage has a magnitude proportional to $(1/\omega_{ref})$. The shorter the reference period, the lower the spur level.

2.4.5 Reference spur magnitude: supply disturbance

Sometimes the most serious disturbances are those coupled by supply or ground lines. These disturbances can even bypass the loop filter, directly reaching the tuning node. The frequency deviation is proportional to the K_{VCO}, but, since the loop filter does not play any role, the closed-loop bandwidth can be very narrow and still the output spectrum can show high spur levels.

Example 2.4 Phase noise estimation in a fourth-order PLL

Let us consider a fourth-order PLL designed by adding a further RC filter to a conventional third-order CP PLL. Data and specifications are listed here:

- Minimum output step = 10 MHz.
- Output tuning range $f_{out} = 5.18$ to 5.805 GHz.
- Division factor $N = 514$ to 585.
- Input phase noise (due to CP, divider, PFD) $S_{\phi,IN}^{OL}(f) \cong -150$ dBc/Hz.
- VCO phase noise $S_{\phi VCO}$ (1 MHz) $\cong -120$ dBc/Hz.
- Noise corner[2] = 200 kHz.
- Power supply = 2.5 V.

In an integer-N PLL, the frequency resolution, 10 MHz, sets the reference frequency. The required frequency range can be covered by choosing $K_{VCO}/2\pi$ equal to 300 MHz/V and tuning over the available 2.5 V voltage range.

The PLL bandwidth is set to the frequency at which the amplified input noise and the VCO noise cross each other. Taking $N = 585$, the input noise is amplified by $20 \cdot \log_{10}(N) \approx$ 55 dB and becomes about −95 dBc/Hz at the output. The VCO phase noise is −120 dBc/Hz

[2] At the noise corner, the slope of the phase noise spectrum changes from f^{-2} to f^{-3}.

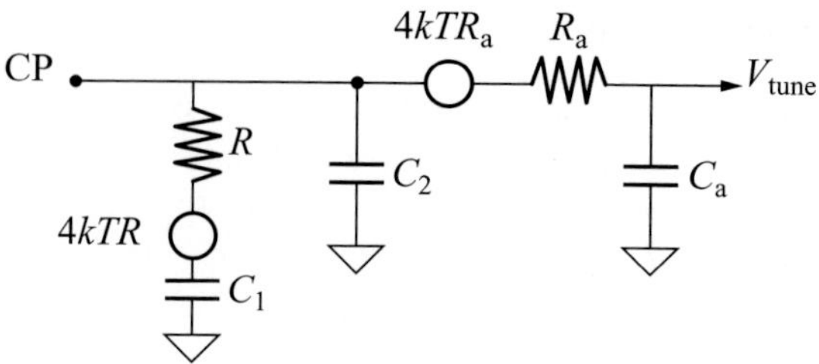

Figure 2.27 Loop filter of Example 2.4, with thermal noise shown

at 1 MHz, reaching up to −106 dBc/Hz at 200 kHz. Below this frequency, the slope changes to −30 dB/decade or −9 dB/octave; thus, the VCO phase noise at 100 kHz is −97 dBc/Hz, i.e., 2 dB lower than the input noise. The PLL bandwidth may, therefore, be chosen to be 100 kHz.

Based on (2.7), the unit-gain frequency may be estimated as $\omega_u/2\pi \approx (K_{VCO}/2\pi) \cdot (I_P R)/(2\pi N)$. Taking the 100 kHz value, it turns out that $I_P R \approx 1.2$ V.

This example is, somehow, oversimplified. For instance, the value of the tuning constant K_{VCO} is too high for a realistic implementation and a reduction of an order of magnitude would be needed. Techniques for reducing K_{VCO} while covering the required tuning range are shown in Chapter 6 (Section 6.5). Moreover, as has already been mentioned, the noise of both input stages and the VCO depends on loop parameters, such as K_{VCO} and the CP current. After these parameters have been set, the noise densities should be evaluated again.

Let us now determine the components of the filter in Figure 2.27. Initially, the presence of the supplementary filter $R_a C_a$ is neglected, since its pole is out of band. The zero and the third pole of the loop can be found by taking their ratio equal to 10, i.e., $b = \tau_Z/\tau_P = 10$, and their geometric mean equal to the unit-gain frequency ω_u, to maximize the phase margin. From these conditions it turns out that $\tau_Z \approx 5$ μs and $\tau_P \approx 0.5$ μs. With a CP current $I_P = 1$ mA, $R = 1.2\,k\Omega$, $C_1 \approx 4.16$ nF and, $C_2 \approx 463$ pF. These high capacitor values require off-chip components.

By placing the additional filtering pole at about 2 MHz, there are no interactions with the other singularities, while a 13 dB extra attenuation of the reference spurs is introduced. It may be taken that $R_a = 500\,\Omega$ and $C_a = 150$ pF. The phase margin is about 48°, and the settling time for an accuracy of 20 ppm obtained from numerical simulations is close to 33 μs.

The thermal noise of the filter resistances is a noise term, which can be accounted for after the design choices above have been made. These noise sources $4kTR$ and $4kTR_a$ [V²/Hz] can first be transferred to V_{tune}, then to the output according to (2.13) and (2.15):

$$S_{\phi out}(f) = S_{V_{tune}}(f) \cdot \left(\frac{K_{VCO}}{2\pi f}\right)^2 \cdot |\text{HP}(j2\pi f)|^2 .$$

At frequencies lower than the fourth pole, the capacitor C_a can be considered to be open. At low frequencies, those between d.c. and the third pole, the capacitive division $[C_1/(C_1 + C_2)]^2$ is practically equal to one and the voltage noise from R is directly fed to the tuning node. In this frequency range, the transfer function HP is given by (2.15). The output phase

Table 2.5 *Design of Example 2.4*

f_{ref}	10 MHz	R	1150 Ω
N	514–585	R_{a}	500 Ω
$K_{\mathrm{VCO}}/2\pi$	300 MHz/V	C_{a}	150 pF
I_{P}	1 mA	φ_m	48°
C_2	0.48 nF	$t_{\mathrm{s}}^{20\mathrm{ppm}}$	33 μs
C_1	4.3 nF	$\varphi_{\mathrm{r.m.s.}}$	1° r.m.s.

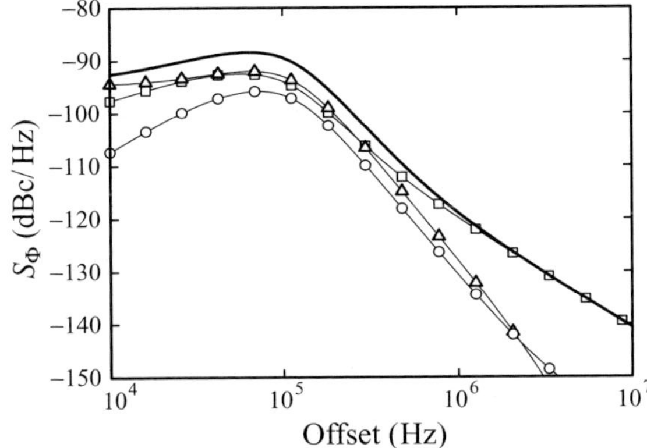

Figure 2.28 Output phase noise of Example 2.4 from the various noise sources: thermal noise from the loop filter (circles), input white noise from CP, PFD and divider (triangles) and VCO noise (squares). The total spectrum is represented by the solid thicker line

noise has a +40 dB/decade slope. Instead, at frequencies higher than the PLL bandwidth, the noise density at V_{tune} decreases as −20 dB/decade and HP is equal to one. The output noise has, therefore, a −20-dB/decade slope. A similar argument can be followed to study the noise from R_{a}.

Figure 2.28 shows the phase noise spectrum of the designed PLL, along with the contributions of its building blocks. The phase noise, integrated between 10 kHz and 1 MHz, is about 1° r.m.s. The selected parameters for the PLL are summarized in Table 2.5.

2.5 References

[1] H. De Bellescize, La réception synchrone, *Onde Electr.*, **11**, 1932, 225–40.

[2] F. M. Gardner, *Phaselock Techniques*, NewYork, NY: J. Wiley & Sons, Inc., 1966, 3rd edn, 2005.

[3] W. C. Lindsey, *Synchronization Systems in Communication and Control*, Englewood Cliffs, NJ: Prentice-Hall, Inc., 1972.

[4] V. Kroupa, *Frequency Synthesis Theory, Design and Applications*, New York, NY: J. Wiley & Sons, Inc., 1973.

[5] A. Blanchard, *Phase-Locked Loops: Application to Coherent Receiver Design*. New York, NY: J. Wiley & Sons, Inc., 1976.

[6] V. Manassewitsch, *Frequency Synthesizers: Theory and Design*, New York, NY: J. Wiley & Sons, Inc., 1976, 3rd edn, 2005.

[7] R. E. Best, *Phase-locked Loops: Design, Simulation, and Applications*, New York, NY: McGraw-Hill, 1984, 5th edn, 2003.

[8] W. F. Egan, *Frequency Synthesis by Phase Lock*, New York, NY: J. Wiley & Sons, Inc., 1990, 2nd edn, 2000.

[9] H. Meyr and G. Ascheid, *Synchronization in Digital Systems: Phase-, Frequency-Locked Loops and Amplitude Control*, New York, NY: J. Wiley & Sons, Inc., 1990.

[10] D. H. Wolaver, *Phase-Locked Loop Circuit Design*, Upper Saddle River, NJ: Prentice-Hall, Inc., 1991.

[11] J. A. Crawford, *Frequency Synthesizer Design Handbook*, Boston, MA: Artech House Inc., 1994.

[12] J. L. Stensby, *Phase-Locked Loops: Theory and Applications*, Cleveland, OH: CRC Press, 1997.

[13] U. L. Rohde, *Microwave and Wireless Synthesizers – Theory and Design*, New York, NY: J. Wiley & Sons, Inc., 1997.

[14] C. Vaucher, *Architectures for RF Frequency Synthesizers*, Dordrecht, Netherlands: Kluwer Academic, 2002.

[15] V. Kroupa, *Phase Lock Loops and Frequency Synthesis*, New York, NY: J. Wiley & Sons, Inc., 2003.

[16] W. C. Lindsey and M. K. Simon, eds, *Phase-Locked Loops and Their Applications*, New York, NY: IEEE Press, 1978.

[17] W. C. Lindsey and C. M. Chie, eds, *Phase-Locked Loops*, New York, NY: IEEE Press, 1986.

[18] B. Razavi, ed., *Design of Monolithic Phase-Locked Loops and Clock Recovery Circuits*, New York, NY: IEEE Press, 1996.

[19] B. Razavi, ed., *Phase-Locking in High Performance Systems*, New York, NY: IEEE Press and J. Wiley & Sons, Inc., 2003.

[20] W. C. Lindsey and C. M. Chie, eds, Special issue of *IEEE T. Commun.*, **30**, Oct. 1982.

[21] M. H. Perrot and G. Y. Wie, eds, Special issue of *IEEE T. Circuits Syst.-II*, **50**, Nov. 2003.

[22] C. A. Sharpe, A 3-state phase detector can improve your next PLL design, *EDN*, Sept. 1976, 55–9.

[23] F. M. Gardner, Charge-pump phase-lock loops, *IEEE T. Commun.*, **28**, Nov. 1980, 1849–58.

[24] B. Y. T. Kamath, R. G. Meyer and P. R. Gray, Relationship between frequency response and settling time of operational amplifiers, *IEEE J. Solid-St. Circ.*, **9**, Dec. 1974, 347–52.

[25] J. P. Hein, *z*-domain model for discrete-time PLL's, *IEEE T. Circuits Syst.*, **35**, Nov. 1988, 1393–1400.

[26] S. Levantino, M. Milani, C. Samori and A. L. Lacaita, Fast-switching analog PLL with finite-impulse response, *IEEE T. Circuits Syst. I*, **51**, Sept. 2004, 1697–1701.

3 Fractional-*N* PLLs

3.1 Beyond the integer-*N* approach

The output frequency of an integer-*N* PLL changes by integer multiples of the input reference frequency ω_{ref}. This simple fact makes it almost impossible to use the architecture in some radio standards.

A critical case is the ubiquitous GSM system. The channel spacing is about 200 kHz while the GSM spectrum is placed near 1 GHz. In an integer-*N* PLL architecture ω_{ref} cannot exceed the channel spacing (200 kHz), and the division factor *N* must be of the order of 1 GHz/200 kHz $\sim$ 5000. The adoption of such a large division factor *N* is detrimental, essentially for three reasons, which are restated here.

Input noise amplification: any input noise superimposed on the reference signal is amplified by N^2. Taking $N = 5000$, the factor N^2 corresponds to a 74 dB increment of the noise floor within the PLL bandwidth. This is why high *N* values are usually not compatible with the in-band noise requirements.

High power consumption: *N* divides G_{loop}, thus it indirectly affects other parameters of the loop and even the power consumption. As an example, taking (2.7) as an approximation of the PLL -3dB bandwidth, the pump current is

$$\omega_{\text{u}} \cong \frac{I_{\text{P}} R}{2\pi} \cdot \frac{K_{\text{VCO}}}{N} \Rightarrow I_{\text{P}} \cong \frac{2\pi \cdot N\omega_{\text{u}}}{K_{\text{VCO}} R},$$

which scales as *N*. Assuming $N = 5000$, $\omega_{\text{u}} = 2\pi \cdot 40\,\text{krad/s}$, $k_{\text{VCO}} = 2\pi \cdot 50\,\text{Mrad/(Vs)}$, $R = 10\,\text{k}\Omega$, the current is $I_{\text{P}} = 2.5\,\text{mA}$, which may exceed the power consumption budget. The current can be reduced, of course, by increasing both the filter resistance *R* and K_{VCO}. However, in this way, the voltage noise from *R* increases, as well as the VCO sensitivity to any disturbances or low-frequency noise.

Narrow loop bandwidth: because of the Gardner limit, the low ω_{ref} imposes a narrow loop bandwidth.

Various solutions have been proposed to circumvent these problems; this chapter is devoted to the discussion of fractional-*N* architectures, where the output frequency changes by a fraction of ω_{ref}. In this way, the reference frequency can be much larger than the channel spacing, thus removing some of the tight trade-offs pointed out.

3.2 Fractional-*N* division

3.2.1 Operating principle

The basic idea behind a fractional-*N* architecture is simple: when the PLL is locked, the division factor is modulated between two integer values, let us say N and $N + 1$, to obtain an 'average' ratio $(N + x)$, with $0 < x < 1$. In many circuits, more than two integer levels are used, but the principle is identical. The problem is to find a more suitable time sequence to change the division factor or, in more technical language, to choose the control pattern for the programmable divider. The following analysis will be performed with reference to a type-II PLL.

Figure 3.1(a) shows the most straightforward way to obtain a fractional division. Consider F cycles of the *input* reference (REF), with a period $T_{ref} = 2\pi/\omega_{ref}$. For the first L cycles the VCO frequency is divided by $N + 1$; for the remaining $(F - L)$ cycles the division factor becomes N; then the pattern is repeated. Figure 3.1(b) shows the signals. Note that during

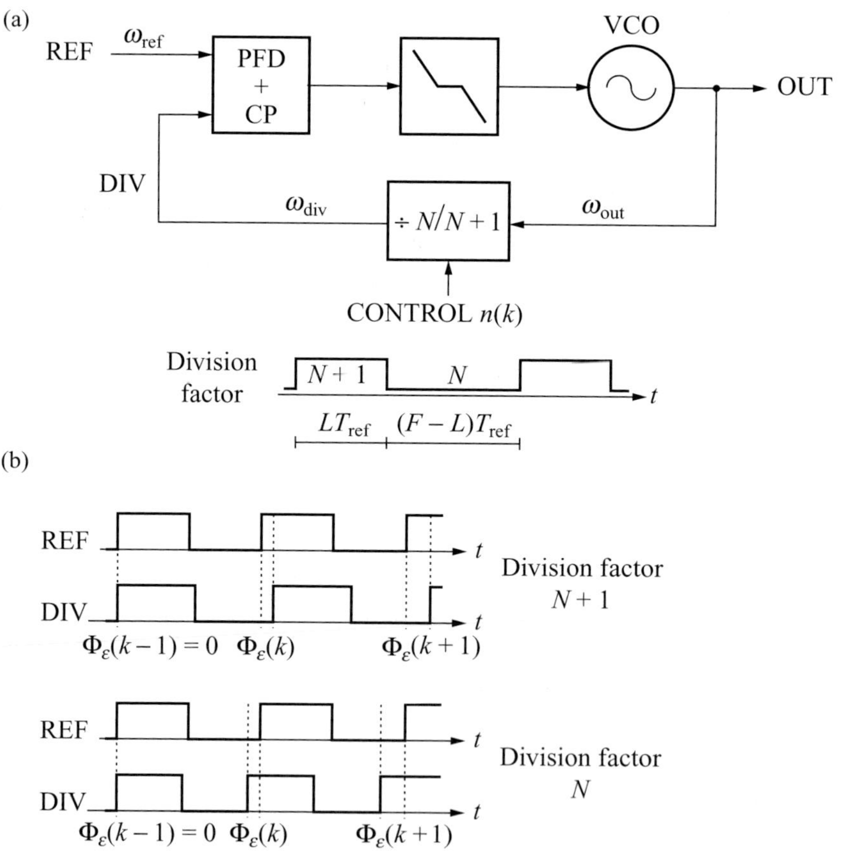

Figure 3.1 (a) The principle of fractional division; (b) sequence of the phase error for division factors equal to $N + 1$ and N

the first L cycles a negative phase error builds up, since the DIV signal has cycles longer than REF. On the contrary, during the remaining $(F - L)$ cycles, when the division factor is N, the DIV signal speeds up, thus recovering the phase lag.

A peculiar feature of the fractional PLL is that the edges of the DIV signal always wander around the edges of REF and, strictly speaking, the PLL never locks, i.e., the two signals never show perfectly synchronous edges. However, what matters for the PLL operation is that:

- The average phase error between REF and DIV remains zero.
- A perfect phase match is recovered after an interval FT_{ref}.

The first condition corresponds to saying that the divider output must make a precise integer number of cycles within the time interval FT_{ref}, even if the division factor changes. In a more quantitative way:

$$F \cdot T_{\text{ref}} = T_{\text{out}} \cdot (N + 1) \cdot L + T_{\text{out}} \cdot N \cdot (F - L),$$

where $T_{\text{out}} = 2\pi/\omega_{\text{out}}$ is the VCO period. The equation can be arranged as

$$\omega_{\text{out}} = \omega_{\text{ref}} \left(N + \frac{L}{F} \right), \tag{3.1}$$

which gives the average division factor, $\overline{N} = (N + L/F)$. By changing L, the frequency can be stepped by ω_{ref}/F, which is the PLL frequency resolution. Note that for (3.1) to hold, the division factor must be changed synchronously with the edges of the divider output, DIV, and not with the reference signal, otherwise N would change before the divider cycle is over. The clock for the logic controlling the division factor, not shown in Figure 3.1, should, therefore, be derived from the signal DIV.

The second issue is, instead, related to the wandering of the signal edges around their average lock condition. This effect makes the time dependence of the phase error ϕ_ε periodic with a period FT_{ref}. The drawback is a periodic modulation of the tuning voltage, which generates spurious tones at the PLL output: the so-called fractional spurs. The spectral distribution of these tones depends on the control pattern, but usually they lie at $\pm m\omega_{\text{ref}}/F$ from the carrier, where m is an integer value. The longer FT_{ref} is, the finer is the PLL resolution, ω_{ref}/F, but the closer the spurs are to the carrier.

From this standpoint it may be pointed out that: (i) as in the integer-N PLL, the spurs may occur at frequency offsets that are multiples of the PLL resolution (e.g., ω_{ref}/F in this case); (ii) while in the integer-N loop the spurs are generated by mismatches or non-idealities, in the fractional-N PLL they are mainly caused by the loop operation. The link between spurs and periodic modulation of the control pattern suggests that the effect could be alleviated by properly scrambling the pattern generation. These topics will be addressed in the following sections.

3.2.2 Control pattern generation and phase error

Let us first discuss the generation of the modulus control pattern for the divider. [1] Denoting the integer sequence $n(k)$ as the control sequence, the loop division factor at the generic kth instant is $N + n(k)$. In the following, $n(k)$ will be either 0 or 1, even if this approach

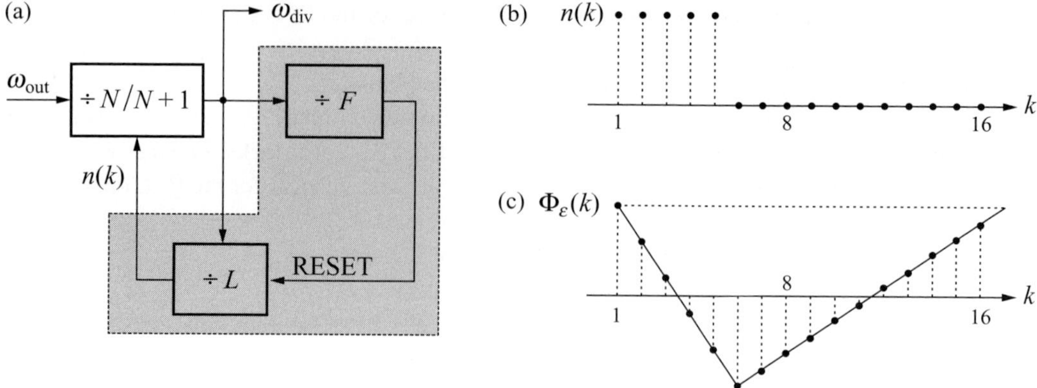

Figure 3.2 (a) Dual-counter generator of $n(k)$ (inside grey box); (b) sequence $n(k)$ for $F = 16$ and $L = 5$; (c) the corresponding phase error

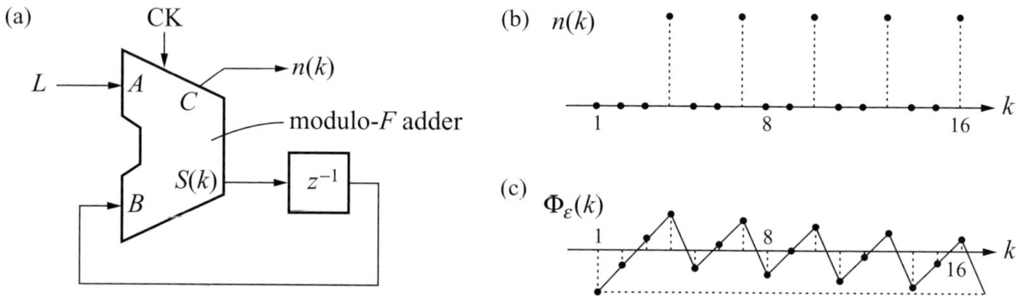

Figure 3.3 (a) Digital phase accumulator (DPA); (b) sequence $n(k)$ for $F = 16$ and $L = 5$; (c) corresponding phase error

may be easily generalized by considering a division factor switching between more than two levels.

The generation of a periodically changing division factor, such as the one shown in Figure 3.1, requires us to synthesize L 1s followed by $(F - L)$ 0s. This is trivially obtained by cascading two programmable counters, as in Figure 3.2(a), one programmed to F and the second programmed to L, both driven by the same clock. After the first L input periods, $n(k)$ changes its status from 1 to 0 and disables the L-counter until the F-counter reaches the final value of F and resets both stages. It can easily be verified that $n(k)$ is the desired sequence. Figure 3.2(b) represents the control sequence when $F = 16$ and $L = 5$; all 1s are grouped together. Figure 3.2(c) shows the corresponding dependence of the phase error. This arrangement is called a dual counter, even if it is just a different use of the pulse-swallow counter considered in Section 2.2.1.

Figure 3.3(a) shows an alternative solution, where a modulo-F accumulator with an input value L is used. [1–2] The accumulated value $S(k)$ is given by

$$S(k) = \begin{cases} S(k-1) + L & \text{if } S(k-1) + L < F \\ S(k-1) + L - F & \text{if } S(k-1) + L \geq F \end{cases} \tag{3.2}$$

and $n(k)$ is given by the overflow, or carry-out, of the accumulator:

$$n(k) = \begin{cases} 0 & \text{if } S(k-1) + L < F \\ 1 & \text{if } S(k-1) + L \geq F. \end{cases} \tag{3.3}$$

The control sequence and the consequent phase error are shown in Figure 3.3(b) and (c), again for $L = 5$ and $F = 16$.

While the dual counter 'collects' all the 1s together, the accumulator distributes them along the control period. This feature guarantees, in most cases, that the spurious tones are more widely spread over the frequency spectrum.

The accumulator exhibits another property, which can be appreciated once the phase error is expressed as a function of the control sequence $n(k)$.

In Section 2.2 the phase error ϕ_ε has been approximated by a continuous-time signal. More precisely, however, the phase error should be described by a discrete-time sequence $\phi_\varepsilon(k)$, giving the phase value at the kT_{ref} time instants. In the same framework, $T_{\text{div}}(k)$ may indicate the period of the signal DIV in Figure 3.1(b). It is $T_{\text{div}}(k) = 2\pi \cdot (N + n(k)) / \omega_{\text{out}}$, where $N + n(k)$ is the value of the division factor in the kth period of the reference. The increment of the phase error at the end of a reference period may be written as:

$$\frac{\phi_\varepsilon(k) - \phi_\varepsilon(k-1)}{2\pi} = \frac{T_{\text{ref}} - T_{\text{div}}(k)}{T_{\text{ref}}} = 1 - \frac{N + n(k)}{T_{\text{ref}} \cdot \omega_{\text{out}}} 2\pi.$$

Since the VCO frequency is

$$\omega_{\text{out}} = \overline{N} \cdot \omega_{\text{ref}} = 2\pi \cdot (N + L/F)/T_{\text{ref}},$$

it turns out that:

$$\phi_\varepsilon(k) - \phi_\varepsilon(k-1) = 2\pi \cdot \frac{L/F - n(k)}{N + L/F}.$$

By summing on both sides, the phase error sequence is finally given by:

$$\phi_\varepsilon(k) = \frac{2\pi}{\overline{N}} \cdot \sum_{j=1}^{k} \left[\frac{L}{F} - n(j) \right] + \phi_\varepsilon^0. \tag{3.4}$$

The ϕ_ε^0 constant should be chosen to set to zero the d.c. value of the sequence. Equation (3.4) quantitatively shows that the phase error in the kth period of the reference results from a sum over the sequence $n(k)$, which is not surprising at all.

By comparing (3.4) with (3.2) and (3.3), it turns out that for the accumulator controller, the value $S(k)$ stored in the accumulator is proportional to the phase error itself. In fact, at any input period the value L is added. When the accumulator overflows, n becomes 1 and F is subtracted from $S(k)$. In a quantitative way, it is:

$$S(k) = \sum_{j=1}^{k} [L - Fn(j)].$$

The phase error in (3.4) may be therefore written as:

$$\phi_\varepsilon(k) = \frac{2\pi}{\overline{N} \cdot F} \cdot S(k) + \phi_\varepsilon^0.$$

The circuit in Figure 3.3 is sometimes called a digital phase accumulator (DPA).

3.2.3 Fractional spurs

The fractional spur at m/FT_{ref} is caused by the mth harmonic of the current signal delivered by the charge pump. The charge pump current can be approximated as small rectangular pulses with amplitude I_{P} and duration $\Delta t(k) = T_{\mathrm{ref}} \cdot \phi_{\varepsilon}(k)/2\pi$. To be consistent with Figure 2.14, the single pulse is written as $I_{\mathrm{P}} \cdot \mathrm{rect}[t/\Delta t(k)]$. This assumes that whenever $\Delta t(k)$ is negative, the current pulse is also negative. The periodic pulse train is then given by a convolution in the time domain (see also [1]),

$$I_{\mathrm{P}}(t) = I_{\mathrm{P}} \left[\sum_{k=1}^{F} \mathrm{rect}\left(\frac{t - kT_{\mathrm{ref}}}{\Delta t(k)} \right) \right] * \left[\sum_{l=-\infty}^{+\infty} \delta(t - l \cdot FT_{\mathrm{ref}}) \right].$$

The sum in the first square brackets is a series of F pulses, each spaced by T_{ref} seconds. The period of the pattern, every FT_{ref} seconds, comes out from the convolution with the train of δ-functions. In the frequency domain the spectrum of the current signal is given by the product:

$$I_{\mathrm{P}} \left[\sum_{k=1}^{F} \left(\mathrm{e}^{-\mathrm{j}2\pi f \cdot kT_{\mathrm{ref}}} \cdot \int_{0}^{\Delta t(k)} \mathrm{e}^{-\mathrm{j}2\pi f \cdot t} \mathrm{d}t \right) \right] \cdot \left[\frac{1}{FT_{\mathrm{ref}}} \sum_{l=-\infty}^{+\infty} \delta\left(f - l \cdot \frac{1}{FT_{\mathrm{ref}}} \right) \right].$$

The fractional spur at m/FT_{ref} is produced by the mth harmonic of $I_{\mathrm{P}}(t)$. Once the integral is solved, the mth harmonic is found to be

$$I_{\mathrm{P},m} = \frac{I_{\mathrm{P}}}{2\pi} \cdot \frac{1}{m} \sum_{k=1}^{F} \frac{\mathrm{e}^{-\mathrm{j}2\pi k \frac{m}{F}}}{\mathrm{j}} \cdot \left(1 - \mathrm{e}^{-\mathrm{j}\frac{m}{F}\phi_{\varepsilon}(k)} \right), \tag{3.5}$$

where the link between $\Delta t(k)$ and the phase error, $\Delta t(k) = T_{\mathrm{ref}}\phi_{\varepsilon}(k)/2\pi$, has been used again. Since, in general, $\phi_{\varepsilon}(k)m/F \ll 1$, the expression can be simplified using $\mathrm{e}^{x} \cong 1 + x$:

$$I_{\mathrm{P},m} \cong \frac{I_{\mathrm{P}}}{2\pi} \cdot \frac{1}{F} \sum_{k=1}^{F} \phi_{\varepsilon}(k) \mathrm{e}^{-\mathrm{j}2\pi k \frac{m}{F}}. \tag{3.6}$$

As expected, it turns out that the spectrum of the charge-pump current at $m(1/FT_{\mathrm{ref}})$ is given by the spectrum of $\phi_{\varepsilon}(k)$ (a discrete Fourier transform) multiplied by the phase-to-current gain $I_{\mathrm{P}}/2\pi$. The output phase signal is eventually:

$$\phi_{\mathrm{out}}(\omega) = I_{\mathrm{P},m} \cdot \left(\frac{Z(\mathrm{j}\omega) \cdot K_{\mathrm{VCO}}/\mathrm{j}\omega}{1 + G_{\mathrm{loop}}(\mathrm{j}\omega)} \right)_{\omega = m\frac{\omega_{\mathrm{ref}}}{F}}. \tag{3.7}$$

The above results provide a reasonable estimate of the fractional spurs. Note that, within the PLL bandwidth, where G_{loop} is reasonably large, the expression inside the parenthesis of (3.7) reduces to $2\pi N/I_{\mathrm{P}}$, which is proportional to N. On the other hand, the phase error in (3.4) is inversely proportional to N and, through (3.6), the same dependence is transferred to $I_{\mathrm{P},m}$. It turns out that the fractional spur given by (3.7) is not dependent on N within the PLL bandwidth.

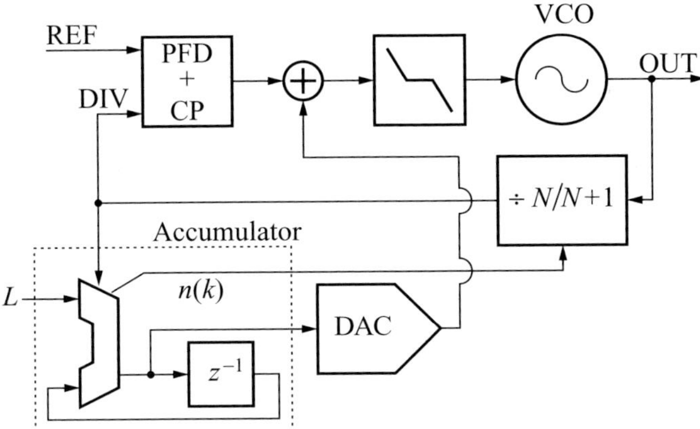

Figure 3.4 Digiphase technique for spur compensation. The division factor is changed synchronously with the divider output

3.2.4 Fractional-spur compensation

It has already been observed that, when the DPA is used, the value $S(k)$ is proportional to the phase error, and thus to the time interval $\Delta t(k)$:

$$\Delta t(k) = \frac{T_{\text{ref}}}{2\pi}\phi_{\varepsilon}(k) = T_{\text{ref}}\frac{S(k)}{NF+L} \cong T_{\text{ref}}\frac{S(k)}{NF}.$$

Therefore, $S(k)$ may be used to compensate the spurious tones originating from the phase error. For example, a partial compensation could be achieved by spilling out of the loop filter a current pulse with a duration of T_{ref}/N and amplitude proportional to $S(k)/F$. A value close to T_{ref}/N is available in the circuit; this is the VCO period. In practical cases, however, this time interval may be too short.

An alternative is to increase the pulse width and proportionally reduce the current. For this purpose, an additional frequency divider may derive a longer time period from the VCO signal. If P is the division factor the interval would be PT_{ref}/N. The amplitude of the current spilled out of the filter should then be proportional to $S(k)/PF$. Figure 3.4 shows a scheme for the implementation of this technique: a digital-to-analogue converter (DAC) generates the current pulse. This solution is sometimes called digiphase, even though this term was originally used to indicate any fractional division technique.

Even if the DAC were ideal, the compensation of fractional tones would not be perfect. The charge-pump current has constant amplitude and it is pulse-width modulated (PWM), while the compensation signal is modulated in amplitude (PAM). Using the same procedure leading to (3.5), the mth harmonic of the compensation current generated by the DAC can be derived as:

$$\begin{aligned} I_{\text{DAC},m} &= \frac{I_{\text{P}}}{\text{j}2\pi m}\cdot\left(1-\text{e}^{-\text{j}2\pi\frac{m}{FN}P}\right)\cdot\sum_{k=1}^{F}\text{e}^{-\text{j}2\pi k\frac{m}{F}}\cdot\frac{S(k)}{PF} \\ &\cong \frac{I_{\text{P}}}{\text{j}2\pi m}\cdot\left(1-\text{e}^{-\text{j}2\pi\frac{m}{FN}P}\right)\cdot\sum_{k=1}^{F}\text{e}^{-\text{j}2\pi k\frac{m}{F}}\cdot\frac{N\phi_{\varepsilon}(k)}{2\pi P}. \end{aligned}$$

The result, as expected, is different from (3.5) even if it reduces to (3.6) within the same approximations.

Digital-to-analogue converter non-linearities and truncation errors arising from the limited number of bits make the compensation even less accurate. Since a DAC with adequate performance is not easy to design and is power consuming, this technique has only been recently adopted in the design of silicon IC synthesizers. [3, 4]

This section ends with a key observation taken from [5]: one could note that the phase error $\phi_\varepsilon(k)$ is always available, without any need for the accumulator. After all, the phase error is simply proportional to the time difference between the REF and DIV edges. A 'perfect' spur compensation could, therefore, be achieved by using an additional charge pump that subtracts a current proportional to $\phi_\varepsilon(k)$. Note, however, that such a solution is equivalent to doing nothing. It would make the PLL insensitive to any phase error arising between the REF and the DIV signal. It merely corresponds to removing the whole PLL and using the VCO in an open loop. The point is that the 'perfect' compensation technique would make the PLL insensitive not only to the deterministic phase error evaluated in (3.4), but also to the random phase fluctuations arising from VCO phase noise, parameters' drift and so forth.

Example 3.1 Dual counter versus DPA

The difference between the dual counter and the DPA can be better appreciated with the help of a few examples. In all cases it will be taken that $F = 128$. This value is much larger in practice, but 128 is useful for illustrative purposes.

To summarize the general characteristic of the signal $n(k)$ and $\phi_\varepsilon(k)$: they are discrete-time sequences, sampled at $f_{\text{ref}} = 1/T_{\text{ref}}$, thus it is sufficient to know their periodic spectra between 0 and $f_{\text{ref}}/2$. Moreover, $n(k)$ and $\phi_\varepsilon(k)$ are also periodic; these signals repeat after FT_{ref} samples, so their spectra are composed by discrete tones at mf_{ref}/F. Finally, $\phi_\varepsilon(k)$ is obtained from $n(k)$ by performing a proper sum.

Let us first consider two very peculiar cases, $L = 1$ and $L = 64$, with $F = 128$.

Taking $L = 1$ corresponds to setting the fractional part of the output frequency to $f_{\text{ref}}/128$. The output sequence in both circuits is the same, that is, $n(k)$ has just one sample equal to 1 followed by 127 0s.

The spectrum of the signal $n(k)$ is, therefore, white. It shows tones separated by $f_{\text{ref}}/128$, all with the same amplitude. The sequence $\phi_\varepsilon(k)$ is a ramp with 127 samples. It is similar to Figure 3.2(c), with the negative slope lasting for only one single sample. The corresponding spectrum of the phase error decreases as -20 dB/decade, from d.c. to $f_{\text{ref}}/2$. According to (3.7), the PLL low-pass filters the spectrum of the signal $\phi_\varepsilon(k)$, thus the output spurs are concentrated at a low-frequency offset from the carrier. This case is very peculiar since there is no difference between the output of the dual counter and the output of the DPA.

The other peculiar case is $L = 64$, $F = 128$. Now the outputs of the two circuits are completely different.

The dual counter produces a square wave $n(k)$, with a 50% duty cycle: 64 1s are followed by 64 0s. The spectrum of $n(k)$ thus features tones at $mf_{\text{ref}}/128$ with $m = 1, 3, 5\ldots$, decreasing as -20 dB/decade. Approaching $f_{\text{ref}}/2$, the spectrum first flattens and then rises again, because of its periodic dependence. Since $\phi_\varepsilon(k)$ follows from a summation on $n(k)$, its

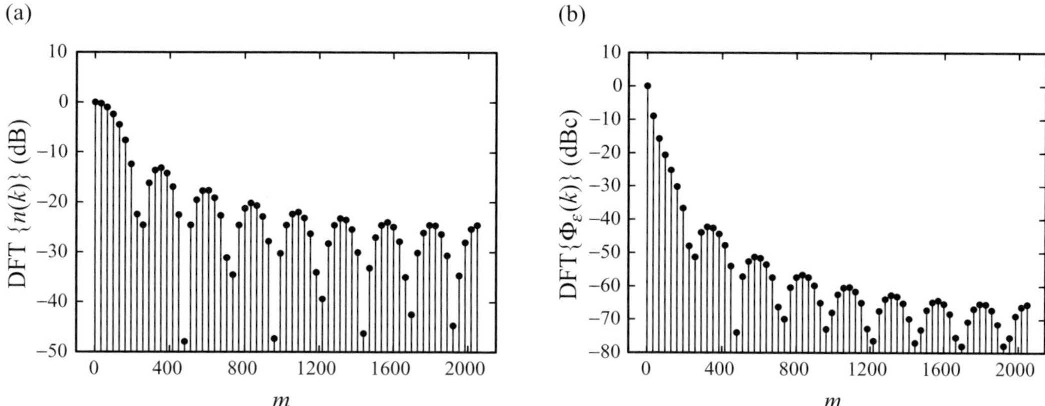

Figure 3.5 Dual-counter generator: (a) discrete spectrum of $n(k)$; (b) spectrum of $\phi_\varepsilon(k)$. $F = 128$, $L = 17$. Index $m = 2048$ corresponds to $f/2$

spectrum shows the same tones, but the low-frequency components are amplified and their amplitude decreases as −40 dB/decade. The spurs' energy is therefore mainly concentrated close to the carrier.

The DPA output is characterized, instead, by the sequence 0, 1, 0, 1 . . ., featuring a very short period, i.e., $2T_{\text{ref}}$. The spectrum of $n(k)$ shows, in addition to the d.c. component, only a single tone at $f_{\text{ref}}/2$. The same happens in the spectrum of the phase error. The output spurs can, therefore, be more efficiently filtered by the low-pass cut-off provided by the PLL.

In the more general case, the $n(k)$ sequence generated by the dual counter is a square wave with a duty cycle depending on L. The $n(k)$ spectrum, therefore, features a cardinal sine (sinc) shape. The spectral components generated by a DPA cannot instead be easily determined, forcing the designer to resort to numerical simulations. However, some further considerations may be helpful.

The period of the DPA output is given by $FT_{\text{ref}}/\text{GCD}(F, L)$, where $\text{GCD}(F, L)$ is the greatest common divisor of F and L. Since F is, in practice, always a power of two ($2^7 = 128$ in this example), any time L is odd, and the period of the DPA output lasts F samples, i.e., the largest possible period. This is the case for $L = 1$. For $L = 64$, the DPA output has, instead, a period of only two samples.

Let us now consider the case $F = 128$, $L = 17$. Figure 3.5 shows the spectra of $n(k)$ and $\phi_\varepsilon(k)$, respectively, as generated by the dual counter. The spectra have been computed by a simulation in the time domain followed by a fast Fourier transform. The FFT was performed over $2^{12} = 4096$ points and the spectra, normalized to the power of the d.c. component, are plotted against the order number of the FFT component. The interval from 0 to 2048 corresponds to the frequency range from 0 to $f_{\text{ref}}/2$. The harmonics are spaced by $f_{\text{ref}}/128$, since FT_{ref} is the period of the $n(k)$ sequence. As expected, the spectrum of $n(k)$ shows a sinc shape. The sequence $\phi_\varepsilon(k)$ is obtained from $n(k)$ by an integration, thus its spectrum shows higher harmonic content at low frequency.

When the DPA is adopted, the outputs are different, see Figure 3.6. The harmonics are still spaced by $f_{\text{ref}}/128$, but the $n(k)$ spectrum shows a high-pass shape, with some additional

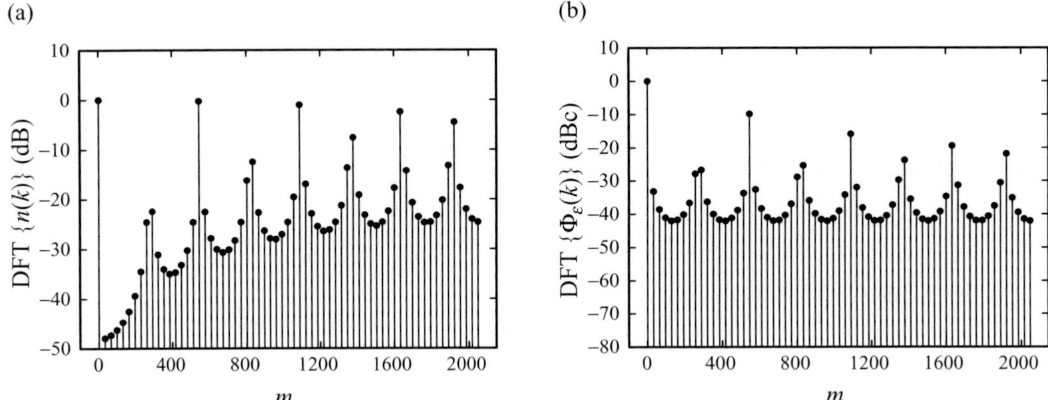

Figure 3.6 Digital phase accumulator: (a) discrete spectrum of $n(k)$; (b) spectrum of $\phi_\varepsilon(k)$, for $F = 128$, $L = 17$

tones superimposed. In fact, in a log–log scale, the harmonics show an amplitude rising at 20 dB/decade; a clear signature of a first-order high-pass filtering. This dependence will be clarified in the following, where it will be shown that the DPA is, in practice, a first-order delta-sigma ($\Delta\Sigma$) modulator with the constant value L as input. Apart from some very particular values of L, the DPA pushes the harmonics at high frequency, where they can be more effectively filtered by the PLL. Figure 3.6(b) shows the corresponding $\phi_\varepsilon(k)$ spectrum, which is approximately white.

It is also interesting to note that the spurious tone with the largest amplitude in Figure 3.6 corresponds to the frequency $f_{\text{ref}}\,(17/128)$. The presence of a larger tone at $f_{\text{ref}}\,(L/F)$ can be intuitively justified by noting that the DPA tries to distribute the L 1s uniformly in the periodic sequence lasting F. In some cases this tone is called a fractional spur, even if this term is not really correct, since all the spurious tones are a result of the fractional division.

3.3 $\Delta\Sigma$ control of division factor

3.3.1 Shaping the spur spectrum

The use of a delta-sigma ($\Delta\Sigma$) modulator as a controller for the division ratio was proposed by Miller and Conley in 1991 [6] and by Riley *et al.* in 1993, [7] albeit a similar arrangement can be found in a European patent dating back to 1984. [8]

The $\Delta\Sigma$ modulation technique is able to shift the quantization noise components to high frequencies. The adoption of the modulator as a division ratio controller is, therefore, expected to emphasize the high-pass shape of the spur spectrum that is already achievable, at least in some cases, by using the digital phase accumulator.

A discussion of this solution needs, however, to recall the fundamentals of the $\Delta\Sigma$ modulation technique. For the sake of completeness, some basics will be illustrated in the next two sections but the interested reader should refer, for a more in-depth discussion, to the literature available on the topic. For instance, reference [9] collects many of the earliest fundamental papers on the subject, while [10] is a more recent tutorial. The reader who is already familiar with the principles of $\Delta\Sigma$ modulation may skip these sections without any loss of continuity.

3.3.2 Quantization noise and oversampling

The speed and the number of bits are the two key figures of an ADC, the latter performance being also expressed in terms of the quantization noise level. In the design of a $\Delta\Sigma$ converter, these figures are traded one against the other.

It is well known that a base-band signal, with f_0 bandwidth, must be sampled at the Nyquist rate $f_S = 1/T_S = 2f_0$ [Hz], or higher. After sampling, each signal sample is translated into a word of B bits. The finite number of bits gives rise to truncation or rounding errors.

The impact of these quantization errors on signal degradation is made following some approximations; these are very coarse indeed.

A signal-to-noise ratio (SNR) may be defined as the ratio between the power of the largest harmonic signal at the ADC input and the quantization noise power. Since the input signal may range from 0 to V_{ref}, the maximum signal power is $V_{\text{ref}}^2/8$. The first approximation is to consider the error amplitude *uniformly* distributed between $-\Delta/2$ and $+\Delta/2$, with $\Delta = V_{\text{ref}}/2^B$. Under this assumption, it is simple to derive the mean square value of these errors, i.e., the *power* of this quantization noise, $\Delta^2/12$. The resulting SNR, when expressed in dB, is given by:

$$(\text{SNR})_{\text{dB}} \cong 6.02 \cdot B + 1.76 \quad [\text{dB}]. \tag{3.8}$$

Designers of data converters use (3.8) in the opposite way. Once the SNR is given, adding to the noise power also the power due to distortion, disturbances and whatever can degrade the signal quality, the number of significant bits of the converter is derived from (3.8).

For instance, if the thermal noise power integrated over the signal band is 75 dB below the maximum signal, the preceding expression says that the useful number of bits is 12. It is useless to further increase B; the quantization error would be smaller than the thermal noise or, equivalently, the extra bits would digitize only the noise and not the useful signal.

A second approximation, often made, regards the power spectral density of the quantization noise S_q. It is assumed to be constant between 0 and f_S. Since the noise power is the integral of S_q, it is $S_q = \Delta^2/(12f_S)$. The quantization process adds white noise over the signal spectrum (Figure 3.7).

In some cases, these two hypotheses are incorrect. Consider a constant signal. In this case, the quantization error is constant, not uniformly distributed in amplitude, and the quantization 'noise' power spectrum is concentrated at d.c. However, the approximations reasonably hold when the signal sweeps all the quantization intervals.

Figure 3.8 indicates a way to increase the number of the bits. If the sampling frequency increases beyond the Nyquist rate, the power spectrum $S_q = \Delta^2/(12f_S)$ decreases. The power of the noise falling *within the signal band* is $(\Delta^2/12)(2f_0/f_S)$, see Figure 3.8. The higher is the so-called oversampling ratio $(f_S/2f_0)$, the lower is the quantization noise

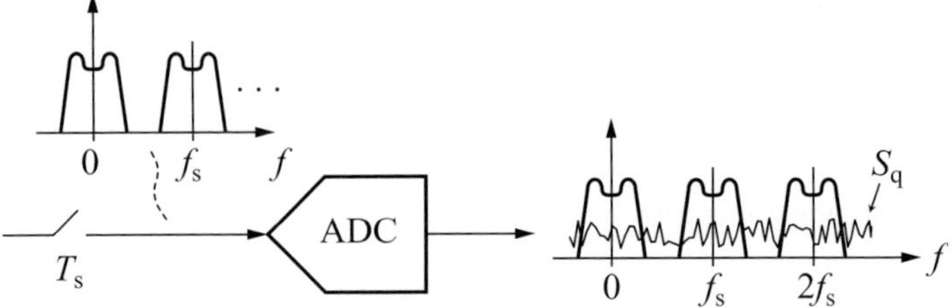

Figure 3.7 The A/D conversion adds a white noise to the signal, with power spectral density S_q

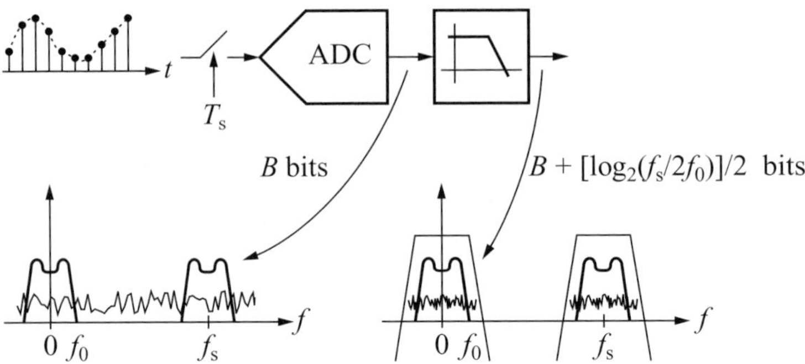

Figure 3.8 Bit gain through oversampling

power. Every time the sampling ratio is multiplied by four, the quantization noise power is reduced by 6 dB. According to (3.8) an additional bit has been gained in some way. In general, since the noise power is halved when the sampling ratio is doubled, the 'bit gain' is $0.5 \cdot \log_2(f_S/2f_0)$ bits.

As a matter of fact, this property is exploited to increase the number of bits of an ADC in some applications where very narrow-band signals are sampled at a rate much higher than the Nyquist one. In these cases, the signal from the ADC can be filtered with a narrow-band digital filter whose output would have more significant bits than the ADC (Figure 3.8). For instance, if the ADC has 12 bits and the signal is oversampled by 16 times, it can be filtered with a narrow-band digital filter, whose output will have 14 significant bits. Two more bits are gained. If the binary words at the filter output are longer than 14 bits, the extra bits do not provide useful information on the signal.

In the limit case even a single-bit ADC may be used. In practice, the ADC output will be a stream of 1s and 0s. When the bit stream is low-pass filtered, i.e., taking some sort of time average, the original signal is recovered up to a certain accuracy.

Note the similarity between the operating principle of such a single-bit converter, and the fractional divider described before. Also, in that case the fractional division factor is obtained with a sequence of N and $N + 1$, whose average is the desired value.

3.3.3 $\Delta\Sigma$ modulation

The next step towards $\Delta\Sigma$ modulation is to shape the quantization noise according to a high-pass function, thus reducing its power in the signal band, at the expense of a higher

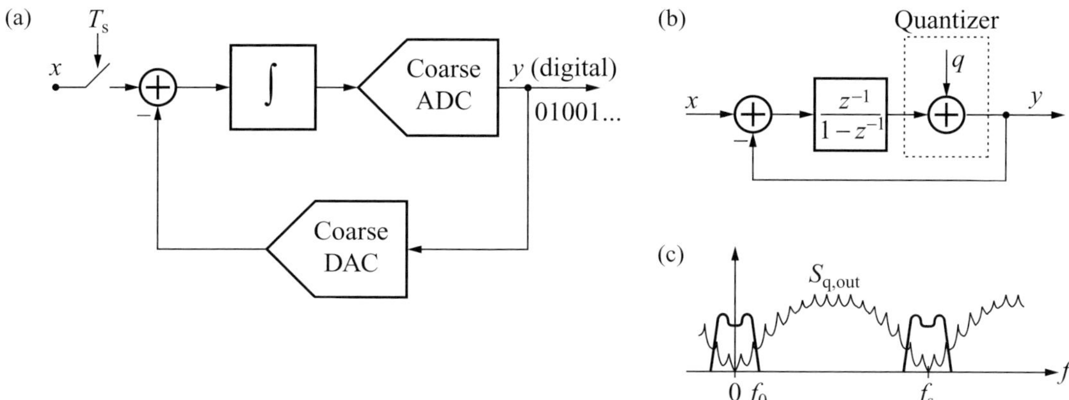

Figure 3.9 (a) First-order delta-sigma modulator; (b) linear model; (c) spectra of signal and noise

out-of-band noise component. The feedback system in Figure 3.9(a) is composed of a sampled and discrete-time integrator, usually implemented with a switched capacitor circuit, a coarse ADC and a DAC. Intuitively, the integrator in the feedback loop forces the average value of the input signal, x, and of the output sequence to be the same. The transfer function from the quantization noise to the output has, instead, the integrator in the feedback path. The noise will, therefore, be high-pass shaped.

In the linear discrete-time model in Figure 3.9(b), the ADC is simply represented as a node in which the quantization noise q is added. The transfer functions for the signal x and the quantization noise q are easily obtained:

$$y = x \cdot z^{-1} + q \cdot (1 - z^{-1}).$$

The signal is delayed but passes unfiltered. The quantization noise is differentiated. Under the assumption that the quantization noise can be described as a white-noise spectrum, $S_q = \Delta^2/(12 f_S)$, the output noise spectrum is derived by using $z = e^{j2\pi f T_S}$ in the preceding expression, and taking the squared magnitude of the result. It turns out that:

$$S_{q,out} = S_q \cdot 4 \sin^2(\pi f T_S).$$

Figure 3.9(c) shows the noise and signal spectra. If the oversampling ratio is high enough, the amount of noise in the signal band can be substantially reduced with respect to the case of simple oversampling.

This system is a first-order delta-sigma ($\Delta\Sigma$) modulator; the term first-order comes from the presence of one integrator in the loop. It is not yet a complete ADC since a digital low-pass filter is necessary to remove the quantization noise. The converter resolution is obtained by calculating the maximum SNR and then using (3.8) to define how many bits are significant at the filter output. Whenever the sampling frequency f_S is much higher than f_0, the squared sine function can be approximated as $(\pi f T_S)^2$ in the signal band.

In many applications the ADC has one bit, i.e., it is a comparator. This choice also simplifies the DAC design, which becomes a simple switch between two voltage values. Also, in this case, the assumption of a white spectrum for the quantization noise may turn out to be inadequate, even if, with respect to a simple ADC, the problem is partially alleviated

(a)

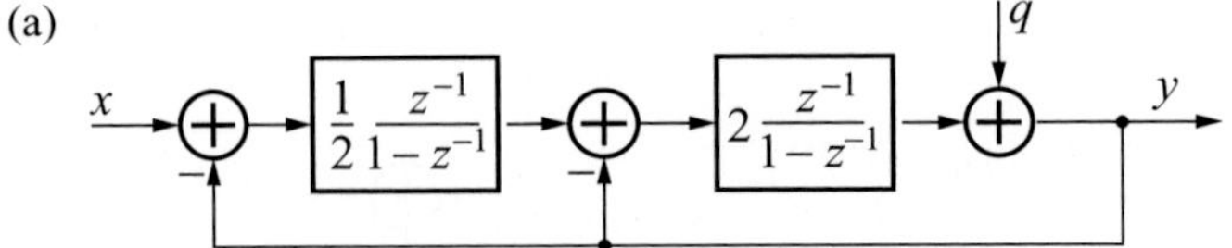

(b)

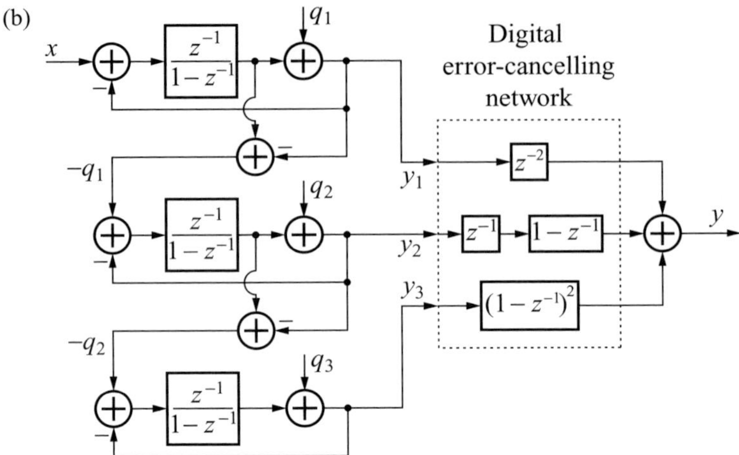

Figure 3.10 Linear models of (a) second-order delta-sigma modulator; (b) third-order 1–1–1 MASH modulator

by the feedback loop that somehow scrambles the signal at the ADC input. Still, for some classes of input signals, in particular d.c. signals, the output noise spectrum can show high tones, sometimes called *pattern noise* or *idle tones*. This effect is caused by the presence of limit cycles at the modulator output and by the correlation of the quantization noise, and it is not predicted by the simple linear model in Figure 3.9(b). To further reduce the amount of in-band noise, and also to alleviate the pattern noise, the order of the modulator can be increased by adding more integrators in the loop. A linear model of a second-order modulator is shown in Figure 3.10(a). The input–output transfer functions for the signal and for the quantization noise are:

$$y = x \cdot z^{-2} + q \cdot (1 - z^{-1})^2.$$

The key point is that the quantization noise is now shaped by a second-order high-pass function. The transfer function for q is now a squared sine that, in term of power spectrum, becomes:

$$S_{\mathrm{q,out}} = S_{\mathrm{q}} \cdot 16 \sin^4 (\pi f T_{\mathrm{S}}),$$

where $S_{\mathrm{q}} = \Delta^2/(12 f_{\mathrm{S}})$. With respect to a first-order $\Delta\Sigma$ modulator, less noise is left in the signal band. Moreover, the presence of two loops reduces the noise correlations, making the assumption of white quantization noise more adequate.

In many cases, the noise correlation is additionally lowered by adding a *dither* to the input signal, that is, a small-amplitude tone of frequency much higher than the signal band.

It does not affect the SNR, since it is out of band, but has the effect of reducing the idle tones.

The order of the modulator can be higher than two, in this case the loop stability becomes a serious issue. Many solutions have been presented in literature, one of the most effective being the adoption of a cascade of first-order modulators. This architecture is also called a multi-stage, or MASH, $\Delta\Sigma$ modulator, a model of which is shown in Figure 3.10(b). The figure shows that the input of the second stage is the quantization noise of the first stage, q_1, with a negative sign, and that the same happens for the third modulator. The error cancellation network combines its three digital inputs as follows:

$$y = y_1 \cdot z^{-2} + y_2 \cdot z^{-1} \cdot (1 - z^{-1}) + y_3 \cdot (1 - z^{-1})^2.$$

Using the input–output transfer function for the first-order modulator obtained above, the output is readily derived:

$$y = x \cdot z^{-3} + q_1 \cdot (1 - z^{-1})^3.$$

The cascade is thus equivalent to a third-order modulator. The noise q_1 is now shaped as a $\sin^3$ wave, and the output power spectrum is $S_q = 64 \cdot \sin^6(\pi f T_S)$. Since this circuit is realized with three first-order modulators, it is also called MASH 1-1-1. Usually, the three first-order modulators have a single-bit quantizer; in this case the output y is a three-bit word. The MASH is always stable, being composed of first-order loops, and is also a modular circuit, since the order can be increased by simply adding another loop.

The $\Delta\Sigma$ concept can also be used to implement a DAC. In this case, the input signal x is a digital word of, let us say, B bits. The analogue building blocks are replaced by digital stages. An accumulator is inserted in place of the switched-capacitors integrator, while the comparator is replaced by a circuit reading only the most significant bit of the input word. The equivalent of an analogue quantizer is a digital circuit that takes more than one bit, starting from the most significant one.

An additional difference between a digital-to-analogue (D/A) and an analogue-to-digital (A/D) $\Delta\Sigma$ modulator is that in a practical application the rate of the B-bit input words is usually close to the Nyquist rate, twice the bandwidth of the digitized signal. For this reason an interpolator is placed before the D/A modulator. The circuit increases the rate of the input word without changing the number of bits. For instance, if the input rate is f_1, the interpolator can first place, between two words of B bits, $N-1$ 0s (*zero padding*). The sequence is then passed through a low-pass digital filter, clocked at Nf_1. The output rate is therefore increased by a factor N.

The output of the D/A single-bit modulator is a stream of 1s and 0s at a rate much higher that the Nyquist one. This stream can be filtered by an analogue low-pass reconstruction filter.

3.3.4 Frequency divider control by a $\Delta\Sigma$ modulator

The adoption of a $\Delta\Sigma$ modulator in a fractional PLL is supposed to more effectively scramble the division factor sequence, shaping the spur spectrum, originated by the quantization noise, according to a high-pass function.

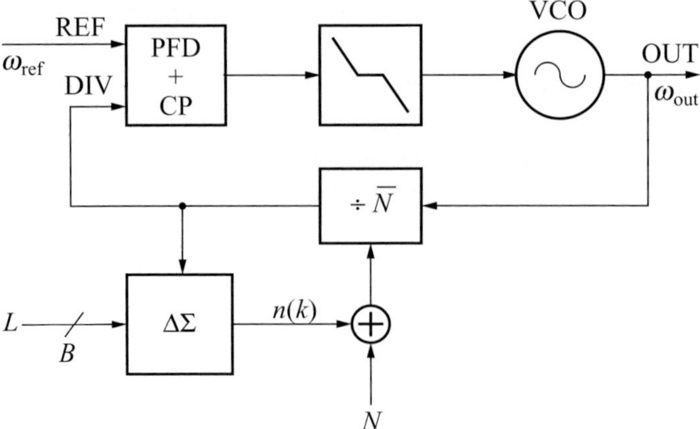

Figure 3.11 A fractional-*N* PLL, in which a $\Delta\Sigma$ modulator controls the divider

Figure 3.11 shows a fractional-*N* PLL controlled by a $\Delta\Sigma$ modulator. Note that now the $\Delta\Sigma$ modulator is digital. Its input signal is a word of B bits giving the fractional part of the division factor. This makes the design relatively simple, the modulator being an ensemble of accumulators and delays. The divider controls the frequency, hence the quantization noise is a frequency noise. The resultant phase noise affecting the PLL output is thus obtained by integrating the frequency fluctuations. Such an integration provides part of the quantization noise filtering; the rest is carried out by the loop filter.

The clock of the modulator is the output of the divider, which is almost equal to the reference frequency. The input of the modulator is a d.c. digital signal, a constant word of B bits setting the fractional part of the division factor. If the input value is L, the PLL output frequency, see (3.1), is given by:

$$\omega_{out} = \omega_{ref}\left(N + \frac{L}{2^B}\right).$$

With respect to (3.1) we have 2^B instead of F, and the minimum frequency step at the PLL output is $\omega_{ref}/2^B$. The number B, as we will see, is determined by the length of the accumulators, so the resolution can easily be very high. For $B = 16$ and a 10 MHz reference input, the minimum step is below 200 Hz. This approach also makes it possible to adjust the frequency to the accuracy required by such standards as the GSM.

In many instances, the output of the $\Delta\Sigma$ modulator is not a single-bit word. For instance, a third-order 1–1–1 MASH is often used, therefore the modulator output has $2^3 = 8$ levels and the division factor can be switched between all the integers in the range $N - 3$ to $N + 4$. The m bits from the modulator are added to the word controlling the programmable divider.

The impact of the $\Delta\Sigma$ modulator on the PLL output spectrum can be evaluated in two steps: first, the power spectrum of its quantization noise should be computed and expressed as a *phase noise*. Then, the noise should be transferred to the PLL output using the linear model.

Equation (3.4) links the time-domain sequences $n(k)$ and $\phi_\varepsilon(k)$. Since, in this specific case, the sequence $n(k)$ is generated by a $\Delta\Sigma$ modulator, its power spectrum, denoted as $S_n(f)$, is high-pass shaped:

$$S_n(f) = \left(\frac{\Delta^2}{12 \cdot f_{ref}}\right) \cdot [2\sin(\pi f T_{ref})]^{2l},$$

where l is the order of the modulator. The large bracket represents the white quantization noise over the bandwidth defined by the modulator sampling frequency, $f_{ref} = 1/T_{ref}$. Since $n(k)$ may change by 1 at each step, the quantization step is $\Delta = 1$. The transfer function between the sequences $m(k)$ and $\phi_\epsilon(k)$ is obtained by transforming (3.4) in the z domain. It turns out that:

$$\phi_\varepsilon(z) = -\frac{2\pi}{N + L/F} \cdot \frac{n(z)}{1 - z^{-1}} = -\frac{2\pi}{N} \cdot \frac{n(z)}{1 - z^{-1}}.$$

The phase error is derived by integrating the sequence $n(k)$. By introducing $z = e^{j2\pi f T_{ref}}$, the power spectrum of the phase error is found to be: [11]

$$S_{\phi_\epsilon}(f) = \left(\frac{2\pi}{N}\right)^2 \cdot \frac{1}{12} \cdot \frac{1}{f_{ref}} \cdot [2\sin(\pi f T_{ref})]^{2(l-1)}, \tag{3.9}$$

where the integer N has been used in place of the average division factor; the difference is negligible. The key point is that the phase spectrum is still high-pass shaped, but because of the integration linking the sequence $n(k)$ and the phase error, the order of the high-pass function is reduced by one (a factor of two in terms of power spectrum). Once such an input-referred spectrum has been derived, the PLL output spectrum is eventually obtained by multiplying (3.9) by the squared magnitude of the PLL transfer function.

3.4 $\Delta\Sigma$ fractional-*N* PLL

3.4.1 Output phase spectrum versus modulator order

Let us now briefly discuss the ideal shape of the output phase spectrum as a function of the modulator order. As an example, let us start with a third-order type-II PLL, with $\omega_{ref} = 2\pi \cdot 40$ Mrad/s, a closed-loop bandwidth $\omega_{(-3dB)} = 2\pi \cdot 500$ krad/s equivalent to $\omega_{ref}/80$, and a division factor $N = 125$. Figure 3.12 shows the PLL output spectra for various modulator orders. In each case the dashed line represents the quantization-induced *input* phase noise and the solid line represents the corresponding PLL *output* phase noise. The input phase noise has been obtained from (3.9) while the transfer function is given by (3.7). Since the latter reduces to N^2 for frequencies within the PLL bandwidth, the in-band noise is simply the quantization-induced phase noise given by (3.9) without the factor N^2 at the denominator.

Figure 3.12(a) shows the results for a first-order modulator, $l = 1$ in (3.9). The input noise is white. It is cut off at -40 dB/decade after the PLL -3 dB bandwidth. The resulting noise level is usually too high to be acceptable. Another drawback of a first-order modulator is the presence of idle tones [9]. If a constant input is fed into a first-order modulator, the spectrum

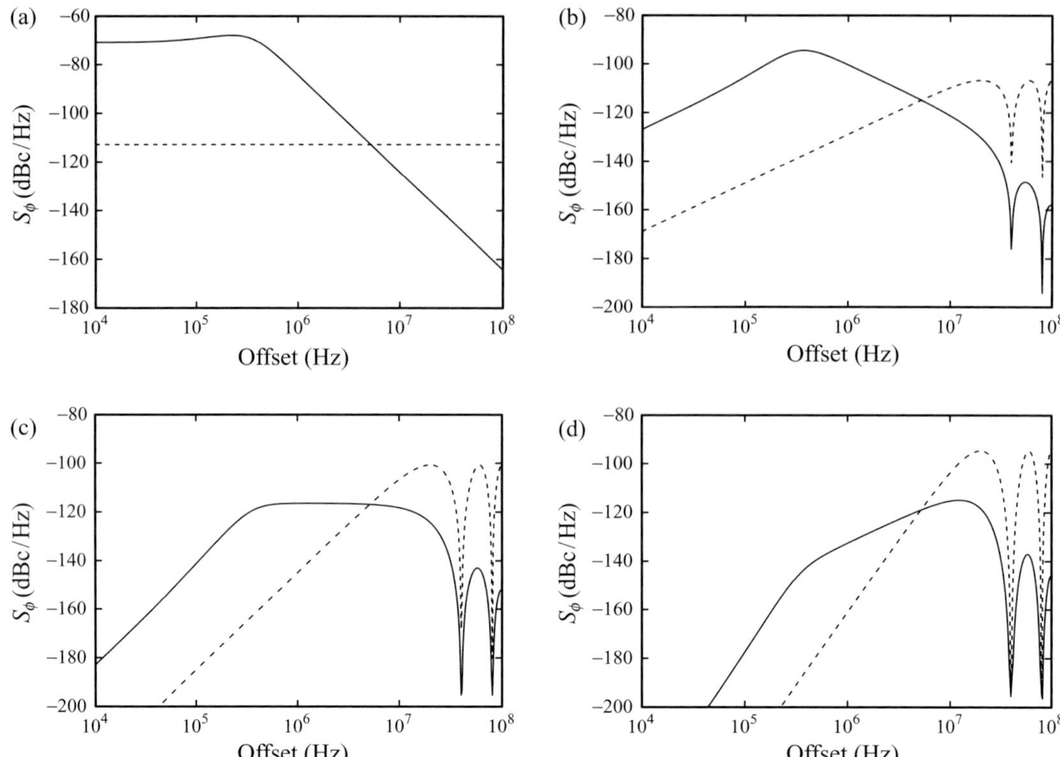

Figure 3.12 Quantization-induced phase noise, input-referred (dashed lines) and output-referred (solid lines) according to linear model, in a third-order type-II PLL, for various orders of the $\Delta\Sigma$ modulator: (a) first, (b) second, (c) third, (d) fourth order

of the quantizer can be very different from the one obtained with the linear approximation and, in general, high spurious tones appear, that were not predicted at all by the linear model. As a matter of fact, it will be shown that the DPA is equivalent to a first-order $\Delta\Sigma$ modulator. With this in mind, note that Figure 3.6(b), referred to in Example 3.1, shows an input-referred phase noise that is almost white but has some high tones. Moreover, for the cases $L = 1$ and $L = 64$ the noise is not white at all. The results can be considered illustrative of the deviation of the performance achievable using a first-order $\Delta\Sigma$ modulator from what is expected based on the linear model.

The linear model usually works better with higher-order modulators. Figure 3.12(b) shows the spectra for a second-order modulator, $l = 2$. The in-band noise rises at 20 dB/decade within the PLL band, and then decreases at −20 dB/decade. The notches lie at integer multiples of reference frequency, $n\omega_{\text{ref}}$ with $n = 1, 2, \ldots$ With a third-order $\Delta\Sigma$ modulator (Figure 3.12(c)) the noise spectrum first rises as f^4 (+ 40 dB/decade), then it flattens after $\omega_{(-3\text{dB})}$, owing to the second-order cut-off provided by the PLL. After $\omega_{\text{ref}}/2$ the noise falls and, following the first notch, the amplitude of the successive noise peaks decreases as −40 dB/decade. A further noise reduction is obtained using a fourth-order modulator (Figure 3.12(d)).

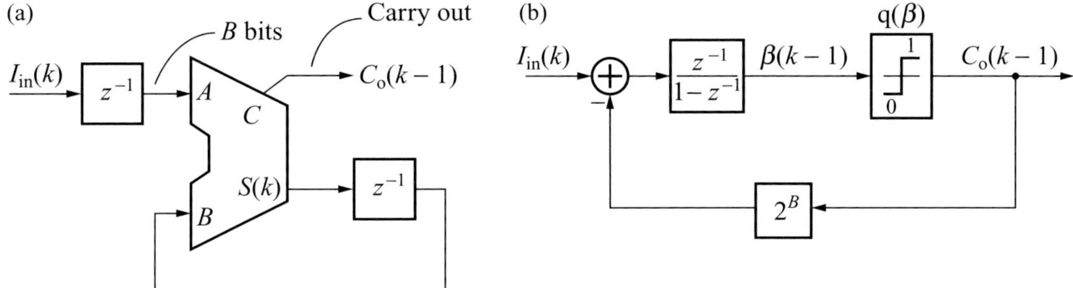

Figure 3.13 Equivalence between (a) the digital phase accumulator and (b) a first-order $\Delta\Sigma$ modulator

In the last three cases, the input-referred and output-referred noise decreases after $\omega_{ref}/2$, where the factor $\sin(\pi f T_{ref})$ reaches its peak value (Figure 3.9(c)).

3.4.2 A $\Delta\Sigma$ modulator for fractional-*N* PLL

Let us now demonstrate that the DPA in Figure 3.13(a) is equivalent to a digital first-order $\Delta\Sigma$ modulator. [6, 12] Referring to Figure 3.13(a), let us write the link between the input $I_{in}(k)$ and accumulated value $S(k)$, at the kth instant:

$$S(k) = I_{in}(k-1) + S(k-1) - 2^B \cdot \underbrace{\text{CARRY}\left[I_{in}(k-1) + S(k-1)\right]}_{C_o(k-1)}, \tag{3.10}$$

where B is the number of the bits in the accumulator and C_o is the carry out of the accumulator. The function CARRY is 1 when the accumulator overflows, otherwise it is 0. It can be written with the help of the function 'sign' as follows:

$$\text{CARRY}(x) = \frac{1}{2} \cdot [1 + \text{sign}(x - 2^B)].$$

It will be useful to introduce the bit quantizer:

$$q(x) = \frac{1}{2} \cdot [1 + \text{sign}(x - 1/2)],$$

whose output is 0 or 1, depending on whether the input is below or above a threshold placed at $\frac{1}{2}$.

By adding $I_{in}(k) - 2^B$ to both sides of the equation (3.10), it is:

$$S(k) + I_{in}(k) - 2^B = \left[I_{in}(k-1) + S(k-1) - 2^B\right] + \left[I_{in}(k) - 2^{B-1}\right]$$
$$-2^{B-1}\text{sign}\left[I_{in}(k-1) + S(k-1) - 2^B\right].$$

Finally, denoting $\beta(k) = I_{in}(k) + S(k) - 2^B + 1/2$, it turns out that:

$$\beta(k) = \beta(k-1) + I_{in}(k) - 2^B \cdot q\left[\beta(k-1)\right],$$

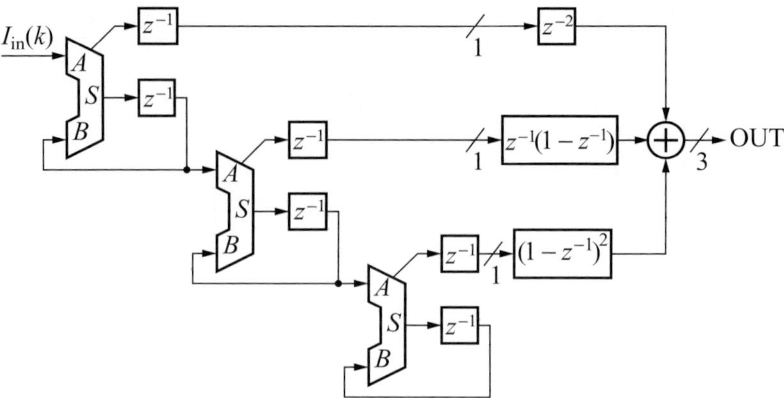

Figure 3.14 An all-digital 1-1-1 MASH modulator

which is the equation describing the scheme in Figure 3.13(b), a first-order modulator. The output of the quantizer is exactly the carry output of the accumulator:

$$\begin{aligned} \mathrm{q}[\beta(k-1)] &= \frac{1}{2}\cdot[1+\mathrm{sign}\,(\beta(k-1){-}1/2)] \\ &= \frac{1}{2}\cdot[1+\mathrm{sign}(I_{in}(\mathrm{k}-1){+}S(k-1){-}2^B)] \\ &= \mathrm{CARRY}\,[I_{in}(\mathrm{k}-1){+}S(k-1)] = C_o\,(k-1)\,. \end{aligned}$$

It is therefore expected that the spectrum of the carry output sequence generated by the DPA in Figure 3.13(a) should be high-pass shaped like the first-order modulator output in Figure 3.13(b).

On the other hand, a digital 1-1-1 MASH can be implemented as in Figure 3.14, using a cascade of three DPAs and by combining the carry-out sequences. In the MASH $\Delta\Sigma$ ADC shown in Figure 3.10(b), the input of the second and the third stage is the difference between the quantizer input and output. In the same fashion in the digital MASH, the input of one DPA is given by the sequence $S(k-1)$ of the previous stage, which can easily be demonstrated to be equal to the difference between the quantizer input $\beta(k-1)$ and the output $2^B\,C_o(k-1)$ in the model in Figure 3.13(b). The output of the digital 1-1-1 MASH is a 3 bit word and the division factor switches among 8 levels from $(N-3)$ to $(N+4)$. [13]

3.4.3 Bandwidth versus reference frequency

A key observation can now be drawn after the discussion of the noise shaping. As pointed out in the introduction to this chapter the fractional-*N* PLL promises to decouple the choice of the reference frequency and the loop frequency resolution. In principle, the loop has the capacity to retain a high reference frequency, and therefore a potentially wide bandwidth, even providing a fine frequency resolution. However, the presence of the quantization noise and the need to filter it out limit the practical performance. It turns out that the bandwidth of a fractional-*N* PLL cannot be as large as one could imagine in principle.

Table 3.1 *Bandwidth versus reference frequency in some published fractional-N PLLs*

Reference	$\omega_{out}/2\pi$	$\omega_{ref}/2\pi$	$\omega_{(-3dB)}/2\pi$	$\omega_{ref}/\omega_{(-3dB)}$
[7]	405 MHz	10 MHz	30 kHz	333
[13]	900 MHz	8 MHz	40 kHz	200
[14]	1.8 GHz	20 MHz	84 kHz	238
[15]	1.8 GHz	26 MHz	35 kHz	742
[16]	1.8 GHz	20 MHz	20 kHz	1000

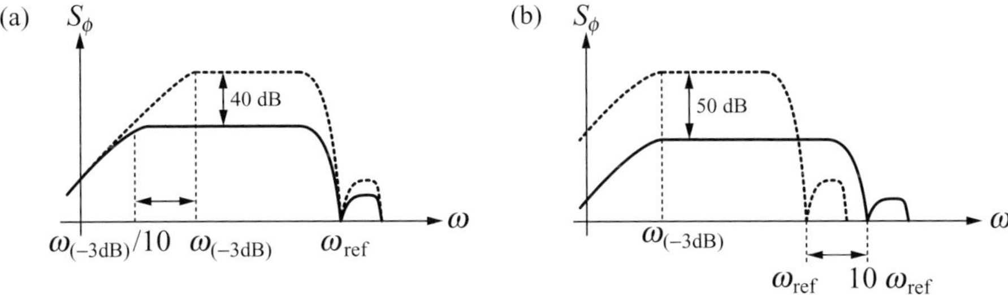

Figure 3.15 Effect of PLL on the $\Delta\Sigma$-induced phase noise: (a) bandwidth variation, and (b) input reference variation

Figure 3.15 refers to a third-order $\Delta\Sigma$ ($l = 3$, Figure 3.12(c)) and shows the change of the output phase noise as a function either of the PLL bandwidth or of its reference. If $\omega_{(-3dB)}$ is reduced by a factor of 10 the noise plateau, Figure 3.15(a), drops by 40 dB. The improvement shown in Figure 3.15(b) is, instead, 50 dB. It is achieved by increasing the reference frequency by a factor of 10 at constant bandwidth. These results can easily be derived from (3.9).

A further concern is the reference spur discussed in the previous chapter. The tone is still present and must be attenuated.

For all these reasons the PLL bandwidth is much lower then the reference frequency, well below the stability limit dictated by Gardner, and also in the fractional-*N* loops an additional low-pass RC filter is sometimes placed before the tuning node, making the PLL a fourth-order system. This filter also helps to substantially reduce the out-of-band noise induced by the $\Delta\Sigma$ modulator.

The examples shown in Figure 3.12 have indeed been selected to highlight the role of the PLL transfer function in the modulator noise. The numbers provided there are not realistic: a 500 kHz bandwidth for a 40 MHz reference is too wide. In many practical implementations the ratio between the reference and the bandwidth is higher than 80. Table 3.1 reports data taken from the literature. The PLLs differ in terms of $\Delta\Sigma$ modulators, the order of the loop filters, and so forth. Nevertheless, it is clear that the design of a wideband fractional-*N* PLL is not straightforward. The main obstacle is the need to cut off the quantization-induced phase noise. A solution could be to widen the bandwidth and compensate the quantization spurs, using a circuit similar to the digiphase shown in Figure 3.4. In [3] the bandwidth has been enlarged up to 460 kHz with a 48 MHz reference frequency, while in [4] the bandwidth attains 700 kHz with a 35 MHz reference frequency.

Example 3.2 Noise in a fourth-order fractional-*N* synthesizer

As in the previous chapter, let us consider the output noise in a type-II PLL. In this case, however, the divider is fractional, and controlled by a third-order $\Delta\Sigma$ modulator. The additional low-pass pole given by R_aC_a, see Figure 2.27, is therefore also useful to attenuate the quantization-induced phase noise. Data and specifications are:

- Reference frequency = 10 MHz;
- Output tuning range $f_{out} = 2.4$–2.5 GHz;
- Integer part of division factor $N = 240$–250;
- Input phase noise (due to CP, divider, PFD) $S_{\phi,\text{In}}^{\text{OL}}(f) \cong -150$ dBc/Hz;
- VCO phase noise $S_{\phi\text{VCO}}$ (1MHz) $\cong -133$ dBc/Hz; $1/f^3$ noise corner frequency = 200 kHz;
- Noise mask = (−50, −80, −105, −125, −145) dBc/Hz respectively at (1 kHz, 10 kHz, 100 kHz, 1 MHz, 10 MHz) offset;
- Power supply = 2.5 V.

The differences with the previous example are the specification of a noise mask and the presence of the $\Delta\Sigma$ quantization noise.

The noise plateau due to the quantization noise *at the output* may be estimated starting from (3.9), multiplying it by N^2, and calculating the resulting expression at the −3 dB bandwidth, where the quantization noise becomes flat. It is:

$$S_{\phi,\text{out}}(f) \cong (2\pi)^2 \cdot \frac{1}{12} \cdot \frac{1}{f_{\text{ref}}} \cdot \left[2\left(\frac{\pi f_{-3\text{dB}}}{f_{\text{ref}}}\right)\right]^4.$$

The approximation $\sin x \cong x$ for small x has been used. Narrow PLL bandwidth would be desirable to filter out the $\Delta\Sigma$ quantization noise. However, wide bandwidth would be useful to filter out the in-band VCO noise. In this example a good compromise is obtained by setting $f_{-3\text{dB}}$ to about 40 kHz and placing the extra pole due to R_aC_a at about 1 MHz.

Proceeding as in Example 2.4, the loop parameters can be chosen as follows: $I_P = 0.6$ mA, $K_{VCO}/2\pi = 100$ MHz/V, $R = 1$ kΩ, $C_1 = 13.2$ nF, $C_2 = 1.5$ nF, $R_a = 500\,\Omega$, $C_a = 0.3$ nF.

Figure 3.16 shows the total output phase noise and its components. Without the low-pass filter R_aC_a, the quantization noise at 5 MHz would be about −125 dBc/Hz.

3.4.4 Dithering the modulator

Idle tones or pattern noises are a typical problem in $\Delta\Sigma$ modulators, in particular when the input is a d.c. signal, which, unfortunately, is the case here. These idle tones clearly affect the output spectrum, as any other spur, and are very difficult to predict. Simulations are, in practice, always necessary.

All the solutions proposed to limit their occurrence try to avoid short-period limit cycles at the $\Delta\Sigma$ output. For instance, it has been noticed in Example 3.1 that in a DPA, the maximum period can be obtained when the input L is odd. When a single DPA is adopted,

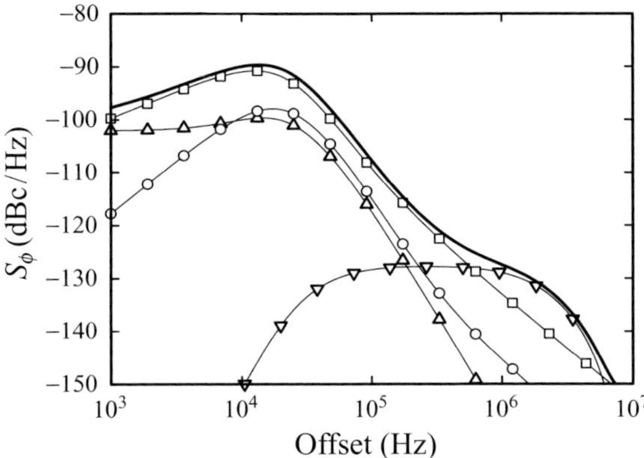

Figure 3.16 Output phase noise due to the different sources. Thermal noise due to the loop filter (circles), input white noise due to CP, PFD and divider (triangles), VCO noise (squares), quantization-induced (upturned triangles). The total spectrum is represented by the solid thicker line

the odd number can be easily synthesized by adding a further bit in the accumulator, and setting the LSB to one.

In general, when higher-order modulators are used, a dithering signal is applied. In practice, it is a high-frequency tone with small amplitude. For instance, in many cases the LSB is toggled between 1 and 0, at the divider output rate (which runs approximately at ω_{ref}). This is equivalent to adding a small tone at $\omega_{\text{ref}}/2$ to the modulator input. A similar approach is reported in [17]. The dithering signal also alters the d.c. value of the input, and therefore the synthesized frequency. Since its duty cycle is 50%, the variation of the input d.c. value is 0.5 LSB. For an accumulator with B bits, the output frequency 'error' is $\omega_{\text{ref}} \cdot 0.5/2^B$, but the minimum output step instead remains $\omega_{\text{ref}}/2^B$. In other cases (e.g., [18]), the dither sequence itself is shaped by a first-order $\Delta\Sigma$ modulator.

3.4.5 Other fractional-*N* techniques

Most of this chapter has been dedicated to the $\Delta\Sigma$ control of the division ratio, mainly because this solution appears as one of the most suitable in silicon ICs. The modulator is relatively simple to design and it is a digital circuit; thus it can take advantage of technology scaling. The resolution can be extremely fine. A minimum frequency resolution below 10 Hz has been reported. [17]

Other solutions have been presented, before and after the introduction of the $\Delta\Sigma$ modulator. It is interesting to remember that [19], published in 1969, was the first work discussing a digiphase, while [20] introduced *random jittering*, i.e., a precursor of the $\Delta\Sigma$ technique. Random jittering is a fractional-spur-compensation technique based on a pseudorandom number generator. A random number $w(k)$ is compared with a number L setting the division ratio. When $w(k) > L$, the division ratio is $N + 1$, otherwise it is N. The average division ratio has the fractional value desired. The problem with this technique was that the spectrum of $w(k)$ is white, thus giving rise to a white frequency noise and to a $1/f^2$ input phase noise. Instead, using the $\Delta\Sigma$ modulator, the spectrum of the randomizing sequence $n(k)$ is high-pass shaped.

If the required resolution is not too high, several other alternative architectures are available. For instance, in [21], the fractional spur is attenuated by averaging the phase error signal before charging or discharging the loop filter. Other architectures perform fractional division by removing or adding pulses at the divider output. For instance, in [22] the output of the divider-by-N consists of short pulses. Then, a logic circuit adds M output between two subsequent divider outputs. In this way, when the PLL is locked, the link between the reference and the output frequency is $\omega_{\mathrm{ref}} = \omega_{\mathrm{out}} \cdot (M+1)/N$. The minimum output step is therefore $\omega_{\mathrm{ref}}/(M+1)$. In [22], it is $M = 8$. Finally, if a multi-phase signal is available, typically when a ring oscillator is adopted, some solutions have been proposed in which the output frequency step is ω_{ref}/M, where M is the number of phases. [23] All these approaches suffer from matching and jitter problems, and apparently are less suited for very large scale integration, with respect to the control of the division factor by a $\Delta\Sigma$ modulator.

3.5 References

[1] D. Butterfield and D. Sun, Prediction of fractional-*n* spurs for UHF PLL frequency synthesizers, *Digest of Technology for Wireless Applications IEEE 1999 MTT-S Symposium*, Feb. 1999, pp. 29–34.

[2] W. F. Egan, *Frequency Synthesis by Phase Lock*, New York, NY: J.Wiley & Sons, Inc., 1990, 2nd edn, 2000.

[3] S. Pamarti, L. Jansson and I. Galton, A wideband 2.4-GHz delta-sigma fractional-*N* PLL with 1-Mb/s in-loop modulation, *IEEE J. Solid-St. Circ.*, **39**, Jan. 2004, 49–62.

[4] E. Temporiti, G. Albasini, I. Bietti, R. Castello and M. Colombo, A 700-kHz bandwidth $\Delta\Sigma$ fractional synthesizer with spurs compensation and linearization techniques for WCDMA applications, *IEEE J. Solid-St. Circ.*, **39**, Sep. 2004, 1446–54.

[5] B. Razavi, *RF Microelectronics*, New York, NY: Prentice-Hall, 1997.

[6] B. Miller and R. J. Conley, A multiple modulator fractional divider, *IEEE T. Instrum. Meas.*, **40**, June. 1991, 578–83.

[7] T. A. Riley, M. A. Copeland and T. A. Kwasnieski, Delta-sigma modulation in fractional-*N* frequency synthesis, *IEEE J. Solid-St. Circ.*, **28**, May 1993, 553–9.

[8] Marconi Instruments, Ltd, *Frequency Synthesisers*, EP Patent 0125790A2, 1984.

[9] J. C. Candy, G. C. Temes, eds, *Oversampling Delta-sigma Data Converters – Theory, Design and Simulations*, New York, NY: IEEE Press, 1992.

[10] I. Galton, Delta-Sigma data conversion in wireless transceiver, *IEEE T. Microw. Theory*, **50**, Jan. 2002, 302–15.

[11] M. H. Perrot, M. D. Trott and C. G. Sodini, A modeling approach for $\Delta\Sigma$ fractional-*N* frequency synthesizers allowing straightforward noise analysis, *IEEE J. Solid-St. Circ.*, **37**, Aug. 2002, 1028–38.

[12] T. P. Kenny, T. A. Riley, N. M. Filiol and M. A. Copeland, Design and realization of a digital $\Delta\Sigma$ modulator for fractional-*N* frequency synthesis, *IEEE T. on Veh. Technol.*, **48**, Mar. 1999, 510–21.

[13] W. Rhee, B. Song and A. Ali, A 1.1-GHz CMOS fractional-*N* frequency synthesizer with a 3-b third-order $\Delta\Sigma$ modulator, *IEEE J. Solid-St. Circ.*, **35**, Oct. 2000, 1453–60.

[14] M. H. Perrot, T. Tewksbury and C. G. Sodini, A 27-mW CMOS fractional-*N* synthesizer using digital compensation for 2.5 Mb/s GFSK modulation, *IEEE J. Solid-St. Circ.*, **32**, Dec. 1997, 2048–60.

[15] B. De Muer and M. Steyaert, A CMOS monolithic $\Delta\Sigma$-controlled fractional-*N* frequency synthesizer or DCS-1800, *IEEE J. Solid-St. Circ.*, **37**, Jul. 2002, 835–44.

[16] C. Heng and B. Song, A 1.8-GHz CMOS fractional-*N* frequency synthesizer with randomized multiphase VCO, *IEEE J. Solid-St. Circ.*, **38**, Jun. 2003, 848–54.

[17] W. Rhee, B. Bisanti and A. Ali, An 18-mW 2.5-GHz/900-MHz BiCMOS dual frequency synthesizer with $<$ 10-Hz carrier resolution, *IEEE J. Solid-St. Circ.*, **37**, Apr. 2002, 515–20.

[18] E. Hegazi and A. Abidi, A 17-mW transmitter and frequency synthesizer for 900-MHz GSM fully integrated in 0.35-μm CMOS, *IEEE J. Solid-St. Circ.*, **38**, May 2003, 782–92.

[19] G. C. Gillette, Digiphase synthesizer, *Proc. of 23rd Frequency Control Symp.*, 1969, 201–10.

[20] V. S. Reinhardt and I. Shahriary, *Spurless Fractional Divider Direct Digital Synthesizer and Method*, US Patent 4815018, 21 Mar., 1989.

[21] S. Pellerano, S. Levantino, C. Samori and A. L. Lacaita, A dual-band frequency synthesizer for 802.11a/b/g with fractional-spur averaging technique, *Digest Tech. Papers of International Solid-State Circuits Conference*, **1**, Feb. 2005, 104–5.

[22] T. Nakagawa and T. Ohira, A phase noise reduction technique for MMIC frequency synthesizers that uses a new pulse generator LSI, *IEEE T. Microw. Theory*, **42**, Dec. 1994, 2579–82.

[23] C. Park, O. Kim and B. Kim, A 1.8-GHz self-calibrated phase-locked loop with precise I/Q matching, *IEEE J. Solid-St. Circ.*, **36**, May 2001, 777–83.

4 Electronic oscillators

4.1 Introduction

Voltage controlled oscillators (VCOs) are used in virtually all wireless systems, where they are key elements of PLLs for frequency synthesis. Oscillators are not the simplest building block to design, simulate or model accurately. Oscillation starts to build up at the power-up, but as the signal amplitude increases, amplifier saturation and other non-linearities stabilize the oscillation amplitude. Large signals and non-linearities cannot be neglected in the prediction of phase noise, tuning range and power dissipation is not trivial at all. Several numerical methods have been developed to simulate VCO topologies rigorously and are nowadays embedded within the most common EDA (electronic design automation) products (see, for instance, [1]). However, quantitative formulae, bridging performance and circuit parameters, are essential to support first-order system partitioning and to provide insight for circuit optimizations.

A second issue is faced when designing silicon ICs. Oscillators for wireless applications have fully integrated LC resonators. External resonators, while offering higher selectivity, introduce other issues, such as the presence of multi-resonances and increasing application costs. However, the integration of all reactive components, such as inductors, capacitors and varactors, makes the VCO design even more challenging. On the other hand, alternative oscillator topologies, avoiding reactive elements, such as ring or relaxation stages, are unable to meet the required phase noise performance.

This chapter is devoted to a discussion of the operation of LC oscillators, introducing the concepts needed to address the phase noise evaluation and its optimization. Phase noise will, instead be covered in Chapters 5 and 7. The integration of reactive components will be discussed in Chapter 6.

4.2 Principles of LC oscillators

4.2.1 Energy balance

In principle a sinusoid can be generated by charging a parallel LC resonant network and leaving it to evolve according to its natural response. The voltage signal across the network

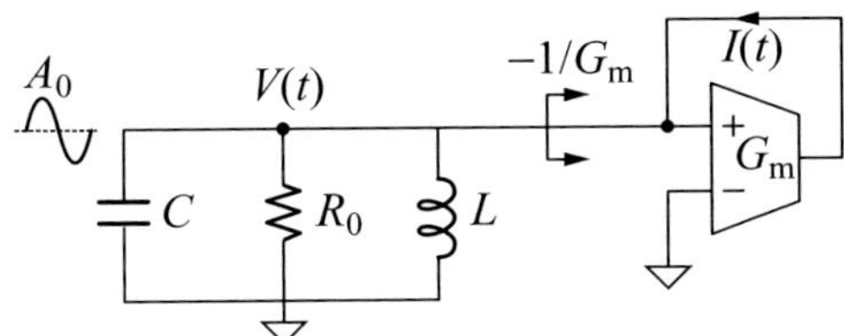

Figure 4.1 Feedback model of a typical LC oscillator

is harmonic with a radial frequency $\omega_0 = 1/\sqrt{LC}$. In practice, however, the amplitude does not remain constant. Energy is lost at each signal cycle, thus leading to an eventually vanishing signal. These losses may be schematically represented by a resistance R_0 placed in parallel with the resonator (Figure 4.1). In this scheme, the resistance is not a physical component purposely placed in the oscillator scheme. It is an equivalent model representing all power losses.

In principle, a harmonic signal, with constant amplitude, may only be obtained by either avoiding all losses or perfectly balancing them by delivering power to the resonant network. In the real world, losses cannot be avoided at all, therefore electrical oscillators are designed using active elements that replace the energy lost by the resonator. Two key points already pop up at this stage: (i) the RLC network in Figure 4.1 must be completed by adding an active element that has to be able to deliver power to the resonator; (ii) there should also be a way to guarantee the perfect balance between the delivered power and the losses, which in turn means that the oscillation amplitude has to remain constant. The first issue will be discussed here, while the amplitude regulation will be described in Section 4.2.4.

In Figure 4.1 a transconductor (e.g., a single transistor) is placed in a positive feedback configuration. The circuit topology makes the transconductor sense the voltage waveform $V(t)$ across the network and deliver a current signal $I(t) = G_m V(t)$. Power is transferred to the resonant tank. Taking $V(t) \approx A_0 \cos(\omega_0 t)$ and $I(t) \approx G_m A_0 \cos(\omega_0 t)$, the average energy delivered per cycle is given by:

$$E = G_m A_0^2 \int_{-T_0/2}^{T_0/2} \cos^2(\omega_0 t)\, dt = \frac{G_m A_0^2 T_0}{2},$$

where $T_0 = (2\pi/\omega_0)$ is the signal period. Perfect balance requires that such energy should balance the energy lost per cycle. Therefore:

$$\frac{G_m A_0^2 T_0}{2} = \frac{A_0^2 T_0}{2R_0}; \quad G_m = \frac{1}{R_0}, \tag{4.1}$$

which sets a condition for self-sustaining oscillation.

Equations (4.1) may also be interpreted in terms of the so-called *impedance cancellation criterion*. Note that the input impedance of the transconductor in Figure 4.1 is $-(1/G_m)$. The negative sign translates the direction of the current flow. As the voltage of the input node rises, the current flows out of the transconductor terminal, being injected into the LC tank. With this in mind, the result (4.1) may be stated by saying that the oscillation takes place when the negative impedance, $-(1/G_m)$, cancels the resistance R_0 and the whole network reduces to an ideal LC resonant tank. This cancellation between the resistive parts adds to a similar cancellation that occurs between the reactances of the LC tank. At resonance, the capacitive and the inductive reactances cancel each other out. The corresponding equation, $\omega_0 C = 1/(\omega_0 L)$, provides the oscillation frequency $\omega_0 = 1/\sqrt{LC}$.

These considerations lead to a more general impedance cancellation criterion: an oscillation occurs whenever a passive network with admittance $Y_p(\omega) = G_p(\omega) + jB_p(\omega)$ is placed in parallel with an active network with an admittance $Y_0(\omega) = G_0(\omega) + jB_0(\omega)$ and there is a frequency ω_0 at which [2]

$$\begin{aligned} G_0(\omega_0) &= -G_p(\omega_0), \\ B_0(\omega_0) &= -B_p(\omega_0). \end{aligned} \tag{4.2}$$

Since an oscillator is a circuit capable of delivering an output signal without being forced by an input, it is also called *autonomous*.

4.2.2 The tank *Q*-factor

The tank losses have been represented so far by the resistance R_0 placed in parallel with the resonator (Figure 4.1). However, the LC losses are usually quoted by using the so-called quality factor, Q. There are indeed several Q-factor definitions. The most general takes the Q-factor as 2π multiplied by the ratio between the maximum energy E_r stored by the reactive components and the energy E_d dissipated per cycle, i.e.:

$$Q = 2\pi \frac{E_r}{E_d}. \tag{4.3}$$

The Q-factor evaluation, therefore, requires knowledge of these two energy terms. The energy E_r can be estimated by taking the peak voltage A_0 across the capacitor. In fact, in so far as the network is linear, when the voltage across the capacitor reaches its peak value, the inductor current is zero. At this time, all the energy is stored in the electric field between the capacitor plates. No residual energy is accumulated in the magnetic field. E_r is therefore given by $E_r = CA_0^2/2$.

As far as the dissipated energy is concerned, its expression may easily be derived by referring to the resonator model in Figure 4.1: $E_d = A_0^2 T_0/2R_0$. The resulting Q factor can, therefore, be written as:

$$Q = 2\pi \frac{E_r}{E_d} = 2\pi \frac{CA_0^2/2}{A_0^2 T_0/2R_0} = \omega_0 R_0 C = \frac{R_0}{\omega_0 L}. \tag{4.4}$$

Let us now put in some numbers. Let us take a fully integrated resonator with $L = 1$ nH and $C = 1$ pF. The oscillation frequency $(\omega_0/2\pi)$ is close to 5 GHz. Assuming a Q factor of 10, which is typical at 5 GHz, the R_0 value would be about 300 Ω.

Note that R_0 is just a model component; it is still necessary, therefore, to link its value to the real losses of the reactive components. Figure 4.2 shows a more realistic model for both the inductor and the capacitor. The components are shown with their series resistances R_{sL} and R_{sC}, which account for the finite conductivity of the current paths. The shunt resistor R_{pC} represents, instead, dielectric losses.

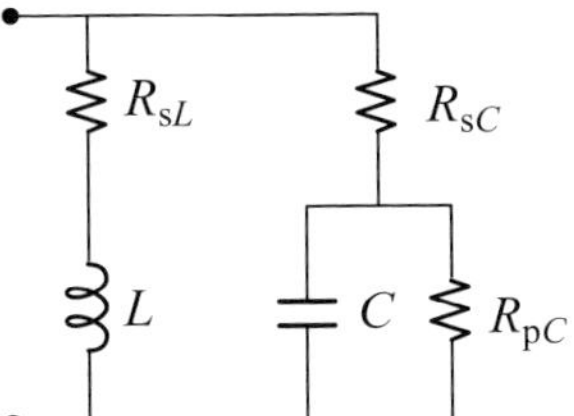

Figure 4.2 Losses of the resonant network

A quality factor can now be defined for each single reactive element following the same energetic definition (4.2). For the inductor, the Q factor may be computed by referring to a test circuit where the component is driven by a current signal at ω_0 with peak value I_0. The magnetic energy is given by $LI_0^2/2$, while the energy dissipated per cycle is $I_0^2 T_0 R_{sL}/2$. It turns out that:

$$Q_{sL} = 2\pi \frac{E_r}{E_d} = 2\pi \frac{LI_0^2/2}{I_0^2 T_0 R_{sL}/2} = \frac{\omega_0 L}{R_{sL}}. \tag{4.5}$$

The final result coincides with another popular definition of the Q factor of a reactive component: the Q factor is the absolute value of the ratio between the imaginary and the real part of the component impedance or admittance.

In a similar way, the Q factor of a capacitor affected by losses due to the resistance R_{sC} is $Q_{sC} = 1/(\omega_0 C R_{sC})$. The Q factor of a capacitor limited by the dielectric losses would instead be $Q_{pC} = \omega_0 C R_{pC}$.

Let us now discuss how all these Q factors combine to give the overall Q factor of the tank. Let us assume the losses to be small at resonance, i.e., $R_{sL} \ll \omega_0 L$, $R_{sC} \ll 1/\omega_0 C$ and $R_{pC} \gg 1/\omega_0 C$, and let us derive the overall quality factor following its energetic definition. Denoting as A_0 the oscillation amplitude, the peak inductor current is approximately $A_0/\omega_0 L$ and the average power dissipated across R_{sL} is given by:

$$P_{sL} \approx \frac{1}{2}\left(\frac{A_0}{\omega_0 L}\right)^2 R_{sL}.$$

On the other hand, the average powers dissipated across R_{sC} and R_{pC} are:

$$P_{sC} \approx \frac{1}{2}(A_0 \omega_0 C)^2 R_{sC}, \quad P_{pC} \approx \frac{1}{2}\frac{A_0^2}{R_{pC}},$$

respectively. By equating $P_{sL} + P_{sC} + P_{pC}$ to the average power $A_0^2/(2R_0)$ dissipated across R_0, it turns out that:

$$\frac{1}{Q} = \frac{1}{Q_{sL}} + \left(\frac{1}{Q_{sC}} + \frac{1}{Q_{pC}}\right) \tag{4.6}$$

where (4.4) has also been used to replace R_0.

Equation (4.6) links the Q factor to the losses of each reactive component. It shows that the overall quality factor is limited by the reactive element with the highest losses, i.e., with

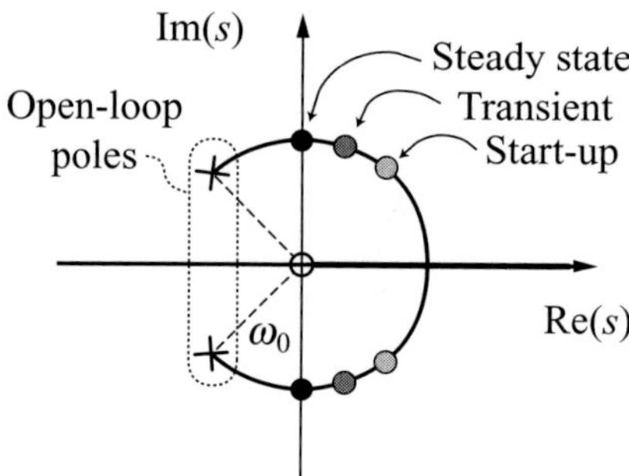

Figure 4.3 Oscillator root locus

lower Q. The term inside the parenthesis is the inverse of the capacitor Q factor, accounting for both the series resistance and the dielectric losses. It should be kept in mind that (4.6) only holds close to ω_0 and for small losses, that is for $R_{sL} \ll \omega_0 L$, $R_{sC} \ll 1/\omega_0 C$ and $R_{pC} \gg 1/\omega_0 C$. It can easily be demonstrated that these conditions translate as saying that Q_{sL}, Q_{sC}, Q_{pC} must be much larger than one.

4.2.3 Barkhausen criterion

The circuit diagram in Figure 4.1 is characterized by a positive feedback. By writing the impedance of the parallel resonant network as

$$Z(s) = \frac{s/C}{\left(s^2 + s\omega_0/Q + \omega_0^2\right)} = R_0 \frac{s\omega_0/Q}{\left(s^2 + s\omega_0/Q + \omega_0^2\right)},$$

the circuit open-loop gain turns out to be:

$$G_{\text{loop}}(s) = G_{\text{m}} R_0 \frac{s\omega_0/Q}{\left(s^2 + s\omega_0/Q + \omega_0^2\right)}.$$

Figure 4.3 shows the corresponding root locus. The locus crosses the imaginary axis, indicating that the network may have imaginary closed-loop poles. Such a network can provide self-sustaining harmonic signals. The condition for closed-loop poles at $\pm \mathrm{j}\omega_0$ is found by solving $G_{\text{loop}}(s) = 1$ after letting $s = \mathrm{j}\omega$. It is:

$$G_{\text{loop}}(\mathrm{j}\omega) = G_{\text{m}} R_0 \frac{\mathrm{j}\omega\omega_0/Q}{\left(\omega_0^2 - \omega^2 + \mathrm{j}\omega\omega_0/Q\right)} = 1. \tag{4.7}$$

The solution of the complex equation leads, again, to the two oscillation conditions $G_m R_0 = 1$ and $\omega = \omega_0$.

The equality $G_{\text{loop}}(\mathrm{j}\omega) = 1$ is also known as the *Barkhausen criterion*. This states that an electronic circuit can sustain a steady-state oscillation whenever a frequency exists at which the open-loop gain is unity. The condition may be intuitively justified by noting that any signal at that frequency, once injected into the loop, returns with the same phase and amplitude after each lap. It is, therefore, self-sustaining or autonomous.

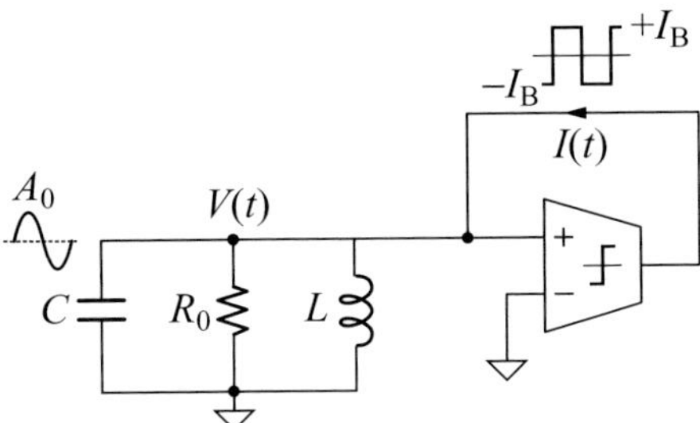

Figure 4.4 Oscillator with the transconductor operating as 'hard limiter' in full switching regime

4.2.4 Start-up and amplitude stability

Based on the Barkhausen criterion two important issues can now be addressed: the condition that should be met to guarantee the oscillation start-up and the mechanism for amplitude regulation.

While the resonance of the bandpass filter defines the oscillation frequency ω_0, the open-loop gain $G_m R_0$ sets the position of the closed-loop poles in the s plane. For $G_m R_0 > 1$, the poles have a positive real part; for $G_m R_0 = 1$, the poles are on the imaginary axis. For $G_m R_0 < 1$, the poles have a negative real part. For the oscillation to start at the power-up, the circuit must have $G_m R_0 > 1$. In this way, a small disturbance or even a fluctuation arising from electronic noise may be able to trigger the harmonic oscillation. Since the poles have a positive real part, the oscillation amplitude rises swiftly. A suitable regulation system should then be introduced, to set the amplitude to a given desired value. In principle, such a regulation has to act on $G_m R_0$, bringing the poles back to the imaginary axis to stop the amplitude rise (Figure 4.3).

There are many methods of controlling the oscillation amplitude. The most straightforward is already present in the basic oscillator scheme in Figure 4.1. Note that, so long as the active element operates in the linear regime, the output current is harmonic. However, as A_0 exceeds the input linear range, the transconductor starts to deliver a distorted current waveform and, for large A_0 values, the output approaches square-wave switching between $+I_B$, i.e., the levels limiting the available output swing. In terms of the circuit model, the transconductor is no longer represented by a linear stage, with its transconductance, but is reduced to a hard limiter (Figure 4.4).

This description is idealized. For example, the delay introduced by the transconductor is neglected. However, the model is quite useful for grasping the consequences of such a fast switching operation. For example, the distorted current waveform has many harmonics, but, if the resonant network has a high Q-factor, its bandpass selectivity may be enough to effectively cut them all off and the voltage across the resonant network remains harmonic at ω_0.

Starting from this observation, the concept of *effective transconductance* can be introduced. It is defined as the ratio between the fundamental harmonic of the output current and the peak amplitude A_0 of the harmonic voltage at the transconductor input. [3] Figure 4.5 shows, schematically, the dependence on A_0 of the effective transconductance $G_{m,eff}$ of

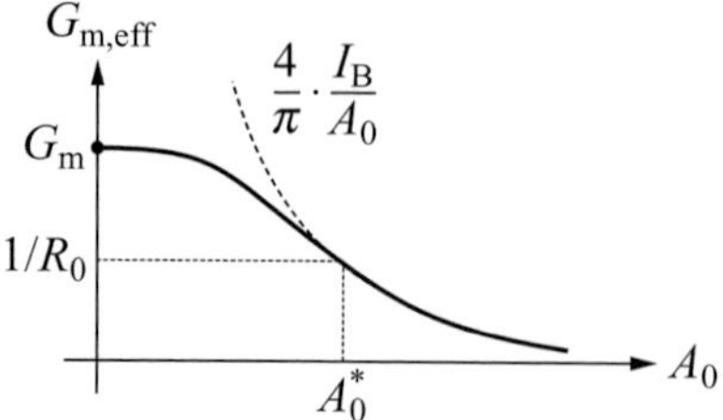

Figure 4.5 Effective transconductance versus oscillation amplitude

the limiter stage in Figure 4.4. So long as A_0 is very small, the transconductor operates in a linear regime, no distortion takes place and $G_{m,eff} = G_m$. When A_0 increases, the effective transconductance decreases. Power is transferred to higher-order harmonics and cut off by the resonant network. When A_0 is so large that the transconductor delivers a square wave, the current component at ω_0 has an amplitude $I_0 = (4/\pi)I_B$ and:

$$G_{m,eff} = \frac{I_0}{A_0} = \frac{4I_B}{\pi A_0}. \tag{4.8}$$

The effective transconductance makes it possible to describe the transconductor operation even beyond the small-signal operation and to extend the use of the open-loop gain concept. Since to a first order approximation, only the fundamental harmonic circulates along the loop. G_m can be replaced by $G_{m,eff}$.

In this frame, however, both the open-loop gain and the position of the closed-loop poles become dependent on the oscillation amplitude. Let us consider the root locus in Figure 4.3. At the power-up, the oscillator voltage gain at resonance, $G_m R_0$, has to be large enough to place the closed-loop poles in the right half of the s plane. In this way, the oscillation amplitude starts to rise exponentially. However, at this stage, the non-linear dependence of $G_{m,eff}$ on A_0 comes into play until non-linearities cause the effective transconductance to decrease. The oscillation amplitude stabilizes when the condition $G_{m,eff}\, R_0 = 1$ is met. From (4.8), it follows that the steady-state oscillation amplitude is given by:

$$A_0 = \frac{4}{\pi} I_B R_0. \tag{4.9}$$

The condition is stable. If for any reason the oscillation amplitude decreases, the effective transconductance increases, the closed-loop poles move to the right-hand side of the s plane, until the oscillation amplitude recovers the nominal value. In a similar way, if A_0 increases, the system reacts to recover the condition set by (4.9).

4.2.5 Effect of the transconductor delay

Up to now the transconductor has been considered as an ideal element, adding no delay in the loop. This condition holds if the oscillator frequency is set well below the transistor cut-off frequency ω_T. In some cases, even if small, the additional loop delay due to the active stages may be relevant. It is therefore worthwhile to address its impact.

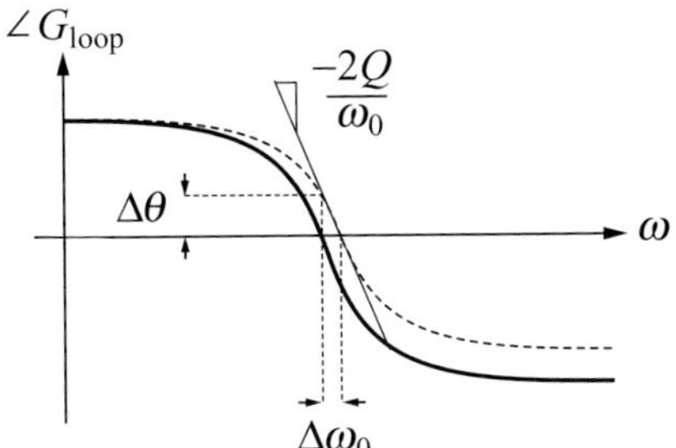

Figure 4.6 Shift of the oscillation frequency due to the transconductor delay

Figure 4.6 shows the Bode diagram for the G_{loop} phase. In the ideal case the phase shift is only created in the LC tank and is zero at $\sqrt{LC}$, where the slope of the phase diagram is $(\mathrm{d}\theta/\mathrm{d}\omega) = -(2Q/\omega_0)$. The presence of additional poles of the active element adds a phase delay, $\Delta\theta$, at the resonance frequency, and makes the oscillation frequency shift back from $\sqrt{LC}$. Since the value of the slope at ω_0 is $(\mathrm{d}\theta/\mathrm{d}\omega) = -(2Q/\omega_0)$, it turns out that the frequency shift is:

$$\Delta\omega_0 \approx -\omega_0 \cdot \frac{\Delta\theta}{2Q}. \tag{4.10}$$

4.2.6 Oscillator tuning

A PLL requires a VCO, that is an oscillator whose frequency may be controlled by a tuning voltage V_{tune}. Frequency tuning is achieved, in practice, by using a resonator with a voltage-controlled capacitor or varactor.

The tuning range of the oscillation frequency $\Delta\omega_0$ may be estimated by differentiating $\omega_0 = 1/\sqrt{LC}$ with respect to the varactor capacitance C. It turns out that:

$$\frac{\Delta\omega_0}{\omega_0} \approx -\frac{\Delta C}{2C}. \tag{4.11}$$

The denominators ω_0 and C are the values in the middle of the frequency and the capacitance range, respectively. In practice (4.10) points out that a 20% tuning range calls for about a 40% capacitance variation.

A varactor may be implemented by using a reverse-biased pn junction by exploiting the dependence of the depletion region capacitance on the bias voltage. An alternative is the use of MOS capacitors swept from accumulation to depletion or from depletion to inversion. These devices and their performance will be analyzed in more detail in Chapter 6. At any rate, the tank capacitance C does not consist only of the varactor capacitance C_V. Parasitic capacitances from the inductor coil and the surrounding active element add to C_V. Since the parasitics are not affected by the tuning voltage variations, the tuning range is reduced to:

$$\frac{\Delta\omega_0}{\omega_0} \approx -\frac{\Delta C_V}{2C_V} \cdot \frac{C_V}{C}. \tag{4.12}$$

It follows that the higher the parasitic capacitance, the narrower the tuning range.

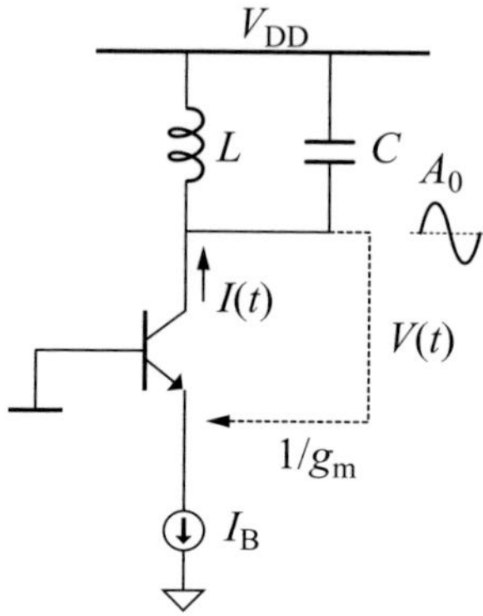

Figure 4.7 Single-transistor oscillator made using a parallel LC resonant load and a collector-to-emitter feedback

4.3 Single-transistor oscillators

Compact oscillators can be built using resonant networks and a single transistor. Cost reasons make these stages the ordinary choice in discrete-component systems such as in crystal oscillators, which provide reference signals to frequency synthesizers. This section is devoted to the discussion of some popular single-transistor LC oscillators, which have also been adopted in RF integrated circuits.

4.3.1 The Colpitts oscillator

There are different ways to link an LC network and a single transistor to make an oscillator. Figure 4.7 shows a solution where a parallel resonant network is placed in series with the transistor collector. Since at resonance the voltage and the current across the resonant network are in phase, a positive feedback loop can be closed around the transistor, feeding a fraction of the collector voltage signal to the emitter.

However, the direct link between collector and emitter, for instance using a large a.c. coupling capacitor, adds losses to the tank. The emitter impedance $1/g_m$ would appear in parallel with the resonant network, impairing the Q factor. This drawback may be tackled using a capacitance divider between collector and emitter. Figure 4.8 shows the typical topology of a Colpitts oscillator, which implements the solution. The resistance R_0 represents the tank losses; a capacitance divider $C_1 - C_2$ connects the tank to the emitter terminal.

Figure 4.9 shows the tank along with the $1/g_m$ impedance of the emitter. The equivalent resistance due to $1/g_m$, in parallel with the tank, may be estimated by writing a power balance. For the series of $C_1 - C_2$ to operate as voltage divider, the impedance of C_2 at the resonant frequency has to be much lower than the emitter impedance: $1/\omega_0 C_2 \ll 1/g_m$. This condition can also be written as a function of the quality factor Q_2 of C_2; $Q_2 = \omega_0 C_2/g_m \gg 1$. Under this assumption, a harmonic signal with A_0 peak amplitude at the collector causes

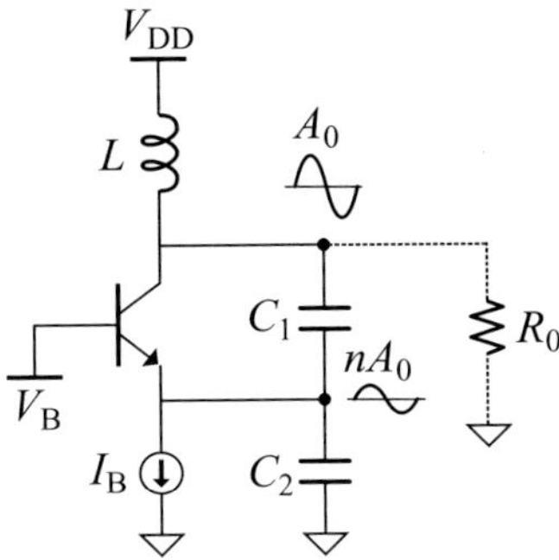

Figure 4.8 Colpitts oscillator

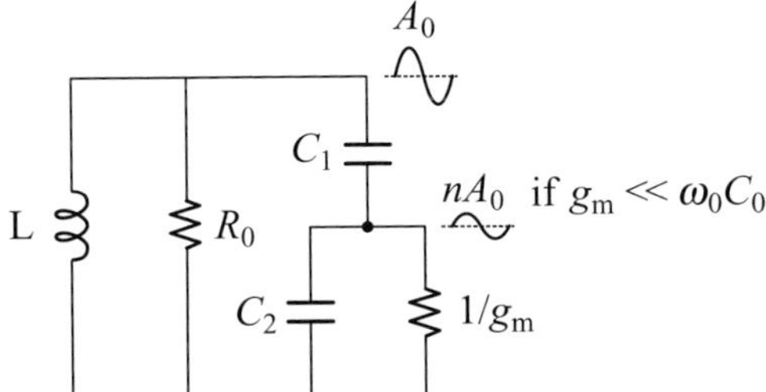

Figure 4.9 Equivalent load at tank terminals of Colpitts oscillator

a voltage signal across C_2 with amplitude:

$$V_{C2} \approx A_0 \frac{C_1}{(C_1 + C_2)} = nA_0,$$

where $n = C_1/(C_1 + C_2)$ is the voltage division factor. The power absorbed by the emitter port is given by:

$$P_{in} \approx g_m \frac{n^2 A_0^2}{2} \approx \frac{A_0^2}{2R_{eq}},$$

where $R_{eq} = 1/(n^2 g_m)$ is the resistance representing the additional losses from the transistor emitter impedance. It appears in parallel with R_0, reducing the overall Q factor. The equivalent parallel resistance of the tank and the equivalent capacitance result:

$$R_T \approx \frac{R_0}{1 + n^2 g_m R_0}, \qquad C \approx \frac{C_1 C_2}{(C_1 + C_2)}. \tag{4.13}$$

Note that the voltage divider is a practical way of realizing an impedance transformation. The capacitive network acts approximately as a transformer with 1:n ratio, so the emitter impedance appears multiplied by $1/n^2 > 1$ at the tank terminals. The same effect can be obtained using a transformer or a voltage partition using two inductors, deriving the so-called Hartley topology. [3] However, the higher number of inductors prevents its adoption in fully integrated schemes.

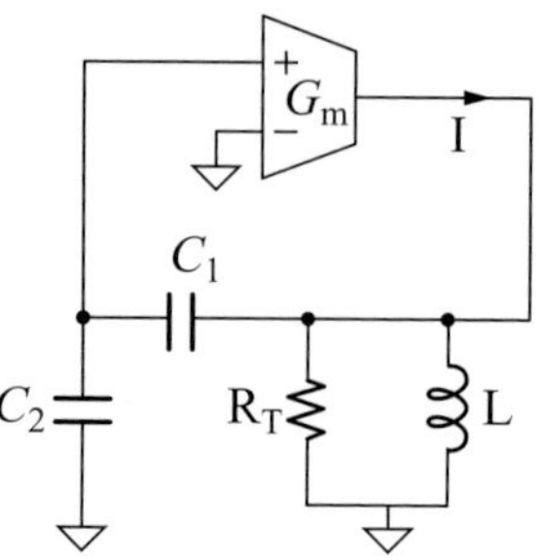

Figure 4.10 Feedback model of Colpitts oscillator

4.3.2 The open-loop gain

The condition $\omega_0 C_2 \gg g_m$ implies that, at resonance, the current signal flowing through C_1 will proceed mainly through C_2 rather than entering the transistor emitter port. The open-loop gain may, therefore, be estimated by neglecting the transistor load in parallel to C_2 and taking the collector current as the product between the transistor g_m and the voltage signal across C_2. Under this assumption:

$$G_{\text{loop}}(s) \approx g_m R_T \cdot n \cdot \frac{s/R_T C}{(s^2 + s/R_T C + 1/LC)}.$$

The expression neglects the current partition between $\omega_0 C_2$ and g_m. In R_T, however, it accounts for the power dissipation across the emitter impedance $1/g_m$ that loads the resonator, see (4.13).

The Barkhausen criterion is fulfilled at $\omega_0 = 1/\sqrt{LC}$ provided that:

$$g_m R_T n = 1.$$

This condition has an obvious meaning; it states that at the oscillation frequency the voltage gain of the common-base stage $g_m R_T$ has to balance the voltage partition between C_1 and C_2. Since R_T depends on g_m through (4.13), the oscillation condition can be rewritten as:

$$n g_m R_T = \frac{n g_m R_0}{1 + n^2 g_m R_0} = 1; \quad g_m = \frac{1}{n(1-n) R_0}. \tag{4.14}$$

To guarantee the start-up, the poles have to be placed in the right-hand side of the s plane. The components and the transistor bias must be chosen so that:

$$g_m > \frac{1}{n(1-n)R_0}. \tag{4.15}$$

Figure 4.10 shows a model equivalent circuit of the Colpitts oscillator. The resistor R_T models the losses. The transistor is regarded as an ideal transconductor, since the voltage signal at the emitter is sensed at high impedance (under the assumption $\omega_0 C_2 \gg g_m$) and the output current is delivered with a high output impedance to the tank.

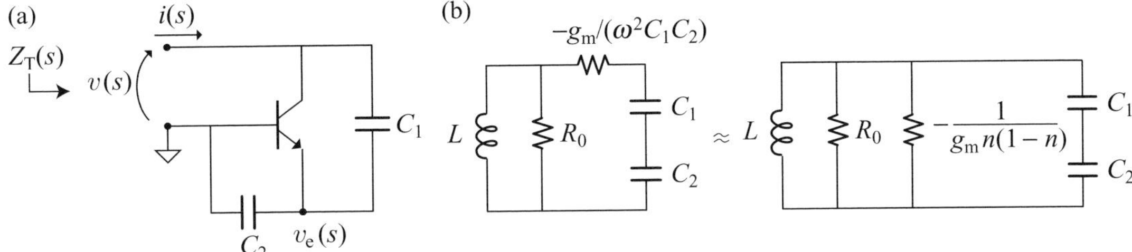

Figure 4.11 Colpitts oscillator: (a) simplified schematic and (b) negative-resistance model

4.3.3 Negative impedance analysis

As an example, let us derive again the oscillation condition, using the impedance cancellation criterion. Figure 4.11(a) shows the Colpitts oscillator without taking accounting of the bias. The capacitance C_2 has been folded around the transistor, since both its bottom terminal and the transistor base are connected to an a.c. ground.

By neglecting the base current, the impedance $Z_T(s)$ at the collector–base port can be easily estimated. A probe signal $v(s)$ between the collector and base terminals causes a signal $v_e(s)$ at the emitter node and an input current signal $i(s)$ (Figure 4.11(a)). It is:

$$\begin{cases} i = sC_2 v_e, \\ i = sC_1(v - v_e) - g_m v_e. \end{cases}$$

By solving these equations, the impedance can be written as $Z_T(s) = v/i = [g_m + s(C_2 + C_1)]/(s^2 C_1 C_2)$, that is:

$$Z_T(j\omega) = -\frac{g_m}{\omega^2 C_1 C_2} + \frac{(C_2 + C_1)}{j\omega C_1 C_2}.$$

This term can be interpreted as the series of a negative resistance and the capacitances C_1 and C_2. Figure 4.11(b) shows a further transformation, which is derived by computing the corresponding admittance. In the shunt-equivalent topology, the two capacitances are now in parallel to a resistance given by:

$$R_p \approx \frac{1}{\dfrac{-g_m}{\omega^2 C_1 C_2} \cdot \left(\omega \dfrac{C_1 C_2}{C_1 + C_2}\right)^2} = -\frac{1}{n(1-n)g_m}.$$

The oscillation frequency is obtained by writing the cancellation among the reactances. The capacitive reactance cancels out the inductive term at ω_0, i.e.:

$$\omega_0 L = \frac{(C_2 + C_1)}{\omega_0 C_1 C_2}.$$

The condition on the loop gain is, instead, obtained by writing the cancellation condition between the positive and the negative resistive terms:

$$R_0 = \frac{1}{n(1-n)g_m}; \quad g_m = \frac{1}{n(1-n)R_0}; \tag{4.16}$$

as already obtained in (4.14).

4.3.4 Oscillation amplitude

Once the oscillation is triggered, the signal amplitude rises. A fraction of this signal is fed back to the emitter, driving the transistor and generating a current in phase with the oscillation. However, because of the exponential *I–V* curve of the bipolar, the collector current is not harmonic at all. Denoting the collector voltage as $V(t) = -A_0 \cos \omega t$ and the average voltage between base and emitter as $\overline{V_{\mathrm{BE}}}$, the base–emitter voltage waveform is $V_{\mathrm{BE}}(t) = \overline{V_{\mathrm{BE}}} + nA_0 \cos(\omega_0 t)$. Therefore, the collector current can be written as:

$$\begin{aligned} I(t) &= I_{\mathrm{S}} \mathrm{e}^{[\overline{V_{\mathrm{BE}}} + nA_0 \cos(\omega_0 t)]/V_{\mathrm{T}}} = I_{\mathrm{R}} \mathrm{e}^{nA_0 \cos(\omega_0 t)/V_{\mathrm{T}}} \\ &= I_{\mathrm{R}} \mathrm{e}^{X \cos(\omega_0 t)}, \end{aligned} \tag{4.17}$$

where I_{S} is the transistor reverse saturation current, $V_{\mathrm{T}} = kT/q$ is the thermal voltage, X is the ratio nA_0/V_{T} and $I_{\mathrm{R}} = I_{\mathrm{S}} \mathrm{e}^{\overline{V_{\mathrm{BE}}}/V_{\mathrm{T}}}$. The collector current, as expected, is not harmonic, and its average value is:

$$\overline{I(t)} = \frac{1}{T_0} \int_{-T_0/2}^{T_0/2} I(t) \mathrm{d}t = \frac{I_{\mathrm{R}}}{2\pi} \int_{-\pi}^{\pi} \mathrm{e}^{X \cos \vartheta} \mathrm{d}\vartheta = I_{\mathrm{R}} \cdot B_0(X),$$

where $\vartheta = \omega_0 t$, $T_0 = 2\pi/\omega_0$ is the oscillation period and $B_0(X)$ is the zeroth-order *modified* Bessel function of the first kind. [4] $B_0(X)$ is an increasing function of X. By noting that $\overline{I(t)}$ has to be equal to the bias current I_{B}, we obtain:

$$I_{\mathrm{R}} = I_{\mathrm{S}} \mathrm{e}^{\overline{V_{\mathrm{BE}}}/V_{\mathrm{T}}} = \frac{I_{\mathrm{B}}}{B_0(X)}, \tag{4.18}$$

which can be inverted into:

$$\overline{V_{\mathrm{BE}}} = V_{\mathrm{T}} \ln \left[\frac{I_{\mathrm{B}}}{I_{\mathrm{S}} B_0(X)} \right] = V_{\mathrm{BE}}|_{X=0} - V_{\mathrm{T}} \ln \left[B_0(X) \right], \tag{4.19}$$

where the voltage $V_{\mathrm{BE}}|_{X=0}$ represents the emitter–base voltage in the absence of the harmonic signal.

Equation (4.19) can be interpreted in the following way. As the oscillation amplitude increases (i.e., as X increases), the average V_{BE} value decreases, to keep the same average current value. The transistor is, therefore, off most of the time, and it is on when the oscillation waveform is close to its minimum. Figure 4.12 shows plots of the output collector voltage and the collector current at increasing bias currents ($I_{\mathrm{B}} = 6$, 7 and 8 mA), as obtained from circuit transient simulations. The larger the amplitude, the sharper the current pulse. Note that the current component superimposed on the periodic current pulses is in with respect to quadrature the collector voltage, denoting that it comes from transistor parasitic capacitances.

Let us now introduce the large-signal effective transconductance of the transistor. It is the ratio between the fundamental harmonic of the collector current and the base–emitter voltage. The former is given by:

$$I_0 = \frac{2}{T_0} \int_{-T_0/2}^{T_0/2} I(t) \cos(\omega_0 t) \, \mathrm{d}t = \frac{I_{\mathrm{R}}}{\pi} \int_{-\pi}^{\pi} \mathrm{e}^{X \cos \vartheta} \cos \vartheta \, \mathrm{d}\vartheta = 2I_{\mathrm{R}} \cdot B_1(X), \tag{4.20}$$

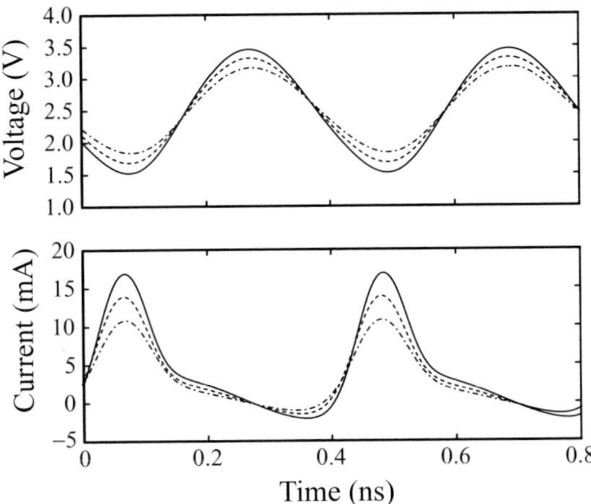

Figure 4.12 Current waveforms in a bipolar Colpitts oscillator for bias currents $I_B = 6$, 7 and 8 mA (dash–dotted, dashed, solid line, respectively)

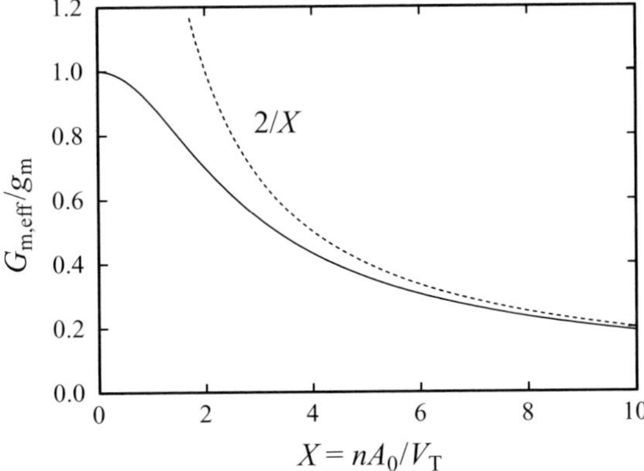

Figure 4.13 Effective transconductance versus oscillation amplitude in bipolar Colpitts oscillators

where $B_1(X)$ is the first-order modified Bessel function of the first kind. [4] The base–emitter voltage is, instead, $nA_0 = X \cdot V_T$. By introducing (4.18) in (4.20), the first harmonic of the collector current can be written as:

$$I_0 = 2I_B \cdot \frac{B_1(X)}{B_0(X)}. \tag{4.21}$$

Thus, the large-signal effective transconductance is given by:

$$G_{m,eff} = \frac{I_0}{nA_0} = \frac{2I_B}{X \cdot V_T} \cdot \frac{B_1(X)}{B_0(X)} = g_m \cdot \frac{2B_1(X)}{X \cdot B_0(X)}, \tag{4.22}$$

which is schematically drawn in Figure 4.13 (solid line). Note that as $X \to 0$, $G_{m,eff} \to g_m$. Instead, when $X \gg 1$, $G_{m,eff}$ can be approximated as:

$$G_{m,eff} \to g_m \cdot \frac{2}{X} = \frac{2I_B}{nA_0}. \tag{4.23}$$

The dashed line in Figure 4.13 shows the latter dependence. By replacing the small-signal transconductance in (4.16) with the effective transconductance, it turns out that:

$$G_{\mathrm{m,eff}} = \frac{1}{n(1-n)R_0}. \tag{4.24}$$

The A_0 dependence on current is derived by using (4.22):

$$\frac{2B_1(X)}{X \cdot B_0(X)} = \frac{1}{n(1-n)}\frac{1}{g_{\mathrm{m}}R_0}. \tag{4.25}$$

Based on (4.23) and (4.24) an estimate can be derived as:

$$A_0 \approx 2(1-n)I_{\mathrm{B}}R_0. \tag{4.26}$$

The approximation is expected to hold well for large emitter voltage amplitudes, i.e., $nA_0 \gg V_{\mathrm{T}}$.

In (4.26), the amplitude is proportional to the bias current. Such a working condition is usually referred to as a *current-limited regime*. The larger the current, the larger the oscillation amplitude. Clearly, if the bias current is too high, the transistor saturates, the oscillation amplitude reaches the limit set by the available voltage dynamics and any further increase of the bias current results only in additional power dissipation. This operation is referred to as a *voltage-limited regime*. The bipolar saturation causes additional delays in the oscillator loop and the oscillation frequency decreases.

Formulae (4.23) and (4.26) may, alternatively, be derived following another path. It should be noted that in the large-amplitude regime the collector current approaches a sequence of narrow δ-like pulses:

$$I(t) \to \sum_{k=-\infty}^{+\infty} q \cdot \delta(t - kT_0).$$

Since the d.c. current component is equal to the bias current I_{B}, the area of the δ-like pulses is $q = I_{\mathrm{B}}T_0$, while the amplitude of the first harmonic of the current is:

$$I_0 = \frac{2}{T_0}\int_{-T_0/2}^{T_0/2} I(t)\cos(\omega_0 t)\,\mathrm{d}t \to 2I_{\mathrm{B}}.$$

The effective transconductance, therefore, approaches (4.23) and the oscillation amplitude is given by (4.26).

The preceding derivations have been obtained under the assumption of $\omega_0 C_2 \gg g_{\mathrm{m}}$. However, when the oscillator operates in a large-signal regime, the transconductance g_{m} is replaced by $G_{\mathrm{m,eff}}$ to the expression of the circuit open-loop gain. The condition for the loop current to flow into the capacitance C_2 and not into the transistor emitter becomes:

$$\omega_0 C_2 \gg G_{\mathrm{m,eff}}; \quad \omega_0 C_2 \gg \frac{1}{R_0 \cdot n \cdot (1-n)},$$

where the last result follows after using (4.23) and (4.26). Since the unloaded tank quality factor is $Q = R_0/\omega_0 L = \omega_0 R_0 \cdot nC_2$, this condition can be further written as:

$$n \ll 1 - \frac{1}{Q}, \tag{4.27}$$

thus linking the capacitor ratio and the tank quality factor.

Example 4.1 Design of a Colpitts oscillator: oscillation amplitude

Let us now consider the design of a 2.4 GHz Colpitts oscillator, in a 2.5 V bipolar technology with 0.4 μm minimum emitter width. The inductor is made using an integrated coil with 1.4 nH inductance and a quality factor of 10 at 2.4 GHz. The capacitors have a Q value of 30 at the same frequency. The target bias current is set to 4 mA.

The resonant frequency of 2.4 GHz is obtained by placing $C = 1/(\omega_0^2 L) = 3$ pF in parallel with the inductor. Neglecting the interconnects and the transistor parasitic capacitances, the capacitance C is given by the series C_1 and C_2, that is, $C = nC_2$. The other capacitance contributions, which offset the resonant frequency, can be taken into account in the second iteration of the design process.

The bipolar transistor should be sized with two key requirements in mind: (i) to minimize the stray capacitances added to the resonator and (ii) to provide minimum additional delay in the loop. In this way the tuning range and the maximum achievable oscillation frequency are increased. In the adopted technology, a current density of about 125 μA/μm^2 guarantees proper transistor operation. The bias current of 4 mA corresponds to an 80×0.4 μm^2 emitter area.

In fast bipolar technologies, the spreading base resistance and the emitter resistance may be some tens of ohms. These parasitics reduce the transconductance and increase the transistor delay. They can be lowered by increasing the number of base and emitter fingers. An emitter finger length of 4 μm is sufficient to reach a cut-off frequency of about 40 GHz. Note also that the peak f_{T} for a bipolar is usually achieved at large current densities (e.g., mA/μm^2) where the exponential dependence of the bipolar current does not hold any more. [5] Even if the formulae obtained above could be modified to address the more general case, in the present example the transistor has always been operated at current densities, where the previous results still hold.

The tank quality factor due to the inductor and the capacitors losses can be derived using (4.6):

$$Q = Q_{\mathrm{C}} Q_{\mathrm{L}}/(Q_{\mathrm{C}} + Q_{\mathrm{L}}) = 7.5.$$

Therefore, (4.27) demands division factors n that are much lower than about 0.87.

To guarantee the oscillation start-up, (4.15) must be satisfied. Hence, the equivalent parallel resistance of the LC tank is $R_0 = \omega_0 L Q = 162\ \Omega$ and the gain $g_{\mathrm{m}} R_0$ is 25.9. From (4.15) it follows that

$$n(1-n) > \frac{1}{g_{\mathrm{m}} R_0}, \tag{4.28}$$

which leads to the n interval:

$$0.04 < n < 0.96. \tag{4.29}$$

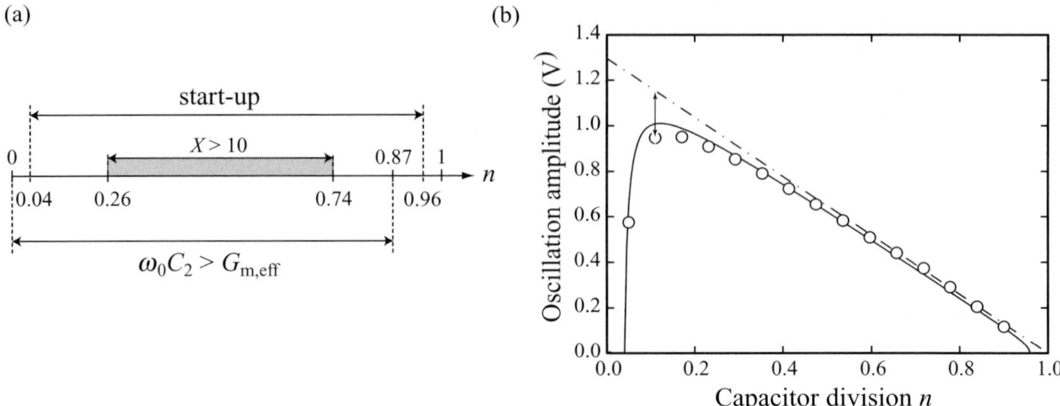

Figure 4.14 (a) Allowed ranges of the voltage division n for the bipolar Colpitts oscillator (b) oscillation amplitude versus n

In practice, low n values reduce the voltage partition and thus the open-loop gain; on the other hand high n values let the emitter impedance decrease the total parallel resistance R_T and in turn degrade the open-loop gain. This explains why the start-up condition generates two bounds for n. The interval also depends on the specific $g_m R_0$ value. Lower $g_m R_0$ values would narrow the permitted range.

Once the start-up has been guaranteed, the oscillation amplitude can be estimated by relying on (4.26), which holds for $X = nA_0/V_T \gg 1$. Figure 4.13 shows that for $X > 10$ the approximated formula fits the more rigorous dependence (4.22) well. Using (4.26), the condition $X > 10$ is translated in terms of n values, leading to:

$$n(1-n) > \frac{10}{g_m R_0},$$

that is

$$0.26 < n < 0.74. \tag{4.30}$$

Figure 4.14(a) summarizes the n ranges identified so far. For $n = 0.26$ the oscillation amplitude is $A_0 = 2I_B R_0 (1-n) = 960\,\text{mV}$.

Figure 4.14(b) shows the A_0 dependence as a function of the capacitive partition. The open circles show the results of detailed transient simulations of the circuit. The solid line is, instead, derived by solving (4.25) numerically. For comparison, the dash–dotted line shows the approximation (4.26). Note that the results from circuit simulations are well accounted for by the curve obtained from (4.30) over the entire n range. The oscillation amplitude A_0 has a maximum of about 1 V at $n = 0.12$. Such an n value corresponds to $C_1 = C/(1-n) = 3.4\,\text{pF}$ and $C_2 = C/n = 25\,\text{pF}$. The approximation (4.26) does not catch the maximum and provides a good approximation down to $n = 0.26$ with a maximum error of about 15%.

These results hold as long as the transistor operates in the forward active region along the whole oscillation waveform. Since the collector voltage is biased at the 2.5 V supply,

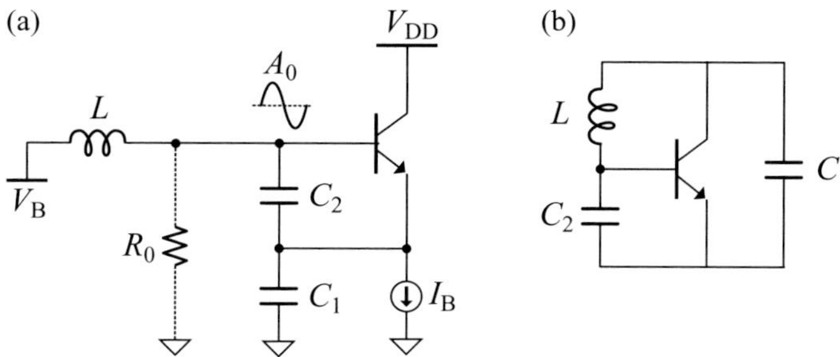

Figure 4.15 Clapp oscillator: (a) schematic and (b) a.c. equivalent circuit

an oscillation amplitude of about 1 V makes the collector voltage swing between 1.5 V and 3.5 V. Transistor saturation is therefore prevented by setting the base bias below 2 V.

4.3.5 Clapp oscillator

Another single-transistor oscillator is the Clapp oscillator shown in Figure 4.15(a). In this scheme, the harmonic signal is generated between the base terminal and ground. Figure 4.15(b) shows the a.c. equivalent of the Clapp oscillator. The circuit turns out to be the same as for the Colpitts oscillator. The design can, therefore, be pursued using the same analytical expressions.

The main advantage of the Clapp configuration is the possibility to derive the voltage signal at the collector output, decoupled from the resonator, simply by placing a resistor on the collector terminal. The solution reduces the resonator loading from the following stage. [6] Another typical variation of the Clapp topology is obtained by placing a capacitor in series to the inductor and biasing the transistor base with a resistor. [7, 8]

4.4 Differential oscillators

Voltage-controlled oscillators for integrated transceivers are usually differential, since they are used to drive differential active mixers. The adoption of oscillator topologies using differential stages is, therefore, the most natural choice. However, differential outputs can also be obtained by coupling two single-transistor oscillators, thus leading to the so-called balanced oscillators. This section is devoted to the discussion of all these circuit solutions.

4.4.1 MOSFET cross-coupled oscillators

Figures 4.16(a) and (b) show two typical MOSFET differential oscillators. [9, 10] The two circuits have the same topology, the only difference being the position of the current generator I_T. They are, therefore, referred to as bottom-biased and top-biased differential schemes, respectively. In practice, there are, indeed, some differences. In the bottom-biased oscillator, the d.c. voltage of the output nodes is set to V_{DD} via the inductors and the non-zero

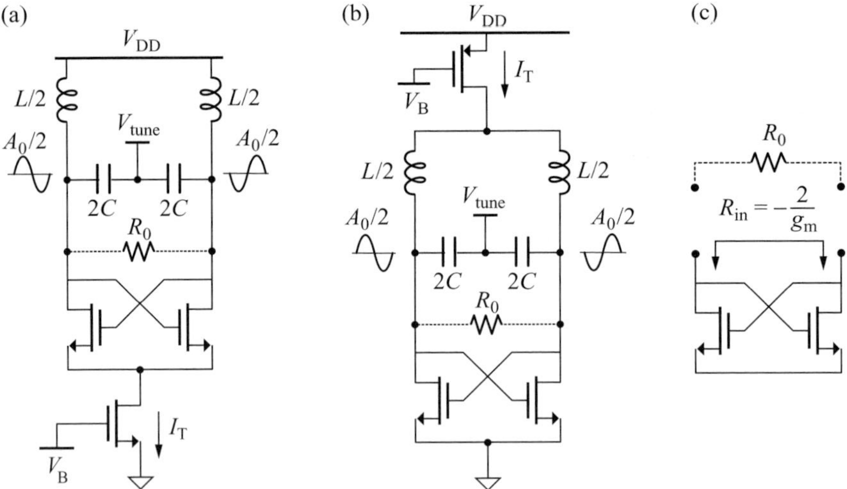

Figure 4.16 MOS differential oscillator: (a) bottom-biased topology, (b) top-biased topology, (c) simplified transconductor circuit

source–bulk voltage of the transistor pairs increases the threshold and makes the oscillator more sensitive to substrate noise. Of course, these drawbacks may be avoided by using a triple-well process or pMOS transistor pairs within their own n-well, and by shortening the bulk and source terminals to limit pick-up from disturbances.

These circuits represent the differential implementation of the prototype oscillator in Figure 4.1. Their analysis is, therefore, straightforward, even if attention should be paid to some features: (i) the amplitude of the differential output is twice the amplitude of each output node swing. Denoting as A_0 the amplitude of the differential signal, each drain node in Figure 4.16(a) swings between $(V_{DD} - A_0/2)$ and $(V_{DD} + A_0/2)$ with opposite sign (a π phase shift). (ii) The resonator inductance L in the prototype circuit shown in Figure 4.1 is here split into two $L/2$ components and the capacitance C is divided into two $2C$ capacitors placed in series. (iii) The tuning voltage V_{tune} is an a.c. ground.

In the small-signal regime, the cross-coupled pair synthesizes a negative input resistance equal to $-2/g_m$. As shown in Figure 4.16(c), this resistance appears in parallel with the positive resistance $R_0 = \omega_0 L \cdot Q$. The start-up requires:

$$g_m R_0/2 > 1. \tag{4.31}$$

The mechanism limiting the oscillation amplitude is the same as has already been described in Section 4.2.4. When the voltage signal is high enough to make the stage fully switching, the current flowing through each half tank is ideally a square wave ranging from 0 to I_T. The differential current delivered to the LC tank is, therefore, a periodic square wave switching between $\pm I_T/2$. In practice, the differential stage acts as a hard limiter. The first harmonic of this square wave is $2I_T/\pi$, and the effective transconductance of the differential pair can be written as:

$$G_{m,eff} = \frac{2I_T}{\pi A_0}.$$

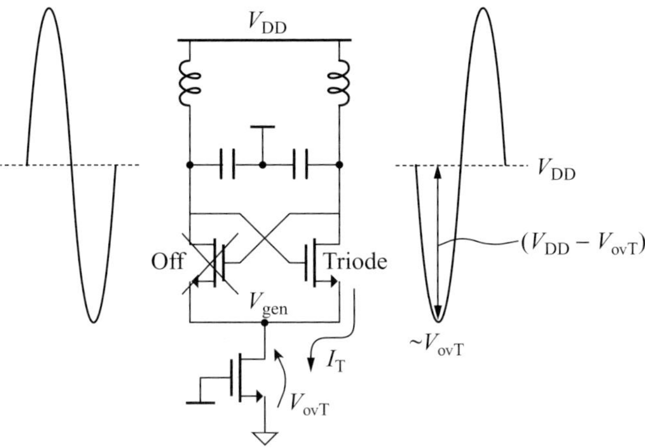

Figure 4.17 Switching operation of a MOS differential oscillator

From the Barkhausen condition on the open-loop gain ($G_{m,eff}R_0 = 1$), the differential oscillation amplitude turns out to be:

$$A_0 = \frac{2}{\pi} I_T R_0. \tag{4.32}$$

This expression holds, so long as the current waveform can be approximated by a square wave, i.e., if the amplitude is much larger than the transistor over-drive voltage $V_{ov} = (V_{GS} - V_T)$ and the transistor delays are negligible.

Note that the oscillation amplitude can be high enough to push the transistor pair alternately in the ohmic region during the negative wave of the drain voltage (Figure 4.17). So long as the tail-current generator is in saturation, nothing changes in the stage operation. The differential pair still behaves as a hard limiter and (4.32) applies. However, it should be considered that a transistor in the ohmic region is, basically, a resistor and, therefore, the negative voltage swing is transferred to the drain of the tail transistor, which may be pushed into the ohmic region too. To retain the transistor in saturation, the negative swing has to be limited to V_{ovT}, the tail transistor over-drive voltage. The maximum single-ended amplitude is, therefore, $(V_{DD} - V_{ovT})$ and the differential amplitude is:

$$A_0 = 2 \cdot (V_{DD} - V_{ovT}).$$

Up to this value the circuit works in a *current-limited regime*, i.e., the oscillation current is set by the bias current. It is worth noting that the differential voltage applies between the gate and the drain of the pair transistors. Therefore, transistor reliability may become an issue when the design is pushed towards large amplitude values.

As the current is increased further, the oscillator enters the *voltage-limited regime.* At the minima of the output voltage signal, the tail transistor becomes ohmic and its current lowers below I_T. The current waveform injected into the resonator departs from the ideal square wave. Its detailed dependence is shown by the solid line in Figure 4.18. Note the valley corresponding to the minimum of the drain voltage.

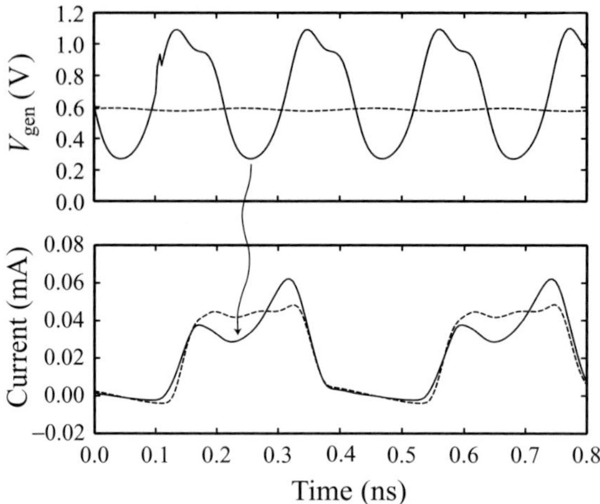

Figure 4.18 Drain voltage V_{gen} of the current generator and current of the transistor pair for the oscillator in Figure 4.17 (solid line) and the oscillator in Figure 4.19 (dashed line)

In VCOs, other mechanisms may limit the oscillation amplitude. For example, in VCOs designed according to Figure 4.16(a), pn junction varactors replace the capacitors. For the diodes to be reverse biased, they should be placed with the cathode linked to the transistor drains, whose d.c. value is V_{DD}. At any rate, they will also limit the maximum achievable peak amplitude. For instance, if $V_{tune} = V_{DD}$, the voltage across each diode is $A_0/2$ and the oscillation amplitude A_0 should be limited to about 1.4–1.6 V, so as not to forward bias the components.

4.4.2 Tail resonator

The oscillation amplitude can be increased even beyond $2V_{DD}$. An extension of the current-limited regime is, for example, made possible by the adoption of a tail resonator at $2\omega_0$ (Figure 4.19). [11] The rationale for this solution may be illustrated by starting from the oscillator in Figure 4.17. Since the transistors of the pair are alternately ohmic, V_S follows the dips of the output voltage waveform (Figure 4.18), oscillating at $2\omega_0$. In Figure 4.19, the drain capacitance C_2 is chosen to be much larger than C_1, to provide a short at $2\omega_0$. In this way, the voltage across the bias generator V_{gen} remains almost constant (Figure 4.18 – dashed line). The tail generator is no longer pushed to become ohmic, even at large amplitudes, and the current injected into the resonator is much closer to the ideal square wave (Figure 4.18). On the other hand, by setting the L_1C_1 resonance at $2\omega_0$ the voltage signal V_S is left free to swing below 0 V, until it is clamped by the source–bulk junctions of the transistor pair. The output differential amplitude can, therefore, increase beyond $2V_{DD}$, as depicted in Figure 4.20 (solid squares). The tail resonator has also beneficial impact on noise filtering, [11] which will be discussed in Chapter 5.

Example 4.2 Design of a MOSFET cross-coupled oscillator

In analogy to Example 4.1, let us discuss the design of the cross-coupled oscillator in Figure 4.16(a) using 2.5 V CMOS technology with 0.25 μm minimum length. As in the previous

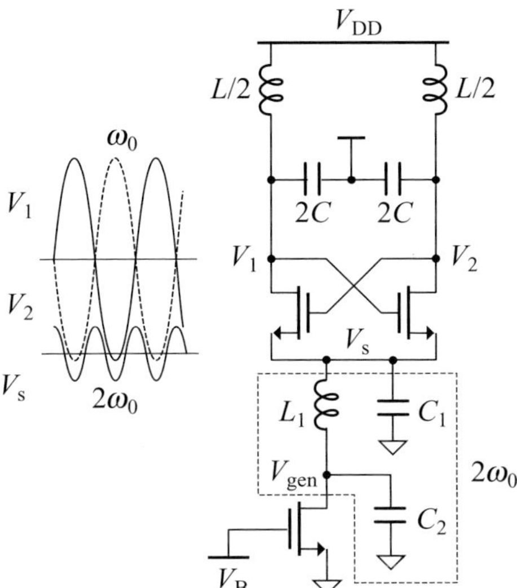

Figure 4.19 MOS differential oscillator with tail inductive peaking

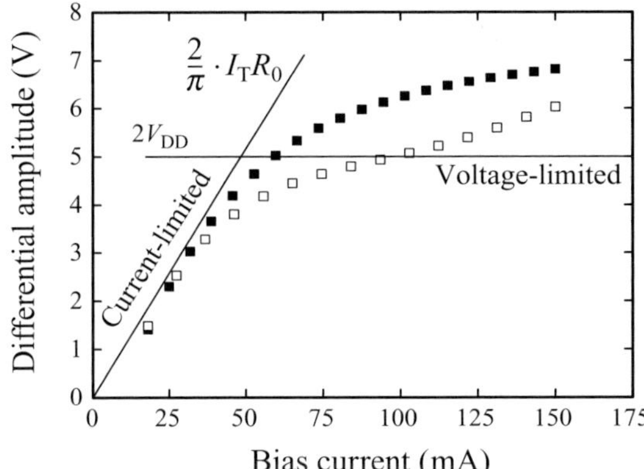

Figure 4.20 Oscillation amplitude versus bias current: the oscillator in Figure 4.16 (b) (hollow squares), the oscillator in Figure 4.19 (filled squares) and current- and voltage-limited behaviour (solid lines)

example, the circuit has to oscillate at 2.4 GHz using capacitors with a Q factor of 30 at 2.4 GHz and an integrated coil with 0.7 nH inductance and Q factor of 10. The voltage supply is applied to a terminal placed in the middle of the integrated inductor.

The overall resonator quality factor is $Q = Q_C Q_L/(Q_C + Q_L) = 7.5$, corresponding to an equivalent parallel loss resistance $R_0 = \omega_0 L Q = 162\ \Omega$. The resonance frequency is set by taking $C = 1/(\omega_0^2 L) = 3$ pF, which means that two capacitors of 6 pF have to be placed in series.

The oscillation start-up condition (4.31) leads to $g_m > 2/R_0 = 12.3$ mS. Unlike bipolar transistors, MOSFETs have a transconductance depending not only on the bias current but also on the transistor aspect ratio. At constant current, the maximum transconductance is achieved by increasing the MOSFET width and moving towards the subthreshold operating region. However, in the design of oscillators, transistor parasitic capacitances reduce the achievable tuning range. The designer should, therefore, aim to achieve sufficient transconductance with the minimum transistor area. These conditions usually prevent the use of MOSFETs in the subthreshold regime.

In the technology adopted for the design, a 10 μm × 0.25 μm MOSFET has a cut-off frequency of 40 GHz at 2 mA and a transconductance of about 3.3 mS. This value is not enough to start up the oscillation. Both the current and the width have to be increased. By using a 45 μm width and increasing the tail current to 18 mA, the g_m reaches 13.5 mS with an over-drive of 1.1 V. The gate polysilicon resistance may raise the transistor delay and thermal noise. Employing a proper number of fingers (nine in this case) limits its contribution to negligible values.

Figure 4.20 (hollow squares) shows the dependence of the oscillation amplitude versus the tail current, as obtained from transient simulations of the circuit. The oscillation starts up for tail currents larger than 18 mA. As the current is increased, the transistors have been widened proportionally, preserving the over-voltage. The amplitude values given by (4.32) are shown by the solid line. This accounts well for the simulation results up to 3 V amplitude. Note that, for current values larger than 100 mA, the oscillator waveform exceeds the $2V_{DD} = 5$ V limit. This is because the graph in Figure 4.20 shows the peak voltage of the differential waveform and not the amplitude of the first harmonic. In this high-current regime, the waveforms are highly distorted and their peaks can be higher than $2V_{DD}$.

4.4.3 MOSFET complementary cross-coupled oscillator

An advantage of CMOS technology is the availability of complementary devices. Figure 4.21(a) shows an LC-tuned oscillator using both transistor types. [12, 13]. The most relevant difference, compared with the differential oscillators seen before, is the re-use of the tail current. Figure 4.21(b) shows that if one nMOS is ohmic, then the pMOS on the opposite side is also ohmic. The topology makes the bias current I_T flow completely through the tank, thus making it possible to reach twice the transconductance and twice the oscillation amplitude for a given I_T.

Denoting as g_{mN} and g_{mP} the transconductance of the nMOS and pMOS transistors, respectively, the small-signal impedance in parallel with the tank is $-2/(g_{mN} + g_{mP})$. The start-up condition becomes:

$$(g_{mN} + g_{mP})R_0/2 > 1.$$

In full switching operation, the current waveform ranges from $+I_T$ to $-I_T$. In the oscillators in Figure 4.16, the differential current was, instead, switching between $\pm I_T/2$. The corresponding voltage amplitude is, therefore:

$$A_0 = (4/\pi)I_T R_0,$$

that is, twice the value of the single stage. Further discussion of the topology can be found in Example 7.2.

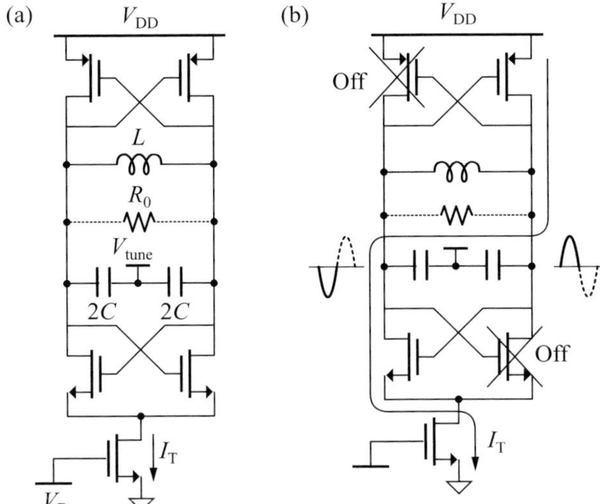

Figure 4.21 MOS differential oscillator with two cross-coupled pairs: (a) schematic and (b) switching behaviour

4.4.4 Bipolar cross-coupled oscillators

The differential oscillator can also be implemented using bipolar transistors. Only the adoption of npn cross-coupled oscillators is feasible, since fast pnp devices are not available in advanced RF bipolar processes. When compared with MOSFETs, bipolar transistors have two main advantages: (i) a lower $1/f$ noise and (ii) lower parasitic capacitances. However, the bipolar transistor may enter saturation for relatively low oscillation amplitudes. In the differential stage in Figure 4.22(a), the output signal amplitude A_0 is also the device's base–collector voltage. Therefore, the oscillator is already voltage limited to $A_0 \approx V_{BCsat}$, which is between 0.8 and 1 V in high-f_T bipolar transistors.

Such a limitation can be extended by decoupling the d.c. voltage of the collector and base terminals. Both capacitors [14] and coupled inductors [15] may be used for this purpose. Figure 4.22(b) shows the approach based on capacitors. The maximum oscillation amplitude now becomes $A_0 = V_{DD} - V_B + V_{BCsat}$, where V_B is the d.c. base voltage. The minimum voltage headroom V_{Gsat} required by the tail-current generator imposes a lower limit to V_B: $V_B > V_{Gsat} + V_{BEon}$, where V_{BEon} is the base–emitter voltage of the bipolar transistors of the differential pair. Therefore, the maximum oscillation amplitude is:

$$A_{0,max} = V_{DD} - V_{Gsat} - V_{BEon} + V_{BCsat}. \tag{4.33}$$

However, an additional margin has to be guaranteed in order not to drive the transistors too close to saturation, where both the current gain and the cut-off frequency are severely reduced.

A further increase of the maximum oscillation amplitude can be obtained by introducing a capacitive voltage divider between the oscillator output nodes and the inputs of the differential transconductor (Figure 4.22(c)). The output signal is attenuated by a factor

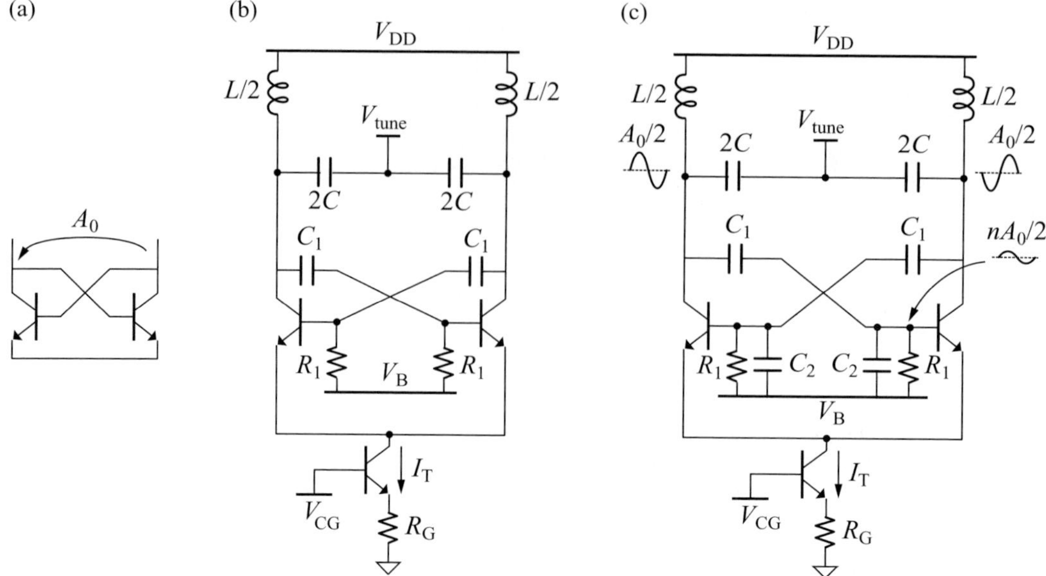

Figure 4.22 Bipolar differential oscillator: (a) simplified transconductor circuit; (b) topology with a.c. coupling between collector and base; (c) topology with capacitive voltage divider between collector and base

of $n = C_1/(C_1 + C_2)$, before feeding the bases of the bipolar transistors. Referring to Figure 4.22(c), the maximum oscillation amplitude becomes:

$$A_{0,\max} = \frac{2}{1+n} \cdot (V_{DD} - V_{Gsat} - V_{BEon} + V_{BCsat}), \tag{4.34}$$

that is, a factor of $[2/(1+n)]$ higher than the value reached by simple a.c. coupling. By choosing small n, the maximum oscillation amplitude can be increased by almost a factor of two. However, the voltage divider reduces the oscillator open-loop gain, leading to a potential start-up problem. The start-up condition becomes

$$\frac{g_m R_0 n}{2} > 1 \Rightarrow n > \frac{8}{\pi} \cdot \frac{V_T}{A_0}, \tag{4.35}$$

since $g_m = I_T/2V_T$ and $A_0 = (2/\pi)I_T R_0$. Using the minimum value of voltage partition given by (4.35), the expression for the maximum A_0 in (4.34) becomes:

$$A_{0,\max} = 2 \cdot (V_{DD} - V_{Gsat} - V_{BEon} + V_{BCsat}) - (8/\pi) \cdot V_T, \tag{4.36}$$

which is about twice the oscillation amplitude achievable without using any voltage partition.

Example 4.3 Design of a bipolar cross-coupled oscillator

This example deals with the design of bipolar cross-coupled oscillators based on the topologies shown in Figure 4.22. The resonator is the same used in Example 4.2, for the MOSFET cross-coupled oscillator.

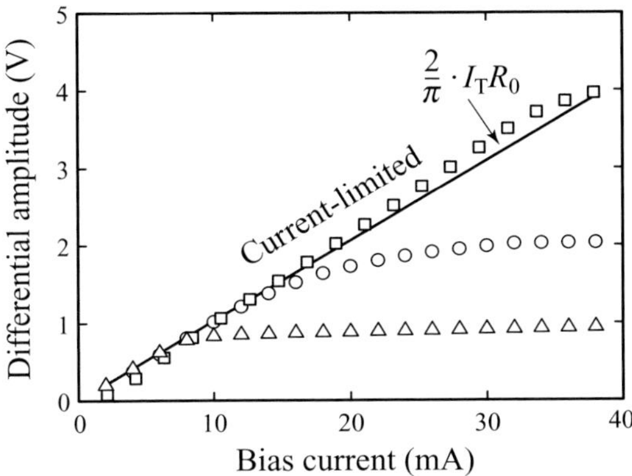

Figure 4.23 Oscillation amplitude versus bias current of bipolar differential oscillators from theory (solid line) and from circuit simulation: (triangles) basic topology, (circles) with a.c. coupling, (squares) with a.c. coupling and capacitive divider

The condition for oscillation start-up requires:

$$g_m R_0/2 > 1 \Rightarrow I_T > \frac{4V_T}{R_0}.$$

Since $R_0 = 162\ \Omega$ (see Example 4.2), the lower boundary to the tail current is 620 μA. Figure 4.23 shows the oscillation amplitude derived from circuit-level transient simulations (triangles) as a function of the tail current. While changing the tail current, the bipolar transistors have been re-sized, to keep the current density constant. According to the square wave model of the current waveform, the oscillation amplitude should follow $A_0 = (2/\pi)\, I_T R_0$. The dependence is plotted as a solid line in Figure 4.23. The theoretical expression is in fair agreement up to about 9 mA, where the amplitude saturates to about $V_{BCsat} = 900$ mV and the base–collector junction starts to be forward biased.

Also, the circuit in Figure 4.22(b) has been simulated to demonstrate the possibility of attaining larger amplitudes. The value of the resistance R_1 must be much larger than the reactance from C_1 at the resonance frequency, $1/\omega_0 C_1$. In this way, the coupling capacitors C_1 act as short circuits, the voltage signal at each collector terminal is transferred to the corresponding base and the open-loop gain remains unhindered. Moreover, the resistance R_1 has to be much larger than the half-tank parallel loss resistance of 81 Ω, to preserve the resonator quality factor. A possible choice is $R_1 = 400\ \Omega$ and $C_1 = 1$ pF (corresponding to a reactance of 33 Ω at 2.4 GHz), which gives negligible degradation to both the open-loop gain and the quality factor.

The d.c. base voltage V_B has been set to 1.3 V, so as to leave about 0.4 V headroom to the current generator. No degeneration resistance is used in the current generator. Therefore, the expected maximum amplitude can be obtained by (4.33): $A_{0,\max} = (2.5 - 0.4 - 0.9 + 0.9)\ \text{V} = 2.1$ V, where $V_{DD} = 2.5\ V$ and $V_{BEon} = 0.9$ V.

The oscillation amplitude versus the tail current obtained from circuit simulations of this case is plotted as circles in Figure 4.23. The simulations confirm the expected saturation of the oscillation amplitude.

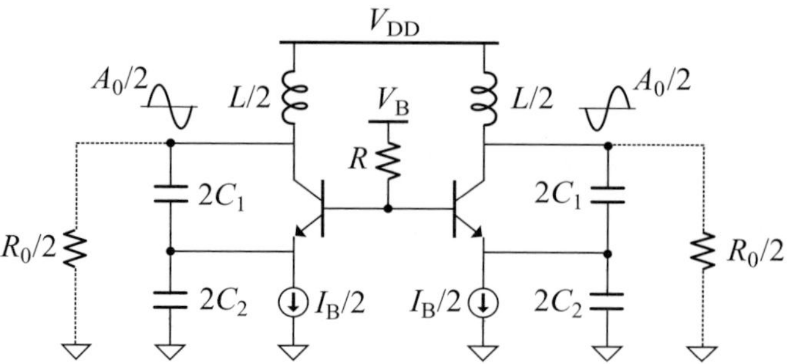

Figure 4.24 Balanced Colpitts configuration

The circuit in Figure 4.22(c) may extend the oscillation amplitude further. At $A_0 \approx V_{DD} = 2.5\,V$, the minimum n value given by (4.35) is of the order of 10^{-2}, which would require a capacitor C_2 about 100 times larger than C_1. A reasonable ratio between C_2 and C_1, limiting the area occupation, is 10. Taking $n = 0.1$ in (4.34), the maximum amplitude is $A_{0,\max} \approx 1.8\,(V_{DD} - V_{Gsat} - V_{BEon} + V_{BCsat}) = 4\,V$. The resistance $R_1 = 400\ \Omega$ is transformed by the voltage divider by a factor of $1/n^2$; therefore, it does not degrade the tank parallel resistance. Taking $C_1 = 1$ pF and $C_2 = 9$ pF, $n = 0.1$ and a capacitance $(1/2)C_1C_2/(C_1 + C_2) = 0.45$ pF is added to the tank capacitance. The latter must, therefore, be lowered to 2.55 pF to meet the 2.4 GHz oscillation frequency. Figure 4.23 shows the simulated oscillation amplitude (squares), which starts to saturate, as predicted, at 4 V.

Example 4.4 Balanced Colpitts oscillator

Colpitts oscillators, as well as other single-transistor oscillators, can be modified to provide two outputs of opposite phase. Figure 4.24 shows two oscillators coupled in a balanced configuration. The reason why the oscillators work with opposite phase can be explained by using an energetic argument. If the oscillation mode is odd, the common base voltage is almost constant and the two circuits work as a single-transistor Colpitts oscillator. If they oscillate in phase (even mode), the common base would oscillate at the same frequency and the resistor R would dissipate more energy. The odd mode, therefore, has fewer power losses and will be triggered more easily. The oscillation of the even mode is prevented by setting R to equal about $100\ \Omega$.

Once the circuit works in odd mode, each oscillator works as a single-ended stage. The oscillation amplitude may be approximated by (4.26) after replacing A_0 with $A_0/2$, I_B with $I_B/2$ and R_0 with $R_0/2$:

$$A_0 \approx (1 - n)\, I_B R_0, \tag{4.37}$$

which is half the amplitude of the single-ended Colpitts oscillator biased at current I_B. This is not surprising, since differential circuits typically double the power consumption compared with their single-ended companions.

Figure 4.25 shows the oscillation amplitude versus the bias current I_B as given from (4.37) for $n = 0.5$ (solid line) and the circuit simulations of the balanced Colpitts oscillator (hollow and filled circles). These are in fair agreement with the theoretical estimation.

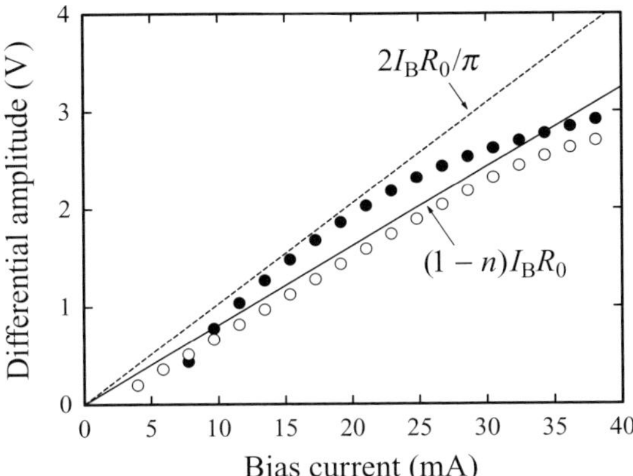

Figure 4.25 Oscillation amplitude versus bias current of the oscillator in Figure 4.24: theoretical dependence for the Colpitts (solid line) and cross-coupled oscillator (dashed line), simulated results for the Colpitts taking $n = 0.5$ (hollow circles) and $n = 0.12$ (filled circles)

Equation (4.37) can be compared with the amplitude of the cross-coupled oscillator with the same bias current I_B and differential resonator resistance R_0:

$$A_0 = (2/\pi)\, I_B R_0.$$

The dependence is also plotted versus I_B as a dashed line in Figure 4.25. It seems that the balanced Colpitts topology, with $n < 0.36$, should feature an oscillation amplitude slightly larger than the cross-coupled topology in the current-limited regime. In practice, (4.37) is no longer accurate for low n values. Simulations for $n = 0.12$ (filled circles in Figure 4.25) show a dependence which does not exceed the theoretical differential limit. In practice, the balanced Colpitts oscillator and the cross-coupled oscillator with identical resonator and power consumption achieve approximately the same maximum oscillation amplitude.

4.5 References

[1] K. S. Kundert, Introduction to RF simulation and its application, *IEEE J. Solid-St. Circ.*, **34**, Sep. 1999, 1298–319.

[2] D. M. Pozar, *Microwave Engineering*, New York, NY: John Wiley and Sons, Inc, 2nd edn, 1997.

[3] K. K. Clarke and D. T. Hess, *Communication Circuits: Analysis and Design*, Reading, MA: Addison-Wesley, 1971.

[4] M. Abramowitz and I. A. Stegun, *Handbook of Mathematical Functions with Formulas, Graphs, and Mathematical Tables*, New York, NY: Dover, 1972, 374–7.

[5] G. A. M. Hurkx, Bipolar transistor physics, in *Bipolar and Bipolar-MOS Integration*, ed. by P. A. H. Hart, Amsterdam, Netherlands: HartElsevier Science, 1994, chapter 3.

[6] L. Dauphinee, M. Copeland and P. Schvan, A balanced 1.5 GHz voltage controlled oscillator with an integrated LC resonator, *Digest Tech. Papers 44th IEEE International Solid-State Circuits Conference*, Feb. 1997, 390–1.

[7] X. Wang, A. Fard and P. Andreani, Phase noise analysis and design of a 3-GHz bipolar differential Colpitts VCO, *Proc. 31st European Solid-State Circuits Conference*, Sep. 2005, 391–4.

[8] A. Bonfanti, S. Levantino, S. Pellerano *et al.*, A voltage-controlled oscillator for IEEE 802.11a and HiperLAN2 application, *Proc. 29th European Solid-State Circuits Conference*, Sep. 2003, 695–8.

[9] J. Craninckx and M. S. J. Steyaert, A 1.8-GHz low-phase-noise CMOS VCO using optimized hollow spiral inductors, *IEEE J. Solid-St. Circ.*, **32**, May 1997, 736–44.

[10] Z. Li and K. K. O, A low-phase-noise and low-power multiband CMOS voltage-controlled oscillator, *IEEE J. Solid-St. Circ.*, **40**, Jun. 2005, 1296–302.

[11] E. Hegazi, H. Sjoland and A. A. Abidi, A filtering technique to lower LC oscillator phase noise, *IEEE J. Solid-St. Circ.*, **36**, Dec. 2001, 1921–30.

[12] A. Hajimiri and T. H. Lee, Design issues in CMOS differential LC oscillators, *IEEE J. Solid-St. Circ.*, **34**, May 1999, 717–24.

[13] H. Wang, Comments on 'Design issues in CMOS differential LC oscillators', *IEEE J. Solid-St. Circ.*, **35**, Feb. 2000, 286–7.

[14] S. Cosentino *et al.*, An integrated RF transceiver for DECT application, *IEEE J. Solid-St. Circ.*, **37**, Mar. 2002, 443–9.

[15] M. Zannoth, B. Kolb, J. Fenk and R. Weigel, A fully integrated VCO at 2 GHz, *IEEE J. Solid-St. Circ.*, **33**, Dec. 1998, 1987–91.

5 Noise in oscillators

5.1 Introduction

The evaluation of the oscillator phase noise is a classical issue. Some of the fundamental papers date back to the 1960s [1–4] and recently the topic has received fresh attention as full integration of RF systems has become the focus of microelectronic design. Designers of VCOs had to rely on noise models either empirically explained or based on highly questionable linear small-signal analysis. Since clear guidelines for circuit optimization were lacking, design was mainly based on a trial-and-error approach. Only at the end of the 1990s was a deeper insight in VCO noise analysis gained. Two frameworks were proposed: one working in the time domain, [5–8] the other in the frequency domain. [9–11] They both succeeded in providing the first quantitative guidelines to noise optimization, linking phase-noise performance to the transfer of the noise sources in the circuit. This grounding is essential for later figuring out proper options and modifications of the circuit topology.

Accurate circuit-level simulators, which have been developed meanwhile, [12–14] simplify and speed up the proper tuning of design parameters and the noise performance evaluation in every operating condition.

This chapter is devoted to describing time-domain and frequency-domain methods for the oscillator noise analysis. Some examples of phase noise calculation based on these theoretical frameworks and verified against the circuit-level simulator results are also shown.

5.2 Linear and time-invariant model

The spectrum of real oscillators is far from being a δ-like function at the oscillation frequency ω_0. It shows, instead, sidebands decaying as a function of the ω_m offset from the carrier frequency. Most of the time, two regions may be clearly defined (Figure 1.4). Very close to the carrier frequency, the sideband power density decays as $1/\omega_\mathrm{m}^3$. At larger offset a $1/\omega_\mathrm{m}^2$ power law better describes the experimental result. The $1/\omega_\mathrm{m}^3$ dependence is originated by up-conversion of $1/f$ (flicker) frequency noise, while the $1/\omega_\mathrm{m}^2$ region depends on the white noise sources in the oscillator components.

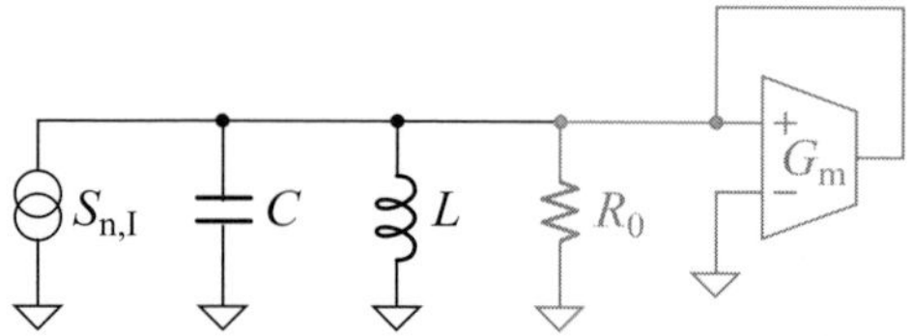

Figure 5.1 Simple model of phase noise: the noise current generator injects noise into an ideal tank

As recalled in Chapter 1, the oscillator quality is measured by the single-sideband-to-carrier ratio (SSCR):

$$\mathcal{L}(\omega_m) = \frac{\text{power in 1 Hz bandwidth}}{\text{carrier power}} = \frac{S_V(\omega_0 \pm \omega_m)}{A_0^2/2} \cong \frac{S_\phi(\omega_m)}{2},$$

the ratio between the noise power in the sideband, measured in 1Hz bandwidth, and the carrier power. Before recent advances, phase noise evaluations were addressed following the semi-empirical model presented in [3]. Let us derive the result by starting with the evaluation of the phase noise contribution arising from tank losses.

Figure 5.1 shows the equivalent loss resistance R_0 and its thermal noise power density $S_{n,I} = 4kT/R_0$. The resulting voltage spectrum at the output node can be derived following the usual noise transfer methodology: by computing the product of the current noise power density, $S_{n,I}$, and the square amplitude of the tank impedance, $Z(\omega)$. The latter is affected by the oscillator feedback. Since the active stage perfectly balances the power dissipated in R_0, this resistive element is cancelled out and $Z(\omega)$ reduces to the impedance of an ideal LC resonant network. For $\omega_m \ll \omega_0$, $Z(\omega)$ may be written as:

$$|Z(\omega_0 \pm \omega_m)|^2 \cong \frac{1}{(2\omega_m C)^2}. \tag{5.1}$$

The argument reduces the overall circuit to a time-invariant resonant network, leading to an output voltage power spectral density, $S_V(\omega_0 \pm \omega_m)$ given by:

$$S_V(\omega_0 \pm \omega_m) = S_V(\omega_m) \cong \frac{4kT}{R_0} \cdot \frac{1}{(2\omega_m C)^2}.$$

From now on, the voltage spectrum at the ω_m offset will be indicated as $S_V(\omega_m)$.

This result should be slightly modified, indeed. The overall power $S_{n,I}$ can be split evenly into amplitude and phase carrier noise. Only half of the noise power, that due to phase fluctuation, is shaped according to (5.1), while the impact of amplitude noise is negligible:

$$S_V(\omega_m) \cong \frac{1}{2} \cdot \frac{4kT}{R_0} \cdot \frac{1}{(2\omega_m C)^2}.$$

The last step may be justified by taking into account the fact that transistors in oscillators operate as switching elements. They have been represented by a hard limiter stage in Figure 4.4. Staying with this equivalent model, note that the current output of the active stage switches on the zero crossing of the input signal. It is, therefore, insensitive to any amplitude fluctuation of the voltage waveform. On the other hand, phase fluctuations of the harmonic signal cause jitter of the zero crossing and shift the switching edges of the current waveform at the transconductor output. In other words, while amplitude modulation is cut off by the limiting transconductor, phase modulation circulates along the loop.

This difference affects the impedance $Z(\omega)$ that should be considered for the resonant network in noise transfer. Since amplitude modulation is cut off, no transconductor feedback is able to cancel out the tank losses. Therefore, for the amplitude noise, $Z(\omega)$ is simply given by the tank parallel resistance R_0. The $1/\omega_{\mathrm{m}}^2$ shaping in (5.1) vanishes close to the carrier, making the amplitude noise contributions to the output voltage spectrum negligible. For phase fluctuations, instead, since phase modulations circulate along the loop, the transconductor is able to cancel out the tank losses for phase noise, shaping the voltage spectrum as $1/\omega_{\mathrm{m}}^2$.

The formula may be semi-empirically extended to account for other noise contributions, such as those related to the active element. By introducing a suitable excess noise factor, F, the current noise spectral density becomes $S_{\mathrm{n,I}} = (1+F)\cdot 4kT/R_0$, and

$$S_{\mathrm{V}}(\omega_{\mathrm{m}} t) \cong \frac{1}{2}\cdot\frac{4kT}{R_0}\cdot\frac{(1+F)}{(2\omega_{\mathrm{m}} C)^2}.$$

The SSCR is

$$\mathcal{L}(\omega_{\mathrm{m}}) = \frac{S_{\mathrm{V}}(\omega_{\mathrm{m}})}{A_0^2/2} = \frac{2kT}{R_0}\cdot\frac{(1+F)}{(2\omega_{\mathrm{m}} C)^2\, A_0^2/2}. \tag{5.2}$$

By using $Q = \omega_0 R_0 C$, it can be written as:

$$\begin{aligned}\mathcal{L}(\omega_{\mathrm{m}}) &= \frac{2kT}{\left(A_0^2/2R_0\right)}\cdot\frac{1}{4Q^2}\cdot\left(\frac{\omega_0}{\omega_{\mathrm{m}}}\right)^2\cdot(1+F)\\ &= \frac{kT}{C}\cdot\left(\frac{\omega_0}{Q}\right)\cdot\frac{(1+F)}{\omega_{\mathrm{m}}^2 A_0^2}.\end{aligned} \tag{5.3}$$

This is essentially the Leeson phase noise result. [3] All the difficulties of noise evaluation are now hidden within F. Since no guidelines were given to link F to the noise sources, the equation was mainly used to fit parameters to measured data. Before proceeding further with the discussion of time-variant techniques for F evaluation, let us highlight some general consequences of (5.2).

5.3 Noise–power trade-off and scaling issues

5.3.1 Figure of merit

Equation (5.3) suggests that phase noise performance can be improved by increasing the tank capacitance C. As an example, if the capacitance C is doubled and the other parameters (Q, F and A_0) remain constant the phase noise power halves, or the phase noise improves by 3 dB.

On the other hand, to keep the same oscillation frequency, the tank inductance L has to be reduced by a factor of two. At constant Q, the equivalent loss resistance $R_0 = Q/\omega_0 C$ decreases by a factor of two and the oscillation amplitude follows the same reduction. To recover the original A_0 value, the bias current should be doubled. In summary, (5.3) suggests that a 3 dB phase noise improvement is obtained by doubling the power consumption.

Such a trade-off between noise and power leads us to consider the figure given by the product Π of $\mathcal{L}$ and the power dissipation $P_\mathrm{d} = V_\mathrm{DD} I_\mathrm{B}$ (V_DD and I_B are the voltage supply and the bias current, respectively). It is:

$$\Pi = \mathcal{L} \cdot P_\mathrm{d} = \frac{kT}{C} \cdot \frac{\omega_0}{Q} \cdot \frac{V_\mathrm{DD} I_\mathrm{B}}{\omega_\mathrm{m}^2 A_0^2} \cdot (1+F).$$

In the current-limited regime the oscillation amplitude is proportional to the bias current and to the tank losses, i.e., $A_0 = \eta R_0 I_\mathrm{B}$, and

$$\begin{aligned}\Pi &= \frac{kT}{C} \cdot \frac{\omega_0}{Q \cdot \omega_\mathrm{m}^2} \cdot \frac{V_\mathrm{DD} I_\mathrm{B}}{\eta I_\mathrm{B} R_0 A_0} \cdot (1+F) \\ &= \frac{kT}{Q^2} \cdot \left(\frac{\omega_0}{\omega_\mathrm{m}}\right)^2 \cdot \frac{1}{\eta} \cdot \frac{V_\mathrm{DD}}{A_0} \cdot (1+F).\end{aligned}$$

In the last step, the term $C R_0$ has been replaced by Q/ω_0. To have a figure whose value increases as the performance gets better, the oscillator figure of merit (FoM) can be defined as proportional to $1/\Pi$. The higher $1/\Pi$, the better the performance trade-off and the lower the $\mathcal{L}(\omega_\mathrm{m})$ value achievable with a given power budget or the lower the dissipated power to get a target $\mathcal{L}(\omega_\mathrm{m})$ value.

It is desirable to desensitize the FoM to the specific $\omega_0/\omega_\mathrm{m}$ ratio, thus enabling comparison between different oscillators operating at different frequencies and having $\mathcal{L}(\omega_\mathrm{m})$ quoted at different carrier offsets. The FoM is, therefore, finally defined as: [15]

$$\mathrm{FoM} = \frac{1}{\mathcal{L} \cdot P_\mathrm{d}} \cdot \left(\frac{\omega_0}{\omega_\mathrm{m}}\right)^2 = \frac{Q^2}{kT} \cdot \eta \cdot \frac{A_0}{V_\mathrm{DD}} \cdot \frac{1}{(1+F)}. \tag{5.4}$$

It is necessary to remember that in the literature the power in (5.4) is always in mW, and the whole FoM is expressed in dB, that is, the number usually provided is $10 \cdot \log_{10}(\mathrm{FoM})$.

Equation (5.4) highlights that the power–phase noise trade-off depends on:

- The tank quality factor.
- The efficiency of the stage in driving the first harmonic of the voltage, as expressed by η.
- The ratio between the supply voltage and the oscillation amplitude.
- The additional noise introduced by the active element.

Usually the designer has in mind that good phase noise performance requires a high tank quality factor. This is surely a prerequisite; however, the choice of the oscillator topology also has an impact, since it limits the achievable oscillation peak amplitude and the A_0/V_DD ratio. In this respect, peaking strategies such as the one implemented in Figure 4.19 may be successfully applied to oscillator circuits to extend the oscillation amplitude.

5.3.2 Limits to the exploitation of noise–power trade-off

Once the figure of merit has been optimized, the only way to reduce the phase noise further is to trade better $\mathcal{L}(\omega_\mathrm{m})$ against higher power consumption at constant FoM. This is obtained by increasing the tank capacitance and decreasing L by the same factor. Unfortunately, feasibility issues impose practical limits to this noise–power trade-off, the limiting factor being the robustness of the LC tank against parasitics.

As an example, let us consider the quality factor of the capacitance due to the series resistance; it is $Q_{sC} = 1/\omega_0 C R_{sC}$. As the capacitance increases, the series parasitic resistances should decrease, to give the same Q factor. However, the larger the capacitance, the longer the interconnections and the higher the parasitic resistances. In other words, as the C/L ratio increases to give lower noise and higher power consumption, parasitics make it difficult to retain a high quality factor of the tank. Furthermore, small inductance values, below some hundreds of pH, are hard to synthesize with good accuracy, since parasitic inductance of the metal interconnects starts to dominate.

A further limit may be posed by thermal effects. As the power consumption becomes large, the increase of the die temperature can be relevant, increasing both the resistance values and the thermal noise. Noise reduction could, therefore, be not as significant as expected.

Finally, the FoM does not address the constraints set by the oscillator tuning range, which is, instead, a fundamental requirement for a VCO. In the next chapter it will be seen that varactors have a trade-off between tuning range and quality factor. A wide tuning range adversely affects the overall Q value.

In conclusion, the FoM is a useful concept but parasitics and other constraints may set limits to performance optimization.

5.3.3 Technology scaling

It is worth noting that the limitations set by the C/L ratio get worse as the technology scales down. The power consumption needed to reach a target phase noise specification $\mathcal{L}(\omega_m)$ can be derived from the FoM expression:

$$P_d = \frac{1}{\mathcal{L} \cdot \mathrm{FoM}} \cdot \left(\frac{\omega_0}{\omega_m}\right)^2.$$

Moreover:

$$P_d = V_{DD} I_B = V_{DD} \cdot \frac{A_0}{\eta R_0} = V_{DD} \cdot \frac{A_0}{\eta Q} \cdot \sqrt{\frac{C}{L}}.$$

In the last step, the loss resistance R_0 has been written as $R_0 = Q/\omega_0 C = Q\sqrt{L/C}$. By combining the two equations it turns out that:

$$\sqrt{\frac{C}{L}} = \frac{1}{\mathcal{L} \cdot \mathrm{FoM}} \cdot \frac{\eta Q}{V_{DD} A_0} \cdot \left(\frac{\omega_0}{\omega_m}\right)^2.$$

This result shows that as V_{DD} scales down as $1/\alpha$, followed by the oscillation amplitude A_0, the same circuit may get the same FoM and phase noise requirements at constant quality factor provided that the C/L ratio increases as α^4.

A simple example could help us to understand the role of the various parameters. By assuming that the supply voltage is reduced by a factor of two, the same power consumption is achieved by doubling the bias current. However, following the supply scaling, the oscillation amplitude also has to be reduced by a factor of two. It turns out that the equivalent tank loss resistance $R_0 = Q\sqrt{L/C}$ should decrease by a factor of four. At constant Q, the C/L ratio should, therefore, increase by a factor of sixteen. In pursuing this target, however, either the Q factor of the tank gets worse because the parasitic series resistances do not properly scale or L gets a minimum feasible value.

The L scaling law may be derived by starting again from (5.3). Taking into account that A_0 is proportional to V_{DD} and that $C = 1/\omega_0^2 L$, it is:

$$L \propto \frac{\mathcal{L}(\omega_{\mathrm{m}}) \cdot \omega_{\mathrm{m}}^2 \cdot Q}{\omega_0^3} \cdot V_{\mathrm{DD}}^2.$$

The expression shows how L should be reduced as the technology scales and, above all, with the rise of operating frequencies. Of course, in making this kind of prediction it is also necessary to take into account how the noise specifications, $\mathcal{L}(\omega_{\mathrm{m}})$, are expected to change in standards operating at higher frequencies.

5.4 Time-variant models

The effects of the noise sources in the active element have been represented so far by the noise factor F, which, in the Leeson framework, should be determined empirically. More recently, a theoretical framework for the F evaluation has been developed. Starting from the concept originally proposed by Kaertner, [16] Hajimiri and Lee studied the transfer of each noise source in the time domain using the impulse sensitivity function (ISF). [5] A second framework considers, instead, the noise transfer in the frequency domain. [9–11] Both methods give the designer guidelines to minimize the phase noise. We will mainly follow the second approach, although the ISF concept and its application will be reviewed.

5.4.1 Time-domain approach: the impulse sensitivity function

An oscillator may be modelled as a system with different inputs, each associated with a noise source, and two outputs: A_0 and ϕ, i.e., the instantaneous amplitude and phase of the oscillator. Noise inputs are represented by current sources forcing the circuit nodes and voltage sources in series with circuit branches. Each input source may cause A_0 and ϕ to change.

Let us now assume that a noise current pulse is injected into a node at time τ. Two responses in the time domain are considered. The function $h_\phi(t,\tau)$ is the impulse response linking the output phase to the driving delta pulse. The same pulse may also change the oscillator amplitude. The function $h_A(t,\tau)$ represents the corresponding impulse response. Since the system is periodic, both functions are expected to be τ-periodic and to explicitly depend on the relative position of the time τ within the oscillation period.

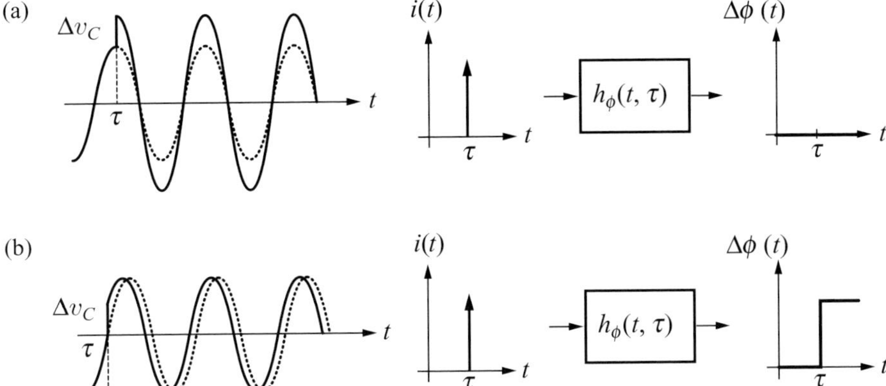

Figure 5.2 Impulse response of the phase depends on time τ: current impulse synchronous with (a) carrier peak and (b) carrier zero crossing

As an example, let us consider the ideal parallel LC network. The voltage across the capacitor and the inductor current are given by:

$$v_C\,(t) = A_0 \cos(\omega_0 t),$$
$$i_L\,(t) = \omega_0 C A_0 \sin(\omega_0 t).$$

The voltage waveform is represented in Figure 5.2. If a current pulse with area Δq suddenly changes the charge across the capacitor, its voltage changes by

$$\Delta v_C = \frac{\Delta q}{C}$$

and the evolution of the oscillator waveform will be affected. The effect is time dependent. If the impulse is applied at the waveform peak, the voltage v_C changes by Δv_C, but no phase shift takes place; only the amplitude will change temporaily, as depicted in Figure 5.2(a). In this case $h_\phi\,(t, \tau) = 0$. Moreover, because of the amplitude-limiting mechanisms, the steady-state oscillation amplitude will be quickly recovered. The response $h_A(t, \tau)$ is, therefore, asymptotically vanishing.

On the other hand, if this impulse is applied at the zero crossings of v_C (Figure 5.2(b)), the phase shift is maximum and the impact on amplitude is nil. Unlike the amplitude, no effect causes the original phase to be recovered. The phase shift persists in the oscillator. This is why the impulse response $h_\phi(t, \tau)$ has been depicted as a step function.

Such a time dependence can also be discussed in the state-variable space. The voltage $v_C(t)$ and the inductor current $i_L(t)$ are the proper state variables of the circuit. Figure 5.3 shows the trajectory over the state-variable plane. The axes are normalized to the peak voltage value, A_0, and to the peak inductor current, $\omega_0 C A_0$. With such a choice, the oscillator limit cycle is circular. The injection of a current pulse into the tank, at the voltage peak, is equivalent to a sudden jump in voltage at point A, which does not affect the phase. Only the amplitude will be changed. The application of an impulse at point B results only in a phase change, without affecting the amplitude. An impulse applied between these two extremes

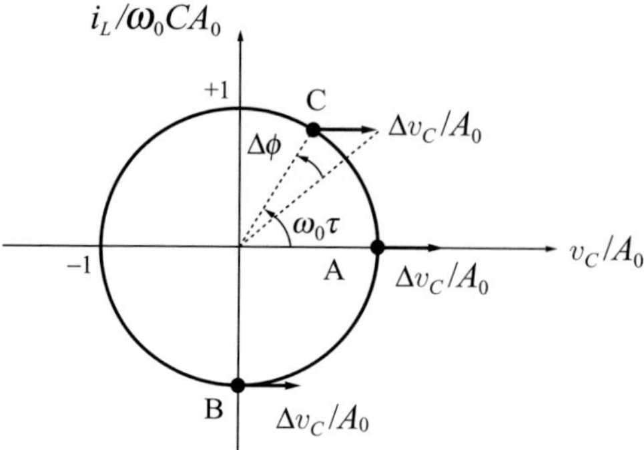

Figure 5.3 Noise effect in phase space

will result in both amplitude and phase changes. From Figure 5.3 at point C the phase shift turns out to be:

$$\Delta\phi = -\frac{\Delta v_C}{A_0}\sin(\omega_0\tau) = -\frac{\Delta q}{A_0 C}\sin(\omega_0\tau) = -\frac{\Delta q}{q_{max}}\sin(\omega_0\tau). \tag{5.5}$$

In the last step we have denoted as $q_{max} = CA_0$ the maximum charge across the capacitor. The corresponding impulse response $h_\phi(t,\tau)$ can, therefore, be written as:

$$h_\phi(t,\tau) = \frac{-\sin(\omega_0\tau)}{q_{max}}u(t-\tau),$$

where $u(t)$ represents the unit step function.

Let us now consider the thermal noise from the tank losses. The noise can be represented as the superposition of uncorrelated harmonic tones, which can be considered to act independently on the circuit. The injection of a noise tone at $\omega_0 + \omega_m$, with amplitude i_n, causes a variation of the output given by:

$$\phi(t) = \int_{-\infty}^{+\infty} h_\phi(t,\tau) i(\tau) d\tau = -\frac{i_n}{q_{max}}\int_{-\infty}^{t} \sin(\omega_0\tau)\cos[(\omega_0+\omega_m)\tau]\, d\tau.$$

That is:

$$\phi(t) = -\frac{i_n}{2q_{max}}\int_{-\infty}^{t} \{\sin[(2\omega_0+\omega_m)\tau] - \sin[\omega_m\cdot\tau]\}\, d\tau.$$

The integration leaves only one significant time-dependent contribution given by:

$$\Delta\phi(t) \approx -\frac{i_n}{2q_{max}\omega_m}\cos(\omega_m t).$$

The noise tone, therefore, causes a phase modulation at ω_m (Section 1.1), with a frequency deviation $\Delta\omega_0 = i_n/2q_{max}$:

$$\frac{\Delta\omega_0}{\omega_m} = \frac{i_n}{2q_{max}\omega_m}.$$

Such a modulation produces two sidebands at $(\omega_0 \pm \omega_m)$, in the voltage spectrum, whose ratio with respect to the carrier power is given by (1.2)

$$\frac{1}{4}\cdot\left(\frac{\Delta\omega_0}{\omega_m}\right)^2 = \frac{i_n^2}{16\cdot q_{max}^2\omega_m^2}.$$

The phase noise from tank losses is obtained by replacing $q_{max} = CA_0$ and $i_n^2/2 = (4kT/R_0)\cdot df$, since the mean square value of the harmonic current tone is $i_n^2/2$. Computing the SSCR, we get:

$$\mathcal{L}(\omega_m) = \frac{1}{2}\cdot\frac{kT}{C}\cdot\frac{\omega_0}{Q}\cdot\frac{1}{\omega_m^2}\cdot\frac{1}{A_0^2}.$$

The result differs from (5.3) by the presence of the factor 1/2 (and of course in F). It is, however, necessary to take into account that noise components symmetrical to the carrier frequency contribute to the same sidebands, with equal weight. So the preceding expression has to be doubled, thus giving the correct amount of noise.

The results can be generalized. The oscillator phase responds to a perturbation caused by the injection of a charge into a circuit node with an impulse response function given by:

$$h_\phi(t,\tau) = \frac{\Gamma(\omega_0\tau)}{q_{max}}u(t-\tau), \tag{5.6}$$

where Γ is the *impulse sensitivity function* (ISF) and is a dimensionless and periodic function. The normalization charge, q_{max}, can be taken as the peak amplitude of the voltage swing at the circuit node times the capacitance load at the same node. Since Γ is periodic, we may write:

$$\Gamma(t) = \frac{a_0}{2} + \sum_{n=1}^{\infty} a_n\cos(n\omega_0 t + \theta_n).$$

Following the same procedure discussed above, it can be demonstrated that each noise tone at $n\omega_0 \pm \omega_m$ contributes to a phase modulation at ω_m with amplitude:

$$\frac{i_n a_n}{2q_{max}\omega_m}.$$

The white-noise components at $n\omega_0 \pm \omega_m$ are folded back around $\omega_0 \pm \omega_m$, weighted by the coefficients $\{a_n\}$ and the sideband-power-to-carrier ratio bandwidth is:

$$\frac{i_n^2 a_n^2}{16\cdot q_{max}^2\omega_m^2}.$$

This frequency translation effect is typical of time-variant systems, like samplers, and it is the basis of mixer up-conversion and down-conversion. Assuming a white-noise current spectrum, $i_n^2/2$ can be replaced by $S_{n,I} \cdot df$. The resulting SSCR is, therefore:

$$\mathcal{L}(\omega_m) = \frac{S_{n,I}}{4q_{max}^2\omega_m^2} \cdot \sum_{n=0}^{+\infty} a_n^2.$$

Note that according to Parseval's theorem:

$$\sum_{n=0}^{+\infty} a_n^2 = \frac{1}{\pi}\int_0^{2\pi} |\Gamma(x)|^2 dx = 2\Gamma_{rms}^2,$$

where Γ_{rms} is the r.m.s. value of $\Gamma(x)$. The final result can, therefore, be written as:

$$\mathcal{L}(\omega_m) = \frac{S_{n,I}}{2q_{max}^2} \cdot \frac{\Gamma_{rms}^2}{\omega_m^2}.$$

The message to the designer is that the contribution to the phase noise can be minimized by reducing the r.m.s. $\Gamma(x)$ value. In the same framework, the close-in $1/f^3$ phase noise derives instead from low-frequency $1/f$ noise up-converted by the coefficient:

$$a_0 = \frac{1}{\pi}\int_0^{2\pi} \Gamma(x) dx.$$

The framework described above has the merit of addressing the time-variant operation of oscillators, overcoming the conceptual limitations of the Leeson model. However, it requires knowledge of $\Gamma(x)$ for each noise source. This means that a transient simulation should be run for every Γ sample. A current pulse should be injected at the desired time instant, and the phase shift has to be derived when the oscillator reaches a steady oscillation. Moreover, the procedure must be repeated for every noise source. The complexity and time duration of the whole process may be higher than simply relying on a phase noise simulator. Nevertheless, first-order estimations can be obtained by approximating the ISF by a simple function (see [5, 8] or the practical cases discussed in Sections 5.5.3 and 5.5.6).

5.4.2 Frequency-domain approach: the harmonic transfer function

In a complementary approach, noise is considered as the superposition of independent harmonic tones, i.e., δ-functions in the frequency domain. Figure 5.4 shows the oscillator model, where the transconductor is represented by a hard limiter, and a generic noise tone. The noise affects the output current delivered to the tank via a transfer function $H[V_0(t)] = H(t)$, which is a periodic function, determined by the harmonic carrier voltage waveform $V_0(t)$, which drives the active element. In the most general case we may write:

$$I(t) = I_U(t) + H[V_0(t)] \cdot i_n \cos(\alpha t + \varphi), \tag{5.7}$$

where $I_U(t)$ is the unperturbed current, while the second term represents the effect due to a single tone. $H(t)$, the *harmonic transfer function* (HTF), accounts for the time-dependent impact of the noise on the output current. Since the amplitude of i_n is negligible with respect

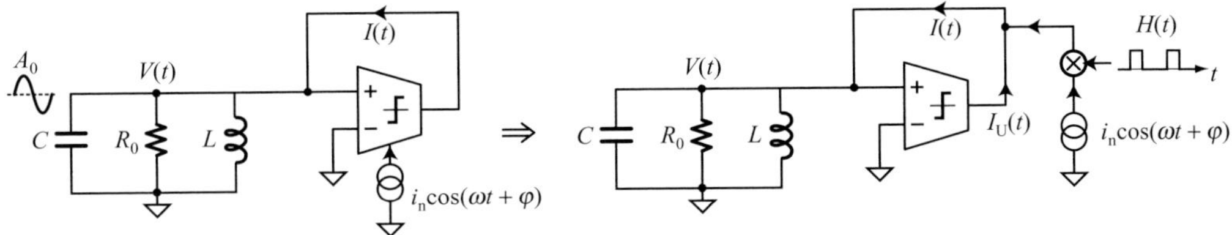

Figure 5.4 Generic noise source affecting the output current

to $I_U(t)$, its influence on $H(t)$ can be neglected. The system remains linear, even if it is time variant.

It is important to remark that for any noise source there is a specific $H(t)$. As a matter of fact, it is obvious that the transfer of the noise to the output current depends on which noise generator is being considered.

The function $H(t)$ is periodic, therefore it can be written as:

$$H(t) = c_0 + \mathrm{Re}\left[\sum_{n=1}^{\infty} 2c_n \mathrm{e}^{\mathrm{j}n\omega_0 t}\right].$$

In general, the coefficients c_n are complex, $c_n = |c_n|\,\mathrm{e}^{\mathrm{j}\theta_n}$. The reason for the factor two in the expansion of $H(t)$ is clarified below.

Figure 5.5, at the top, schematically shows the spectrum of a generic $H(t)$, i.e., the magnitude of the coefficients c_n, the spectrum is indicated as $H(\omega)$. Instead, $I(\omega)$ is the noise spectrum, which is composed by a single tone I_n at $\omega_0 + \omega_m$. Based on (5.7), the perturbation caused by this tone has a spectrum given by the convolution integral $H(\omega) * I(\omega)$. The bottom graph in Figure 5.5 shows a spectrum that presents the characteristic sideband harmonics at $n\omega_0 \pm \omega_m$, that is, the single-noise tone has been folded around the harmonics of the oscillator. It is not surprising that the approach both in the time domain and in the frequency domain lead to the same results, since they are alternative ways to account for the oscillator's time-varying behaviour.

An important assumption now is that the spectral components falling outside the tank bandwidth are filtered out. The only terms leading to excess noise are those at $\omega_0 \pm \omega_m$. These tones can be associated with the phasors:

$$c_0^* I_n = |c_0||I_n|\,\mathrm{e}^{\mathrm{j}(\varphi-\theta_0)}, \qquad c_2 I_n^* = |c_2||I_n|\,\mathrm{e}^{\mathrm{j}(\theta_2-\varphi)}.$$

The factor two has disappeared because of the products between two sinusoids. To get a more general expression, which could be extended to noise contributions arising from higher-order harmonics, it is convenient to keep the complex conjugate of the c_0 coefficient, even if it is coincident with c_0, since c_0 is always real. The corresponding angle θ_0 is either zero or π. Figure 5.6 depicts the phasors in the carrier frame. The phasor weighted by the coefficient c_0^* lies at $\omega_0 + \omega_m$, so in the carrier frame it rotates at $+\omega_m$, while the other phasor turns at $-\omega_m$. Note that the two sidebands, at $\pm\omega_m$ from ω_0, are correlated.

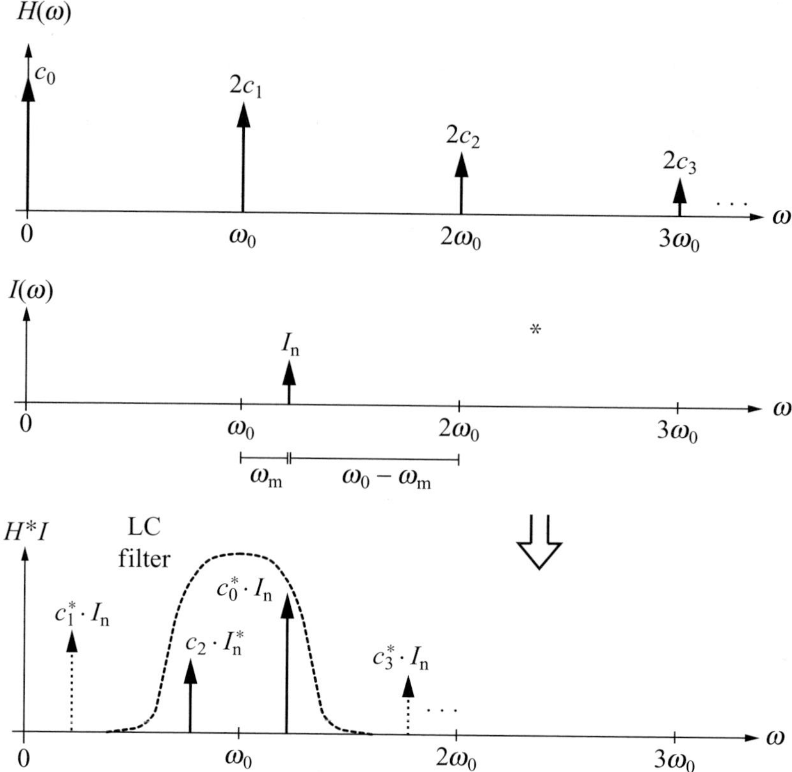

Figure 5.5 Folding of one single-noise tone I_n

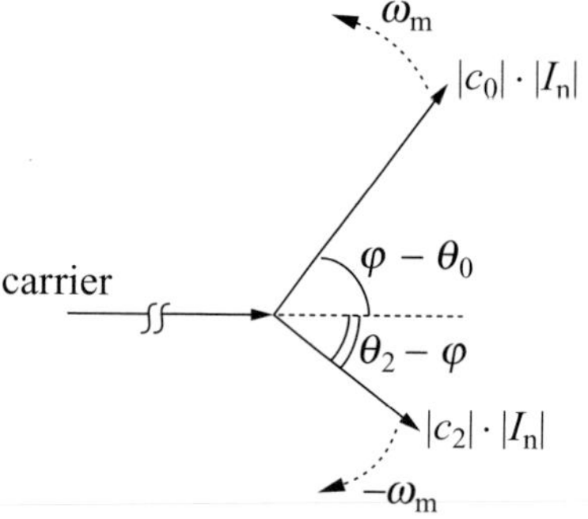

Figure 5.6 Phasors referred to carrier frame

It is now necessary to take into account the fact that only phase fluctuations are of interest. Amplitude variations do not circulate in the oscillator loop. Figure 5.7 shows how the tone at $+\omega_m$ can be separated in a phase modulation of the carrier at ω_m and an amplitude modulation at the same frequency. If only the phase-modulation tones are considered, it emerges that the single-noise tone at $\omega_0 + \omega_m$ gives rise to two correlated side tones, at $\omega_0 \pm \omega_m$.

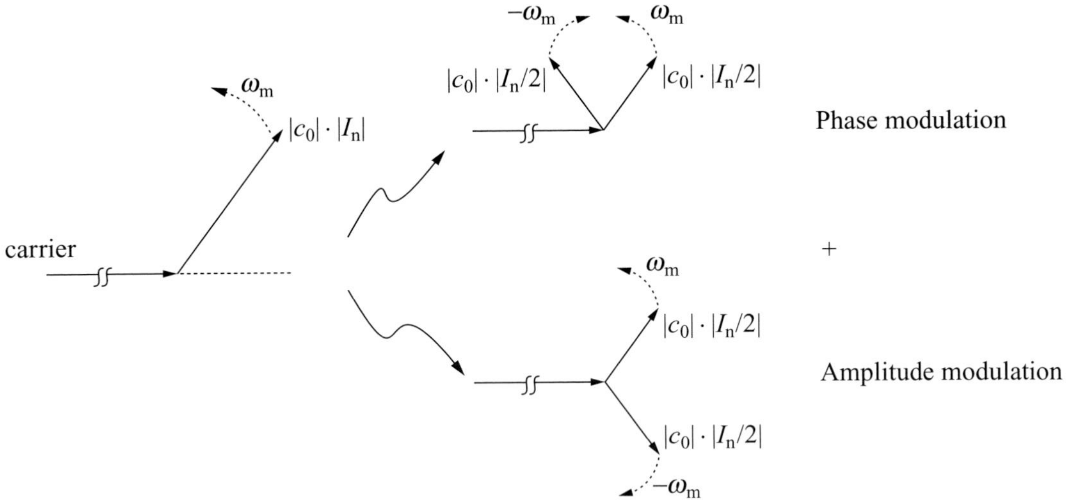

Figure 5.7 Decomposition of one noise tone into phase and amplitude modulation

The same decomposition holds for the tone folded at $\omega_0 - \omega_m$ and weighted by c_2. Combining the two contributions to the higher sideband (the one at $\omega_0 + \omega_m$) we get:

$$\frac{|I_n|}{2}\left[|c_0|\,e^{j(\varphi-\theta_0)} + |c_2|\,e^{j(\pi+\varphi-\theta_2)}\right] = \frac{|I_n|\,e^{j\varphi}}{2}\left[|c_0|\,e^{-j\theta_0} - |c_2|\,e^{-j\theta_2}\right].$$

This corresponds to a power:

$$\frac{|I_n|^2}{8}\left||c_0|\,e^{-j\theta_0} - |c_2|\,e^{-j\theta_2}\right|^2.$$

Assuming that the noise source has a white spectrum, $|I_n|^2/2$ can be replaced by $S_{n,I} \cdot df$. Moreover, an equal contribution exists at $\omega_0 - \omega_m$, and similar terms are determined by noise around all frequencies that are multiples of ω_0, at $n\omega_0 \pm \omega_m$, involving different c_n coefficients. Taking into account all these terms, the total folded current noise, let us say $S_{tot,I}$, is found to be:

$$S_{tot,I} = \frac{S_{n,I}}{2}\left(\left|c_1^* - c_1\right|^2 + \sum_{n=0}^{+\infty}\left||c_n|\,e^{-j\theta_n} - |c_{n+2}|\,e^{-j\theta_{n+2}}\right|^2\right). \tag{5.8}$$

The term $|c_1^* - c_1|^2 = 4\,|c_1|^2 \cdot \sin^2\theta_1$ justifies the up-conversion of low-frequency noise components into the resonator bandwidth. However, it is nil when c_1 is real. The sidebands at $\omega_0 \pm \omega_m$ correspond to current signals injected into the LC tank. The voltage–power spectrum is obtained by multiplying the above current–power spectral density by the impedance of the tank losses at $\omega_0 \pm \omega_m$, that is, $1/(2C\omega_m)$. The voltage spectrum is, therefore:

$$S_V(\omega_m) = \frac{1}{(2C\omega_m)^2} \cdot \frac{S_{n,I}}{2}\left(4\,|c_1|^2 \cdot \sin^2\theta_1 + \sum_{n=0}^{+\infty}\left||c_n|\,e^{-j\theta_n} - |c_{n+2}|\,e^{-j\theta_{n+2}}\right|^2\right).$$

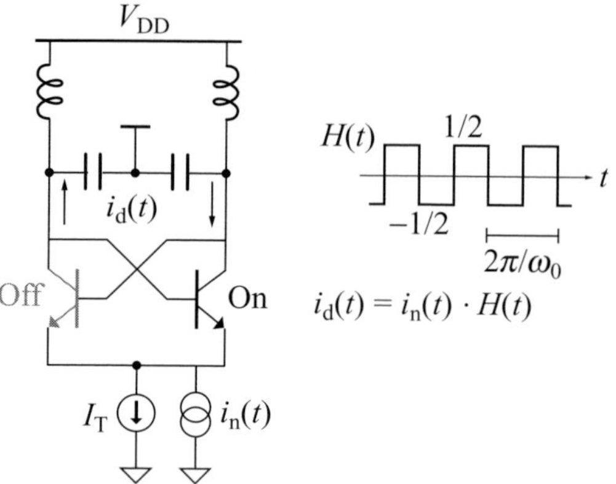

Figure 5.8 Tail noise in cross-coupled oscillator: time-domain window $H(t)$ between tail noise $i_n(t)$ and differential current $i_d(t)$ forced into the tank

The noise factor F is obtained dividing $S_{tot,I}$ by $2kT/R_0$, since only half of the R_0 thermal noise power density contributes to the phase noise:

$$F = \frac{S_{tot,I}}{2kT/R_0} = \frac{S_{n,I}R_0}{4kT}\left(4\,|c_1|^2 \cdot \sin^2\theta_1 + \sum_{n=0}^{+\infty} \left||c_n|\,e^{-j\theta_n} - |c_{n+2}|\,e^{-j\theta_{n+2}}\right|^2\right).$$

The above derivation is quite general. In most practical cases it is not, however, necessary to take into account too many terms of the sum. Noise at a higher frequency is typically cut off by circuit parasitics before any folding may occur.

A final remark is needed on a theoretical aspect that should not be overlooked. The noise source in Figure 5.4 may be stationary. However, its effect on the noise across the tank has been computed based on (5.7), by multiplying each tone for the time-variant function $H(t)$. It turns out that the resulting noise is not stationary any more. It is cyclostationary, since its statistics change periodically, following $H(t)$. [17–19] The output voltage spectrum in (5.8) is recovered because the LC tank is a narrow-band filter, averaging out the noise time variation. However, the cyclostationary noise exhibits correlated spectrum sidebands around the carrier, as highlighted in Figure 5.5. This property is not removed by the narrow-band LC filtering. It would be removed only by a single-sideband passband filter. [17]

5.5 Application to some practical cases

5.5.1 Cross-coupled oscillator: tail noise

Let us estimate the contribution to F arising from the noise of the tail transistor in Figure 5.8. There is no difference between bipolar and CMOS implementation. The cross coupling acts as a single-sideband mixer for the current noise coming from the tail. The differential current noise reaching the tank is, therefore, given by the tail noise, multiplied by the $H(t)$ square wave at ω_0 in Figure 5.8. Its amplitude is $\pm 1/2$.

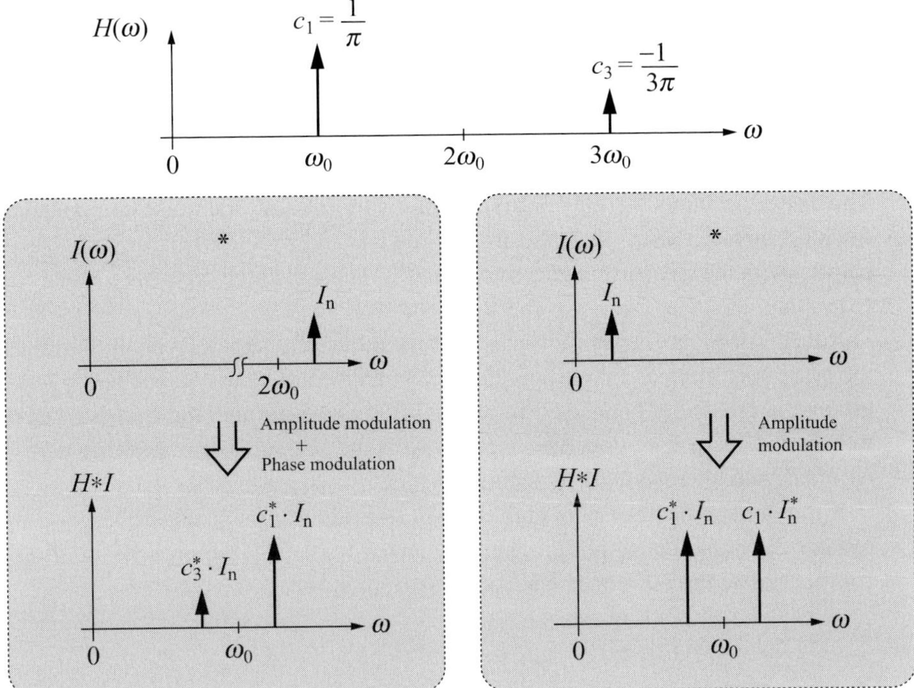

Figure 5.9 Frequency-domain transfer: tone I_{n} close to $2\omega_0$ (left diagram) and close to zero frequency (right diagram)

Figure 5.9 shows some coefficients of the $H(t)$ Fourier expansion. In more general terms, they are:

$$c_{2n} = 0, \quad c_{2n+1} = \frac{(-1)^n}{(2n+1)\pi}, \qquad n = 0, 1, 2, \ldots$$

According to (5.8), the noise component at $2\omega_0 + \omega_{\mathrm{m}}$ will be folded at $\omega_0 + \omega_{\mathrm{m}}$ weighted by the c_1^* coefficient. The same noise component also gives a contribution at $\omega_0 - \omega_{\mathrm{m}}$, weighted by the c_3^* coefficient of the $H(t)$ harmonic expansion. The same coefficients fold the noise component at $2\omega_0 - \omega_{\mathrm{m}}$, and a similar transformation applies to all the noise components at $n\omega_0 \pm \omega_{\mathrm{m}}$, with n even, the only difference being the folding coefficients. Denoting as $S_{\mathrm{n,I}}^{(\mathrm{tail})}$ the tail noise power spectral density, the resulting phase-noise current injected into the tank turns out to be:

$$S_{\mathrm{tot,I}} = \frac{S_{\mathrm{n,I}}^{(\mathrm{tail})}}{2} \cdot \frac{1}{\pi^2} \sum_{n=0}^{+\infty} \left| \frac{1}{2n+1} + \frac{1}{2n+3} \right|^2 = \frac{S_{\mathrm{n,I}}^{(\mathrm{tail})}}{8}.$$

The result can be intuitively justified as follows: the differential current flowing through the tank is only half the current provided by the current generator. This is accounted for by the $1/2$ amplitude of $H(t)$. In terms of noise power, the $1/2$ factor leads to a factor of $1/4$

multiplying the $S_{\rm n,I}^{\rm (tail)}$ term. An additional $1/2$ factor is produced by taking into account that only half of the noise power is contributing to phase fluctuations.[1]

The noise factor is now derived by dividing by $2kT/R_0$:

$$F = \frac{S_{\rm n,I}^{\rm (tail)} R_0}{16kT}. \tag{5.9}$$

This factor accounts for the contributions arising from all the noise components. Most of the time, however, only the noise around $2\omega_0$ has an impact. The large stray capacitance in parallel with the tail-current generator usually shunts the high-frequency noise components to ground.

Note that if a large capacitor is purposely introduced in parallel with the current generator, the noise performance may even degrade. In fact, such a large capacitance acts as an a.c. ground, alternatively connected to the tank terminals through the transistor of the pair in the ohmic operating region. This effect lowers the average loss impedance across the tank terminals and, therefore, the Q factor.

If the voltage headroom is high enough, the tail transistor can be degenerated with a resistor $R_{\rm T}$ reducing its noise. The tail-current noise $S_{\rm n,I}^{\rm (tail)}$ approaches $4kT/R_{\rm T}$ and the contribution to the noise factor becomes:

$$F \approx \frac{4kT}{R_{\rm T}} \cdot \frac{R_0}{16kT} = \frac{R_0}{4R_{\rm T}} \approx \frac{\pi}{8} \cdot \frac{A_0}{V_{R_{\rm T}}},$$

where $V_{R_{\rm T}}$ is the voltage drop across $R_{\rm T}$. A more effective solution is based on the peaking technique shown in Figure 4.19. Note that the tail resonator not only avoids a d.c. voltage drop and extends the maximum oscillation amplitude. It also cuts off the most important noise contribution coming from the tail generator. Since the network resonates at $2\omega_0$, the tail-current noise component at $2\omega_0$ is diverted to ground. [11]

A final remark concerns the low-frequency noise coming from the tail generator, which can be very high, especially in CMOS technology. A single tone at frequency $\omega_{\rm m}$ is up-converted to $\omega_0 \pm \omega_{\rm m}$ by the effect of the pair switching. The current noise injected into the tank can be obtained by multiplying this tone amplitude by the first Fourier coefficient c_1 of $H(t)$ (Figure 5.9). Since only the coefficient c_1 is involved, the two sidebands produced at $\omega_0 \pm \omega_{\rm m}$ are identical and they represent an amplitude noise. This result is intuitive. As long as the oscillator works in the current-limited regime, low-frequency variations of the tail current imply fluctuations of the oscillation amplitude. Unfortunately, even if leading to amplitude modulation, the role of this noise component cannot be neglected. The non-linear effects discussed in Chapter 7, which arise for instance from the presence of voltage-dependent capacitances, convert this amplitude modulation noise into phase noise.

5.5.2 Cross-coupled oscillator: noise of the transistor pair (HTF method)

Let us consider the noise generated by each transistor of a differential pair. Figure 5.10 shows a bipolar oscillator, though the results can easily be extended to a CMOS stage. The

[1] A passband white noise can be decomposed into two components with half of the total noise power. The first component is in phase with a carrier located in the middle of the noise bandwidth, while the second is in quadrature with the same carrier. Only the second component is able to modulate the phase of the carrier.

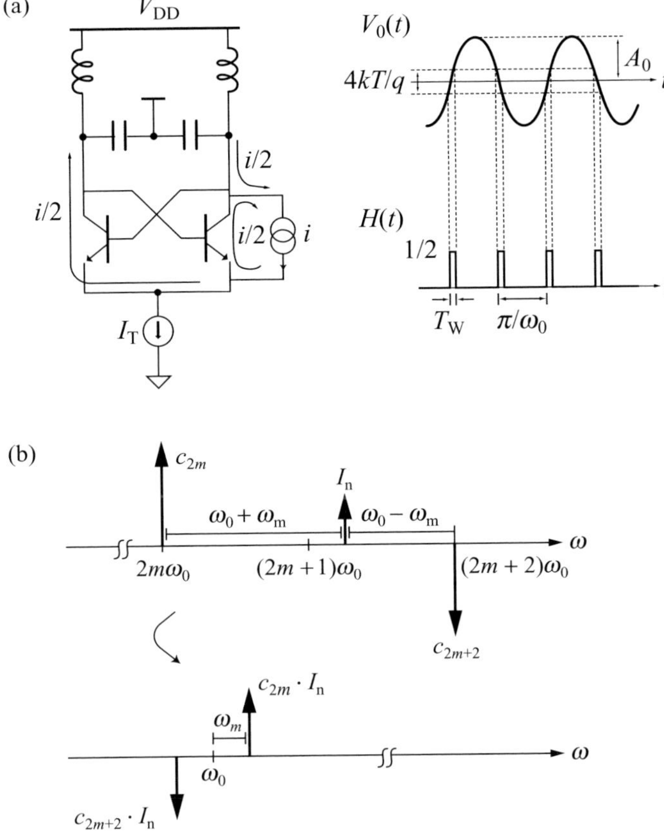

Figure 5.10 Transistor pair noise in cross-coupled oscillator: (a) circuit schematic and time-domain window function $H(t)$; (b) frequency-domain transfer of the tone I_n close to odd harmonics of the carrier

current noise generator of each transistor is switched off for about half the period. At each zero crossing, when the stage is balanced, both the transistors are on. In this condition, half of the noise current of each generator reaches the tank, while the remaining half circulates within the transistor. When the stage is fully switched, the conducting transistor finds itself well degenerated by the tail current source. Therefore, its shot noise does not reach the tank.

Such collector noise is not stationary. It may, however, be depicted as a stationary noise source $2qI_C$ transferred to the tank through the harmonic transfer function depicted in Figure 5.10(a). $H(t)$ is approximately a train of rectangular pulses with half the period of the oscillation, switching between zero (noise off) and $1/2$ (noise on). The value of the pulse duration T_W can be derived by taking into account the fact that the input linear dynamic of a bipolar differential stage is $4kT/q$, while the slope of the voltage waveform close to the zero crossings is given by $\omega_0 A_0$. It follows that:

$$T_W = \frac{4kT}{qA_0\omega_0}. \tag{5.10}$$

Let us now derive the Fourier coefficients of $H(t)$. Since the frequency of $H(t)$ is $2\omega_0$, only even multiples of ω_0 are present in its spectrum. Moreover, if the carrier is written as $A_0 \cdot \cos(\omega_0 t)$, it turns out that all the harmonic components of $H(t)$ are cosines, at $2\omega_0, 4\omega_0, 6\omega_0, \ldots$, with alternate signs. All these cosines add in phase at the zero crossings of the carrier. Referring to Figure 5.5, the Fourier coefficients are:

$$c_{2m} = (-1)^m \frac{T_W}{T_0} \cdot \frac{\sin(m\omega_0 T_W)}{m\omega_0 T_W}; \qquad c_{2m+1} = 0;$$

where $T_0 = 2\pi/\omega_0$ is the period of the carrier. For $m \ll (T_0/2\pi T_W)$, the sine function can be approximated by its argument. Since $T_W \ll T_0$, the condition is true up to very large values of m. In this case, it is $c_{2m} \cong (-1)^m T_W/T_0$; namely, all the coefficients have about the same amplitude. Figure 5.10(b) shows that in this case a single generic noise tone gives rise to phase noise. In fact, the two folding coefficients are real, with the same magnitude and opposite signs. This result is in agreement with the picture derived using the impulse sensitivity function. The current noise considered here is injected at the zero crossing of the carrier, where the sensitivity is maximum and where the noise completely affects the oscillator phase.

Let us now notice that the coefficients c_n and c_{n+2} in (5.8) correspond to the terms c_{2m} and c_{2m+2} just derived above. Therefore, it turns out:

$$S_{\text{tot,I}} = \frac{S_{\text{n,I}}}{2} \sum_{m=0}^{+\infty} \left(\frac{T_W}{T_0}\right)^2 \left| \frac{\sin[m\omega_0 T_W]}{m\omega_0 T_W} + \frac{\sin[(m+1)\,\omega_0 T_W]}{(m+1)\,\omega_0 T_W} \right|^2 .$$

In the limit $T_W \ll T_0$, the series gives:

$$S_{\text{tot,I}} = \frac{S_{\text{n,I}}}{2} \cdot \frac{T_W}{T_0}.$$

Again, this quantitative result can be justified in an intuitive way. The current noise spectral density $S_{\text{n,I}}$ is multiplied by $1/4$, since only half of the current reaches the tank. The result is also weighted by the duty cycle of $H(t)$, which is $2T_W/T_0$. Since the noise injection occurs at the zero crossing, it gives only phase noise.

The noise factor, as usual, is calculated by dividing by $2kT/R_0$. For each bipolar transistor it is:

$$\begin{aligned} F_{\text{single}} &= \frac{S_{\text{n,I}}}{2} \cdot \frac{T_W}{T_0} \cdot \frac{R_0}{2kT} \\ &= \frac{2qI_C}{2} \cdot \left(\frac{4kT}{qA_0\omega_0 T_0}\right) \cdot \frac{R_0}{2kT} = \frac{1}{4} \end{aligned} \qquad (5.11)$$

where I_C is the collector current, half of the bias current of the stage, and $2qI_C$ the corresponding collector shot noise. To obtain the result, it is necessary to remember that the oscillation amplitude is $A_0 = (2I_C)\,R_0 \cdot 2/\pi$. Adding the contribution of both transistors the resulting noise factor is $F = 2F_{\text{single}} = 1/2$. Unlike the tail noise, the present contribution represents a fundamental, unavoidable limit to the noise performance.

The case of a MOSFET oscillator is practically identical. The width of the injection window may be approximated by $T_W = (2V_{\text{ov}}/\omega_0 A_0)$, where V_{ov} is the over-drive voltage.

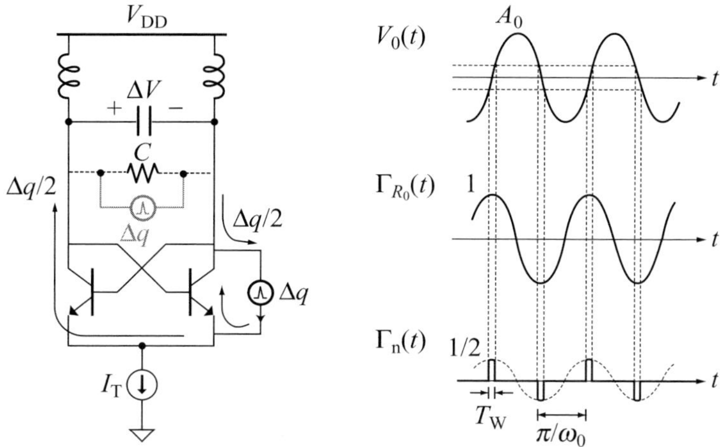

Figure 5.11 Transistor pair noise in cross-coupled oscillator: ISF function $\Gamma_n(t)$

Moreover, the channel noise is written as $4kTg_m\gamma$, where the transconductance $g_m = (2I_D/V_{ov})$. In this case, the noise factor from both transistors, $2F_{single}$, is equal to γ.

5.5.3 Cross-coupled oscillator: noise of the transistor pair (ISF method)

The noise factors from the oscillator noise sources can also be derived by relying on the ISF method. The ISF relative to the tank thermal noise has already been derived in Section 5.4.1. If the voltage waveform across the tank is $V_0(t) = A_0 \cdot \cos(\omega_0 t)$, the ISF for the tank resistance is obtained from (5.5) and (5.6): $\Gamma_{R_0}(t) = -\sin(\omega_0 t)$, i.e., it is a maximum at the zero crossing of the voltage waveform $V_0(t)$.

To obtain the ISF of one switching transistor, the voltage variation across the tank caused by a current pulse in parallel with the transistor has to be evaluated. When only one of the two switching transistors is on, the current pulse has, ideally, no effect. Instead, during the time window T_W around the zero crossings of $V_0(t)$, both transistors drive the same current and half of the current pulse charges the tank capacitance C. The capacitance voltage swing caused by the pulse with charge Δq is half the voltage swing ΔV_{R_0} caused by a Δq pulse in parallel with R_0:

$$\Delta V_n = \frac{1}{2} \cdot \frac{\Delta q}{C} = \frac{\Delta V_{R_0}}{2}.$$

Therefore, the transistor ISF depicted in Figure 5.11 is half the ISF of the tank noise inside the time window and zero outside it. Its root mean square value is:

$$\Gamma^2_{n,rms} = \frac{2}{T_0} \int_{-T_W/2}^{T_W/2} \frac{1}{4} \cos^2(\omega_0 t)\, dt \simeq \frac{T_W}{2T_0}.$$

The factor $2/T_0$ arises from the fact that the period of $\Gamma_n(t)$ is $T_0/2$. Hence, the noise

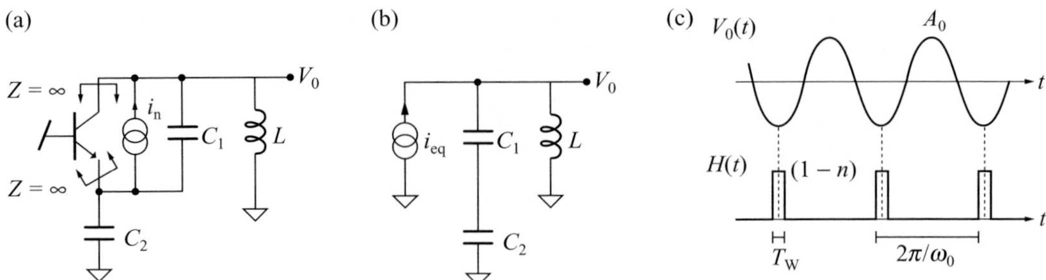

Figure 5.12 Transistor noise in Colpitts oscillator: (a) collector-current generator; (b) equivalent noise generator; (c) time-domain window function $H(t)$

factor is:

$$F_{\text{single}} = \frac{\Gamma^2_{\text{n,rms}} \cdot i_{\text{n}}^2 df}{\Gamma^2_{R_0,\text{rms}} \cdot i^2_{R_0} df} = \frac{T_\text{W}/2T_0}{1/2} \cdot \frac{2qI_\text{C}}{4kT/R_0} = \frac{1}{4},$$

which matches (5.11).

5.5.4 Cross-coupled oscillator: noise of the spreading resistance

An analogous procedure can be followed to evaluate the noise of the spreading resistance $r_{\text{bb}'}$ in a bipolar differential stage (or the gate resistance noise in MOSFET oscillators). [9] Whenever the differential pair is balanced, with both transistors on, the noise voltage causes fluctuations of the current injected into the tank. Whenever a single transistor is on, the collector current is set by the bias source and is not sensitive to voltage fluctuations at the base. The time-variant noise transfer can, therefore, be described using a time-dependent transconductance, at frequency $2\omega_0$, which is non-zero only for short time windows T_W close to the crossing point of the harmonic voltage waveform.

The calculation follows the lines of that just performed for the collector noise. The noise of each spreading resistance can be transformed as a collector-current noise of the corresponding transistor, with noise power given by $S_{\text{n,I}} = g_\text{m}^2 \cdot 4kTr_{\text{bb}'}$. By replacing $S_{\text{n,I}} = g_\text{m}^2 \cdot 4kTr_{\text{bb}'}$ in (5.11) it turns out that $F_{\text{single}} = g_\text{m}r_{\text{bb}'}/2$ and $F = g_\text{m}r_{\text{bb}'}$ for the entire bipolar pair.

5.5.5 Colpitts oscillator: collector current noise (HTF method)

Let us now estimate the noise contribution of the collector shot noise in a Colpitts oscillator. Figure 5.12(a) shows the position of the noise generator with respect to the tank, while it omits the bias generator. The folding effects will be computed using the harmonic transfer function. Since the generator is not referred to ground, the $H(t)$ evaluation requires an intermediate step. For this purpose let us assume that the transconductor is considered to be a voltage-controlled current source, with infinite impedance at both the input and the output (Figure 5.12(a)). As discussed in the previous chapter this is a reasonable approximation since the emitter impedance of the transistor is higher than the C_2 impedance at the resonance frequency. Figure 5.12(b) shows the circuit with an equivalent ground-referred generator. By equating the output voltage produced by the two generators it turns out that:

$$i_{\mathrm{eq}} \cdot \frac{sL}{1+s^2/\omega_0^2} = i_{\mathrm{n}} \cdot \frac{sL}{1+s^2/\omega_0^2} \cdot \frac{C_2}{C_1+C_2}.$$

From the theory of the Colpitts oscillator the capacitance ratio on the right-hand side can be expressed as $(1-n)$, where $n = C_1/(C_1+C_2)$. If $S_{\mathrm{n,I}}$ is the collector power spectral density, the spectrum for the equivalent noise is $S_{\mathrm{n,I}}\,(1-n)^2$.

The current noise delivered by the transistor is multiplied by the function $H(t)$, which is approximately given by a train of rectangular pulses, at frequency ω_0, with T_{W} duration. Since the average current value is set by the bias current I_{B}, the current pulses have a peak of $I_{\mathrm{B}} T_0/T_{\mathrm{W}}$. As shown in Figure 5.12(c), the current injection corresponds with the minima of the output signal. Referring to a carrier $A_0 \cdot \cos(\omega_0 t)$, the coefficients of the $H(t)$ expansion are:

$$c_l = (-1)^l \cdot \frac{T_{\mathrm{W}}}{T_0} \cdot \frac{\sin(l\omega_0 T_{\mathrm{W}}/2)}{(l\omega_0 T_{\mathrm{W}}/2)}.$$

Using (5.8) the total folded current spectrum is given by:

$$\begin{aligned} S_{\mathrm{tot,I}} = {} & S_{\mathrm{n,I}}\,(1-n)^2 \cdot \frac{1}{2} \cdot \left(\frac{T_{\mathrm{W}}}{T_0}\right)^2 \\ & \sum_{l=0}^{+\infty} \left| \frac{\sin[l\omega_0 T_{\mathrm{W}}/2]}{[l\omega_0 T_{\mathrm{W}}/2]} - \frac{\sin[(l+2)\,\omega_0 T_{\mathrm{W}}/2]}{[(l+2)\,\omega_0 T_{\mathrm{W}}/2]} \right|^2. \end{aligned}$$

The noise current is $S_{\mathrm{n,I}} = 2qI_{\mathrm{peak}} = 2qI_{\mathrm{B}} \cdot (T_0/T_{\mathrm{W}})$ and, in the limit $T_{\mathrm{W}} \ll T_0$, the series is equal to $\left(2\pi^2/3\right) \cdot (T_{\mathrm{W}}/T_0)$. Therefore, the total current spectrum reduces to:

$$S_{\mathrm{tot,I}} = \frac{2\pi^2}{3} \cdot (1-n)^2 \cdot qI_{\mathrm{B}} \cdot \left(\frac{T_{\mathrm{W}}}{T_0}\right)^2, \tag{5.12}$$

where

$$\frac{T_{\mathrm{W}}}{T_0} = \frac{I_{\mathrm{B}}}{I_{\mathrm{peak}}}. \tag{5.13}$$

On the basis of (4.17) and (4.18), $I_{\mathrm{peak}} = I_{\mathrm{R}}\mathrm{e}^X$ and $I_{\mathrm{B}} = I_{\mathrm{R}} B_0(X)$. After applying the approximation of the modified Bessel function for large amplitudes (i.e., $X \gg 1$):

$$\frac{T_{\mathrm{W}}}{T_0} = \frac{B_0(X)}{\mathrm{e}^X} \approx \frac{1}{\sqrt{2\pi X}}. \tag{5.14}$$

Combining (5.12) and (5.14):

$$S_{\mathrm{tot,I}} = \frac{2\pi^2}{3} \cdot \frac{(1-n)^2 \cdot qI_{\mathrm{B}}}{2\pi \cdot X} = \frac{\pi}{3} \cdot (1-n)^2 \cdot \frac{qI_{\mathrm{B}} V_{\mathrm{T}}}{nA_0}.$$

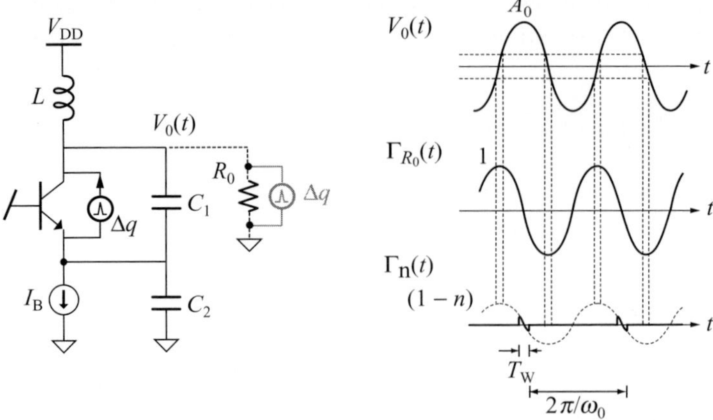

Figure 5.13 Transistor noise in Colpitts oscillator: ISF function $\Gamma_n(t)$. The collector noise can be regarded as a tank noise once scaled by $(1-n)$ (Figure 5.12)

Using the linear relationship (4.26) linking amplitude to current in the current-limited regime, the noise spectrum can be written as:

$$\begin{aligned} S_{\text{tot,I}} &= \frac{\pi}{3}\cdot(1-n)^2\cdot\frac{kT\cdot I_B}{n\cdot 2(1-n)\,I_B R_0} \\ &= \frac{2kT}{R_0}\cdot\underbrace{\left(\frac{\pi}{12}\cdot\frac{1-n}{n}\right)}_{F}. \end{aligned} \tag{5.15}$$

5.5.6 Colpitts oscillator: collector current noise (ISF method)

The same result can be obtained using the ISF method. Assuming that the voltage waveform across the tank is $V_0(t) = A_0\cdot\cos(\omega_0 t)$, the ISF for the collector noise shown in Figure 5.13 is zero when the transistor is off and is equal to $(1-n)$ times the ISF of the tank noise, i.e., $-(1-n)\cdot\sin(\omega_0 t)$, when the transistor is on. Therefore:

$$\Gamma^2_{\text{n,rms}} = \frac{1}{T_0}\int_{-T_W/2}^{T_W/2}(1-n)^2\sin^2(\omega_0 t)\,\mathrm{d}t \simeq \frac{\pi^2}{3}\left(\frac{T_W}{T_0}\right)^3(1-n)^2$$

for $T_W \ll T_0$ and the noise factor is

$$F = \frac{\Gamma^2_{\text{n,rms}}\cdot i_n^2\mathrm{d}f}{\Gamma^2_{R_0,\text{rms}}\cdot i_{R_0}^2\mathrm{d}f} = \frac{\frac{\pi^2}{3}\left(\frac{T_W}{T_0}\right)^3(1-n)^2}{1/2}\cdot\frac{S_{\text{n,I}}}{4kT/R_0}.$$

Substituting the expression for the current spectrum $S_{\text{n,I}} = 2qI_B\cdot(T_0/T_W)$, given above, and using (5.14), we get

$$F = \frac{\pi}{12}\cdot\frac{1-n}{n},$$

which matches the result in (5.15).

So far, the transistor switching has been modelled by a rectangular window function. However, in this case, this approximation underestimates the noise factor. In practice, the collector current pulse is closer to a pulsed triangle wave. Since the area and the peak of the triangle pulse have to be identical to the rectangle pulse so far considered, the time window is twice the previous one:

$$\frac{T'_{\rm W}}{T_0} = 2\frac{I_{\rm B}}{I_{\rm peak}} \simeq \frac{2}{\sqrt{2\pi X}}$$

for $X \gg 1$. The shot noise is periodically modulated by the window function. Its power density is given by $2qI_{\rm peak} \cdot \left(1 - 2|t|/T'_{\rm W}\right)$ inside the window. This can be regarded equivalently by taking account of a constant noise power density $S_{\rm n,I} = 2qI_{\rm peak}$ and for an effective ISF equal to the sinusoid windowed by the square root of the triangle pulse:

$$-(1-n) \cdot \sin(\omega_0 t) \cdot \sqrt{(1 - 2|t|/T_{\rm W})}.$$

The root mean square of this effective ISF is

$$\Gamma^2_{\rm n,rms} \simeq \frac{1}{T_0}\int_{-T_{\rm W}/2}^{T_{\rm W}/2} (1-n)^2 \left(1 - \frac{2|t|}{T'_{\rm W}}\right) \sin^2(\omega_0 t)\,{\rm d}t$$

$$\simeq \frac{2\pi^2}{3}\left(\frac{T_{\rm W}}{2T_0}\right)^3 (1-n)^2$$

for $T_{\rm W} \ll T_0$ and the noise factor is equal to

$$F = \frac{\Gamma^2_{\rm n,rms} \cdot i^2_{\rm n}{\rm d}f}{\Gamma^2_{R_0,\rm rms} \cdot i^2_{R_0}{\rm d}f} = \frac{\frac{2\pi^2}{3}\left(\frac{T'_{\rm W}}{2T_0}\right)^3 (1-n)^2}{1/2} \cdot \frac{S_{\rm n,I}}{4kT/R_0} = \frac{\pi}{6} \cdot \frac{1-n}{n}, \tag{5.16}$$

which is twice the result obtained with the rectangle pulse approximation. Obviously, the wider the time window, the higher the transistor noise factor.

Example 5.1 Phase noise of a bipolar cross-coupled oscillator

This example applies the phase-noise analysis to the bipolar cross-coupled oscillator in Figure 5.10, whose design has been already proposed in Example 4.3. The resonator parameters are the same as in the previous example: $R_0 = 162\ \Omega$, $C = 3$ pF and $L = 1.4$ nH, which correspond to a Q factor of 7.5 at the resonance frequency of 2.4 GHz. The oscillation amplitude is limited by the saturation of the transistors of the differential pair. A maximum A_0 of 900 mV has been achieved; however, at A_0 of 600 mV the bipolar transistors already enter the saturation region.

The tank losses have been represented by a resistor R_0 in parallel with the tank. The phase noise due to the thermal noise of this resistance can be estimated using (5.3) and neglecting

Table 5.1 *Noise factors of bipolar cross-coupled oscillator*

	Noise type	Noise factor, F
Differential pair	Collector shot noise	1/2
	Spreading resistance	$g_m r_{bb'}$
Tail generator		$\frac{S_{n,I}^{tail} R_0}{16kT}$

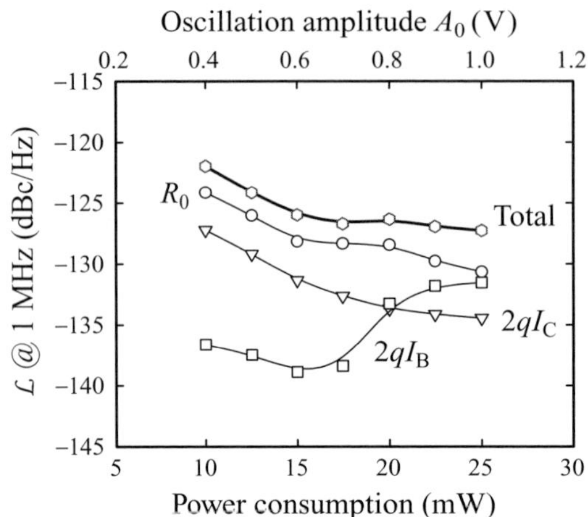

Figure 5.14 Phase noise contributions of bipolar cross-coupled oscillator. The lines are fittings of the numerical results from circuit simulations

the noise factor $(1 + F)$:

$$\mathcal{L}(\omega_m) = \frac{kT}{C} \cdot \left(\frac{\omega_0}{Q}\right) \cdot \frac{1}{\omega_m^2 A_0^2}. \tag{5.17}$$

At $A_0 = 500$ mV, the phase noise at an offset $\omega_m/2\pi = 1$ MHz from the carrier estimated from (5.17) is -125.7 dBc/Hz at a temperature of 25 °C. The tail current is 5 mA and the dissipated power is 12.5 mW. The circuit has been simulated by using a periodic steady-state analysis. [20] For the sake of simplicity, the tail generator has been considered ideal.

Figure 5.14 shows the different phase noise contributions at 1 MHz, while Figure 5.15 shows the noise factors F of the collector noise and of the base noise. Circles in Figure 5.14 highlight the phase noise contribution due to R_0 versus the dissipated power. The estimate given by (5.17) is very close to the simulated results.

The triangles in Figure 5.14 show the impact of the collector noise. The collector shot noise $2qI_C$ of the differential pair has a theoretical noise factor F of 1/2 (Table 5.1). At $A_0 = 500$ mV, the phase noise at 1 MHz is, in fact, 3 dB lower than the pure tank contribution. At amplitudes larger than about 0.7 V, the contribution of the base shot noise, $2qI_B$, which is normally negligible, increases (squares in Figures 5.14 and 5.15), thus becoming comparable to the collector shot noise at 0.8 V. This can be explained by taking into account the fact that the transistors of the pair are in deep saturation. Therefore, their

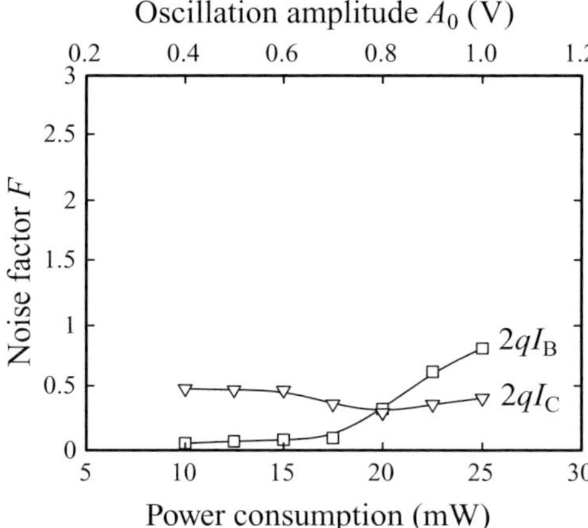

Figure 5.15 Noise factor of various noise sources in bipolar cross-coupled oscillator. Results derived from circuit simulations

base current becomes comparable to the collector current and the effect of the base shot noise becomes of the same order as the collector shot noise.

Example 5.2 Phase noise of a bipolar oscillator with extended amplitude

Low phase noise can be attained by increasing the oscillation amplitude without pushing the bipolar transistor in saturation. The bipolar oscillator with the capacitive divider in Figure 4.22(c) may be an option. It has been described in Example 4.3. The capacitive divider has a division factor $n = C_1/(C_1 + C_2) = 0.2$. The bias resistor $R_1 = 400\ \Omega$ does not degrade the quality factor; therefore, it adds negligible noise. Since $R_0 = 162\ \Omega$, the tank contribution to the phase noise is the same as in the previous example.

The collector shot-noise contribution due to the differential pair changes. The output amplitude improvement is made at the expense of a lower oscillation amplitude at the transistor base terminal. A lower amplitude produces a lower slope at the zero crossings and, in turn, a larger time window T_W during which the transistor injects shot noise into the tank. Since the amplitude at the base terminal is n times lower, the time window is n times larger. Following the same calculation as for the basic topology, the noise factor due to the collector noise is increased by $(1/n)$: $F = 1/(2n)$. Thus, in this specific design, it turns out that $F = 2.5$. The larger amplitude given by the capacitive partition is traded against higher noise of the transistor pair. An optimum value n exists that minimizes the phase noise. It may easily be calculated.

Figures 5.16 and 5.17 show the phase noise at 1 MHz and the noise factor, respectively, for each noise source. The collector noise is 2.5 times larger than the pure tank noise, as estimated. As the bias current is increased, the transistor area gets larger, while the base spreading resistance decreases. The larger the area, the higher the number of base fingers. In practice, the product $g_m r_{bb'}$ is almost constant along the bias current range and is equal to about 0.3. Also, the noise of the spreading resistance is affected by the larger time window T_W. Its contribution to the noise factor is estimated to be $g_m r_{bb'}/n = 1.5$, which is close to the simulated results.

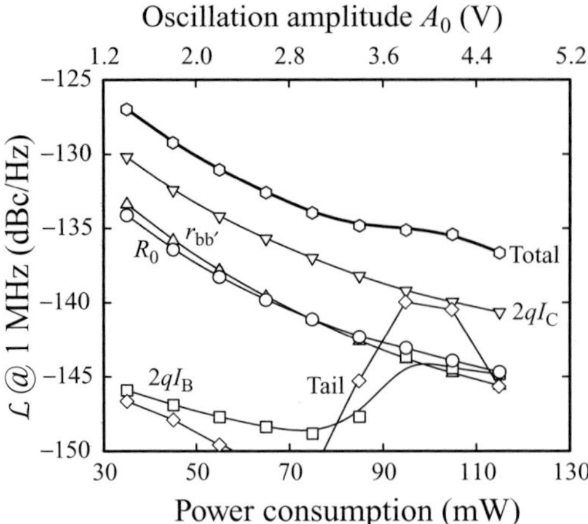

Figure 5.16 Phase-noise contributions of bipolar cross-coupled oscillator with capacitive divider and tail inductor. Results derived from circuit simulations

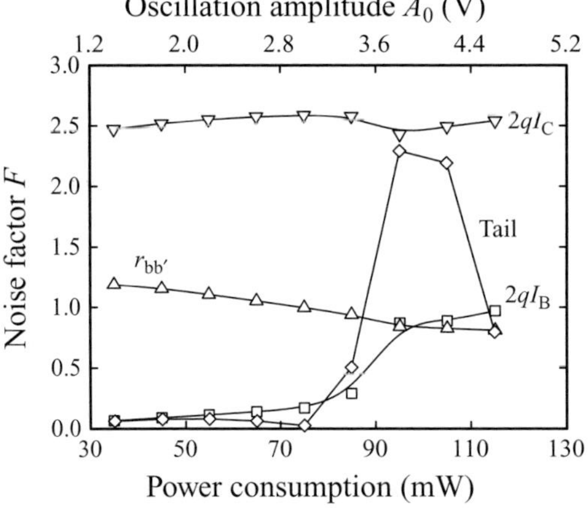

Figure 5.17 Noise factor of various noise sources in bipolar cross-coupled oscillator with capacitive divider and tail inductor. Results derived from circuit simulations

To maximize the voltage headroom, the tail current may be provided by a current mirror without degeneration resistance. In this case, (5.9) suggests a noise factor of $F = g_{m,T} R_0/8$, where $g_{m,T}$ is the tail transistor transconductance. However, numerical simulations do not follow such a theoretical dependence, featuring a noise factor that is almost a factor of two better. This result can be justified by taking into account that the pair switching is not instantaneous, i.e. $H(t)$ is not a square wave but its transition edges have a finite T_W duration. During this window, the tail noise simply produces a common mode signal at the tank terminals with no impact on the phase noise of the differential output (except for amplitude-to-phase modulation conversion). Therefore, the larger n, the lower the impact of the tail current noise.

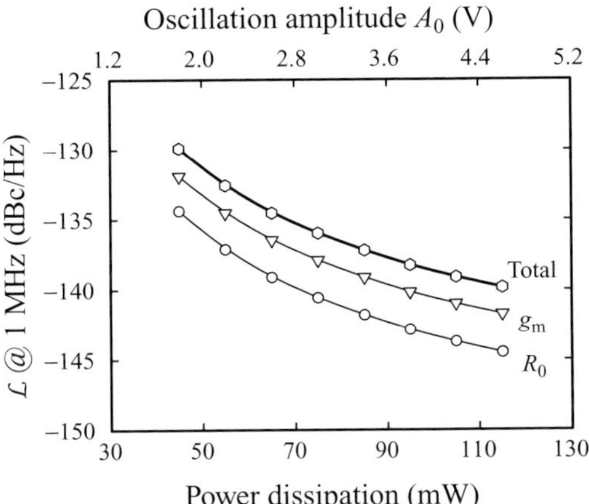

Figure 5.18 Phase-noise contributions of the CMOS cross-coupled oscillator designed in Example 4.2

Nevertheless, at typical values of n, the noise contribution of the tail generator would still be dominant (in this example, it is $F = 10$ at 2.5 V oscillation amplitude). The tail resonator at twice the oscillation frequency can be adopted to filter this noise. Figure 5.17 (diamonds) shows the simulated noise factor from the tail generator after the insertion of the tail resonator. This contribution is negligible, so long as the transistor pair operates in the forward active region.

Even in this circuit, the total phase noise is ultimately limited by the saturation of the transistor pair, which produces an increase in the contributions of tail noise and base shot-noise. Nevertheless, the adoption of the capacitive divider has pushed such a transition to larger amplitudes, thus optimizing the achievable performance.

Example 5.3 Phase noise of a MOSFET cross-coupled oscillator

Let us consider the simulation of the nMOS oscillator designed in Example 4.2. The nMOS noise factor γ extracted from the transistor model is equal to 1.77, along the whole range of bias currents. The tail generator is made of a current mirror. In order to limit the drain-to-source drop across the tail transistor, large transistors are used. However, they add a relevant parasitic capacitance to the source terminals of the switching pair. A tail resonator has been adopted to improve the maximum achievable amplitude, to filter out the tail current noise and, mainly, to reduce the noise contribution of the transistor pair. The presence of a large source capacitance would cause the transistor pair to be connected to an a.c. ground. Therefore, the channel thermal noise of the transistor pair would be injected into the tank during the whole period and not only during the time window T_W.

Figure 5.18 shows the simulated phase noise together with the two dominant contributors, namely the nMOS channel noise of the transistor pair and the tank noise. The noise factor of the transistor pair noise is shown in Figure 5.19 for both the circuits with and without tail resonator. Note that the estimated noise factor F, equal to $\gamma = 1.77$, is obtained only when the tail resonator is adopted. Otherwise, the noise factor increases, following the transistor transconductance.

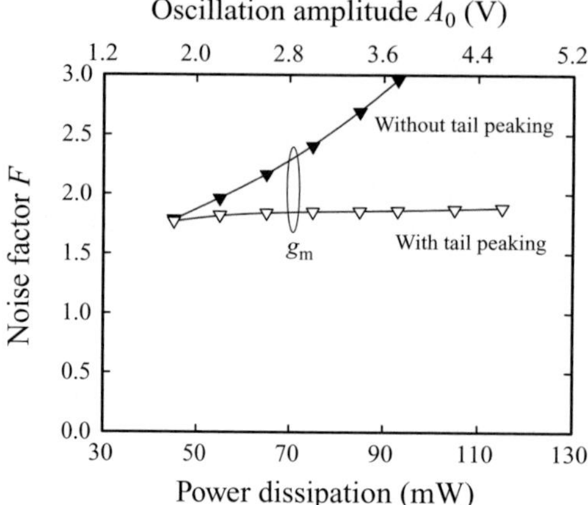

Figure 5.19 Noise factor of the transistor pair in CMOS cross-coupled oscillators with and without the tail resonator

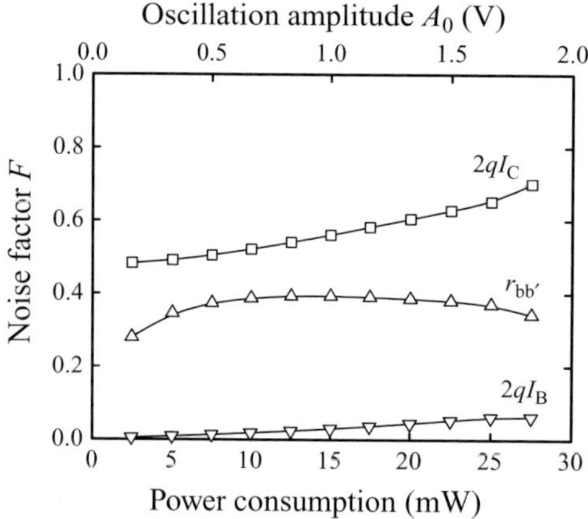

Figure 5.20 Noise factor of various noise sources in the bipolar Colpitts oscillator in Example 4.1. Results from circuit simulations

Example 5.4 Phase noise of a bipolar Colpitts oscillator

Let us now comment on the noise simulations of the single-transistor Colpitts oscillator designed in Example 4.1. The voltage division ratio n is set to 0.5. The resonator parallel resistance is 162 Ω and the bias current has been swept from 1 to 11 mA. So the amplitude varies between 0.16 V and 1.8 V.

In this condition, the noise factor from the transistor shot noise can be estimated from (5.16) as about 0.52. The noise factors of the transistor collector and base shot noise and the spreading resistance noise are shown in Figure 5.20. The base shot noise is negligible, since the bipolar transistor is kept far from the saturation region. The factor of the collector shot noise goes from 0.49 to about 0.7, in fair agreement with the theoretical estimation. This result also demonstrates that the triangle pulse approximation leading to (5.16) is more accurate than the rectangular shape for modelling the transistor noise injection.

5.6 Additional issues in low-phase-noise design

The results shown in this chapter may be summarized as follows. Low phase noise demands careful assessment and minimization of all noise contributions. The discussion has been limited to the main noise contributors. Other sources, such as thermal noise of the polysilicon gate terminals and of the bulk resistance, can always be made negligible by careful design. Gate noise has to be reduced by using short fingers. Bulk noise has to be limited by minimizing the resistance between the local bulk terminal of the transistors and the oscillator ground or voltage supply, depending on whether the transistor is n or p type.

As a final remark, note that the thermal design is another issue of a low-phase-noise oscillator. If the loss resistance is assumed to change proportionally with temperature, the phase noise from the tank losses increases as the fourth power of the absolute temperature:

$$\mathcal{L}(\omega_{\mathrm{m}}) \div \frac{kTR_0}{A_0^2} = \frac{kTR_0^3}{I_{\mathrm{B}}^2} \div T^4.$$

Taking the aluminium thermal coefficient of 5000 ppm/°C, it turns out that a temperature increase of 20 °C can be responsible for 1 dB phase noise degradation, which should be compensated for by a 12% increase in the bias current. A current generator proportional to the temperature (PTAT) may partially compensate for this degradation in phase noise.

5.7 References

[1] M. Lax, Classical noise. V. Noise in self-sustained oscillators, *Phys. Rev.*, **160**, Aug. 1967, 290–307.

[2] W. A. Edson, Noise in oscillators, *Proc. IRE*, **48**, Aug. 1960, 1454–66.

[3] D. B. Leeson, A simple model of feedback oscillator noise spectrum, *Proc. IEEE*, **54**, Feb. 1966, 329–30.

[4] K. Kurokawa, Noise in synchronized oscillators, *IEEE T. Microw. Theory*, **16**, Apr. 1968, 234–40.

[5] A. Hajimiri and T. H. Lee, A general theory of phase noise in electrical oscillators, *IEEE J. Solid-St. Circ.*, **33**, Feb. 1998, 179–94.

[6] A. Hajimiri and T. H. Lee, Design issues in CMOS differential LC oscillators, *IEEE J. Solid-St. Circ.*, **34**, May 1999, 717–24.

[7] A. Hajimiri, S. Limotyrakis and T. H. Lee, Jitter and phase noise in ring oscillators, *IEEE J. Solid-St. Circ.*, **34**, Jun. 1999, 790–804.

[8] P. Andreani, X. Wang, L. Vandi and A. Fard, A study of phase noise in Colpitts and LC-tank CMOS oscillators, *IEEE J. Solid-St. Circ.*, **40**, May 2005, 1107–18.

[9] C. Samori, A. L. Lacaita, F. Villa and F. Zappa, Spectrum folding and phase noise in LC tuned oscillators, *IEEE T. Circuits-I*, **45**, Jul. 1998, 781–90.

[10] J. J. Rael and A. Abidi, Physical processes of phase noise in differential LC oscillators, *Proc. IEEE 2000 Custom Integrated Circuits Conf.*, May 2000, 569–72.

[11] E. Hegazi, H. Sjoland and A. A. Abidi, A filtering technique to lower LC oscillator phase noise, *IEEE J. Solid-St. Circ.*, **36**, Dec. 2001, 1921–30.

[12] K. S. Kundert, Introduction to RF simulation and its application, *IEEE J. Solid-St. Circ.*, **34**, Sep. 1999, 1298–1319.

[13] A. Demir, A. Mehrotra and J. Roychowdhury, Phase noise in oscillators: a unifying theory and numerical methods for characterization, *IEEE T. Circuits-I*, **47**, May 2000, 655–74.

[14] S. K. Magierowski and S. Zukotynski, CMOS LC-oscillator phase-noise analysis using nonlinear models, *IEEE T. Circuits-I*, **51**, Apr. 2004, 664–76.

[15] P. Kinget, Integrated GHz voltage controlled oscillators, in *Analog Circuits Design: (X)DSL and Other Communication Systems; RF MOST Models; Integrated Filters and Oscillators*, ed. W. M. C. Sansen, J. H. Huijsing and R. van der Plassche, Boston, MA: Kluwer, 1999, 351–83.

[16] F. X. Kaertner, Analysis of white and $f^{-\alpha}$ noise in oscillators, *Int. J. Circ. Theor. App.*, **18**, 1990, 485–519.

[17] F. X. Kaertner, Determination of the correlation spectrum of oscillators with low noise, *IEEE T. Microw. Theory*, **37**, Jan. 1989, 90–101.

[18] B. Razavi, A study of phase noise in CMOS oscillators, *IEEE J. Solid-St. Circ.*, **31**, Mar. 1996, 331–43.

[19] J. Roychowdhury, D. Long and P. Feldmann, Cyclostationary noise analysis of large RF circuits with multitone excitations, *IEEE J. Solid-St. Circ.*, **33**, Mar. 1998, 324–36.

[20] Affirma RF Simulator User Guide, Cadence product documentation, version 4.4.6, Apr. 2001.

6 Reactive components in oscillators

6.1 Introduction

The key impact of reactive elements on VCO performance has been highlighted extensively in the preceding chapters. Noise, tuning range, signal amplitude and power dissipation are all strongly dependent on the 'quality' of inductors and variable capacitors. This chapter is, therefore, devoted to discussing the margin left to the designer for judiciously tailoring these components to squeeze out the best performance.

The optimization of reactive elements is not fully under the control of the circuit designer. Some properties depend directly on the process and often the only option left to the designer is to choose the inductor geometry or the varactor type, within a limited library of components that have previously been characterized and modelled.

On the other hand, a larger degree of freedom would call for a full customization of the reactive components, which is a very risky and delicate process, involving some trial-and-error steps, starting from numerical modelling at the device level, and proceeding with its experimental characterization, the fine tuning of the structure and the development of compact models for circuit simulation.

However, the increase in operating frequency is making the optimization of reactive components a key ability of an RF designer. The following sections give some basic guidelines.

6.2 Integrated inductors

6.2.1 On-chip inductors

On-chip inductors are widely used in silicon-based RF integrated circuits as series or shunt elements in resonant VCO tanks, as well as impedance-matching elements in LNAs and choke components. Inductors have traditionally been integrated in GaAs MMICs. However, while III–V compound materials are semi-insulating, the silicon substrate is more conductive. In silicon integrated circuits, the variable magnetic fields induce, therefore, significant currents in the substrate itself, leading to losses and degradation of the inductor quality factor. In this environment, the evaluation of component parameters, such as the inductance value, the losses, the quality factor and the self-resonance frequency, is not simple at all.

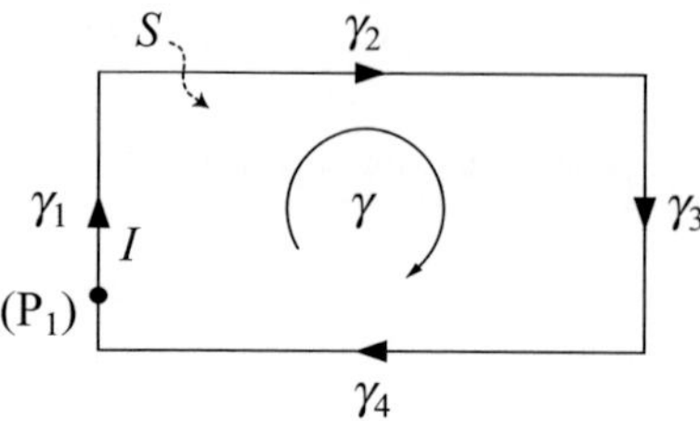

Figure 6.1 Current loop that can be decomposed into partial inductances

The design flow of an inductor can be conceptually divided into the following steps:

- Choice of the component shape.
- Definition of the geometric parameters and of the occupied area.
- Detailed electromagnetic simulations of the component topology.
- Translation of the numerical results obtained from the electromagnetic simulators into a lumped element model.

In terms of component geometry: some options are available depending on the inductance values, circuit topologies, operating frequencies. In the 1–6 GHz range, inductor values usually range from 100 pH to 10 nH and the inductors are designed using conventional planar spiral structures. For frequencies above 10 GHz the inductances have values well below nH. In this case, microstrips are more appropriate.

As far as electromagnetic simulations are concerned, numerical simulators are nowadays available, but the layout refinement is always extremely time consuming. The outcomes of these simulators are the S parameters of the structure over a discrete set of frequency values. Since circuit analysis is performed in the time domain, the S values then have to be translated into a lumped model to be used in the circuit simulator.

Often, the circuit designer does not have access to the whole design flow. He or she can only choose some parameters (e.g., the inductance value, the number of turns): the inductor layout and the corresponding lumped model are then automatically generated. A fully customized design requires instead mastering all the steps above. Let us recall some key concepts that should be properly handled while going through the complete design procedure.

6.2.2 Partial inductance

The first step in the inductor design is the evaluation of the inductive impedance. This task can be performed by following the partial inductance approach.

Let us consider the rectangular loop γ in Figure 6.1. It defines a surface S and a current, I, flows through it. The flux Φ of the magnetic induction field $\vec{B}$ through S is given by $\Phi = LI$, where L is the loop inductance. On the other hand, the field $\vec{B}$ can be written in terms of the magnetic vector potential $\vec{A}$, and Stokes' theorem gives:

$$\begin{aligned} L &= \frac{\Phi}{I} = \frac{1}{I}\int_S \vec{B}\cdot\vec{n}\cdot \mathrm{d}S \\ &= \frac{1}{I}\int_S \left(\nabla\times\vec{A}\right)\cdot\vec{n}\cdot\mathrm{d}S = \frac{1}{I}\int_\gamma \vec{A}\cdot\mathrm{d}\vec{\gamma}. \end{aligned}$$

The subscript S indicates a surface integral, while the line integral in the last step is evaluated along the closed path γ. This last term can, obviously, be divided into the sum of four contributions, one for each side. For instance L_1 may denote the integral of the vector potential along the side γ_1 divided by I, and so on.

Each one of these four integrals can be further decomposed. Considering again the side γ_1, the vector potential $\vec{A}_1$ at (P_1) is the sum of four terms. $\vec{A}_{1,j}$ with $j = 1 \ldots 4$, where $\vec{A}_{1,j}$ is the vector potential at (P_1), due to the current in side j. It turns out that:

$$\begin{aligned} L_1 &= \frac{1}{I}\int_{\gamma_1} \vec{A}_{1,1} \cdot \mathrm{d}\vec{\gamma}_1 + \frac{1}{I}\int_{\gamma_1} \vec{A}_{1,2} \cdot \mathrm{d}\vec{\gamma}_1 + \frac{1}{I}\int_{\gamma_1} \vec{A}_{1,3} \cdot \mathrm{d}\vec{\gamma}_1 \\ &\quad + \frac{1}{I}\int_{\gamma_1} \vec{A}_{1,4} \cdot \mathrm{d}\vec{\gamma}_1 \\ &= L_{1,1} + L_{1,2} + L_{1,3} + L_{1,4}. \end{aligned}$$

The first term $L_{1,1}$ is called the *partial self-inductance* of the segment, while the others are referred to as *partial mutual inductances*. Sometimes, the partial self-inductance is erroneously confused with the internal inductance, which is the contribution to the inductance from the magnetic flux crossing the inner part of the conductor itself, but this is obviously not the case.

Note that the vector potential has the same orientation as the current. For instance, the vector potential generated by the current flowing through the segment γ_2 in Figure 6.1 is orthogonal to side γ_1. The same applies to the vector potential $\vec{A}_{1,4}$ generated along γ_1 by the current flowing through γ_4. It follows that $L_{1,2} = L_{1,4} = 0$, while $L_{1,3} < 0$ because the currents in γ_3 and γ_1 are opposite. The partial self-inductance is, instead, always positive. The total inductance of a loop is the sum of all the terms:

$$L = \sum_{i,j=1}^{N} L_{i,j}.$$

For example, the inductance of the rectangular loop in Figure 6.1, is given by $L = 2 \cdot (L_{1,1} - |L_{1,3}|) + 2 \cdot (L_{2,2} - |L_{2,4}|)$. The symmetry makes it possible to consider only the self and mutual terms arising from side γ_1 and γ_2, and then multiplying the result by two.

The inductance values of complex geometries may be estimated by starting from the partial self-inductances and mutual inductances of elementary configurations, like straight metal segments. It can be shown that the partial self-inductance of a straight wire having length l and featuring a circular cross section of radius, r, is given by:

$$L_{i,i} = \frac{\mu}{2\pi} \cdot l \cdot \left[\ln\left(\frac{l}{r} + \sqrt{\left(\frac{l}{r}\right)^2 + 1} \right) + \frac{r}{l} - \sqrt{\left(\frac{r}{l}\right)^2 + 1} \right], \tag{6.1}$$

where $\mu \cong \mu_0 = 4\pi \cdot 10^{-7}\,\mathrm{H/m}$. For the more usual rectangular cross section it is:

$$L_{i,i} \cong \frac{\mu}{2\pi} \cdot l \cdot \left[\ln\left(\frac{2l}{w+t} \right) + 0.5 + \frac{w+t}{3l} \right], \tag{6.2}$$

where w and t are the width and the thickness of the line, respectively. [1, 2].

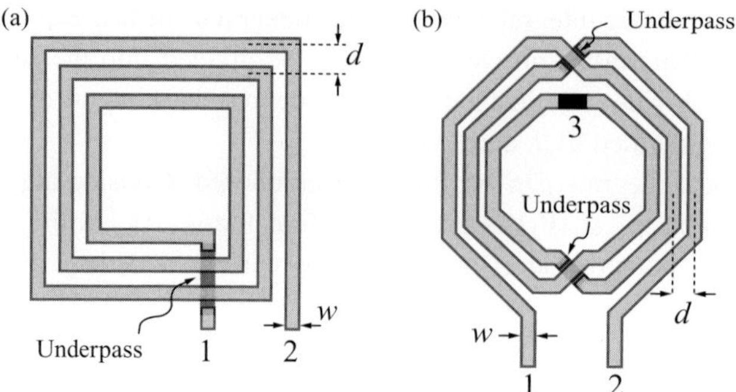

Figure 6.2 Examples of spiral inductors: (a) square and (b) octagonal

For a conductor much longer than r, (6.1) reduces to $L_{i,i} \cong (\mu/2\pi)l \cdot \ln(2l/r)$. In a similar way, for $l \gg w, t$ (6.2) can be approximated by $L_{i,i} \cong (\mu/2\pi)l \cdot \ln[2l/(w+t)]$. Note that the partial self-inductance is not simply proportional to the length l, so it is not possible to quote a partial inductance per unit length rigorously, even if it is done quite often.

The mutual inductance between two parallel conductors is approximately given by:

$$L_{i,j} \cong \frac{\mu}{2\pi} \cdot l \cdot \left\{ \ln\left[\frac{l}{d} + \sqrt{1+\left(\frac{l}{d}\right)^2}\right] - \sqrt{1+\left(\frac{d}{l}\right)^2} + \frac{d}{l} \right\}, \tag{6.3}$$

where d is the geometric mean distance between them. For practical purposes, d can be taken as the distance between the centres of the conductors. Equation (6.3) is always positive; however, if the currents in the segments i and j are opposite, it has to be taken as negative.

The resonator of a differential oscillator may be built using inductors with the square shape in Figure 6.2(a). However, since the parasitic capacitances at ports 1 and 2 are different, symmetry can be achieved by using two inductors, one for each oscillator output. They may be drawn by mirroring the square structure along the right side leading to terminal 2. In this way, the terminals at 1 will be connected to the transistors and the two terminals at 2 will be tied together to the power supply. Since the current signal is differential, the two mirrored sides leading to terminals at 2 will always carry identical currents, but flowing in opposite directions. The resulting mutual inductance is, therefore, negative, and lowers the total inductance. To limit the effect, the two spirals must be spaced apart.

This problem does not arise for symmetrical coils like the one shown in Figure 6.2(b). The component provides a mearly symmetrical load at its terminals. Terminals 1 and 2 are driven by the drain (or collector) terminals of the transistor pair, while the point 3 is linked to the power supply. The inductor shown in Figure 6.2(b) is octagonal but it may also be designed using a square shape. However, the octagonal geometry has a higher inductance value per unit area. In fact, note that in the octagonal shape two successive segments are not orthogonal and their mutual inductance is positive. In the square shape, two successive segments are, instead, orthogonal, leading to a nil mutual term.

The two following analytical examples help to better handle the concept of partial self-inductance, and its link to the loop inductance.

Example 6.1 Partial inductance of two wires of infinite length

Let us consider first a line made by two identical infinite parallel wires of circular cross-section, with radius r, and obviously carrying opposite currents. From classical textbooks, [3] it turns out that the inductance *per unit length* of this line is given by:

$$L' = \frac{\mu}{\pi} \cdot \cosh^{-1}\left(\frac{d}{2r}\right),$$

where d is the distance between the centres of the wires. When $d \gg r$ the approximation $\cosh^{-1}(x) \cong \ln(2x)$ leads to $L' \cong (\mu/\pi) \cdot \ln(d/r)$. It would now be interesting to obtain the same results using the notion of partial inductance.

The partial self-inductance of the wire is given by (6.1). For $l \gg r$, it reduces to $L_{i,i} \cong (\mu/2\pi)l \cdot \ln(2l/r)$. The partial mutual inductance $L_{i,j}$ between the two parallel cylindrical conductors spaced by d is given by (6.3), which, for $l \gg d$, becomes $L_{i,j} \cong (\mu/2\pi)l \cdot \ln(2l/d)$. Since the currents are opposite, this contribution has to be taken with a negative sign. The total inductance for a length l turns out to be:

$$2(L_{i,i} - L_{i,j}) = (\mu/\pi)l \cdot \ln(d/r).$$

This result is consistent with the expression for the inductance per unit length given above. Note that the inductance increases both as d gets larger and as r reduces at constant d. This corresponds to increasing the surface between the two conductors. As the inductance increases, the capacitance per unit length decreases, since the product LC must remain constant and equal to $\varepsilon\mu$.

Example 6.2 Partial inductance and loop inductance

A second example deals with the dependence of a coil inductance on the number of turns, n. It will be shown that the inductive impedance of the coil increases almost as n^2. The result may be derived in two ways. The classical argument starts by considering the flux Φ of the magnetic induction field B through the loop. The field strength, and, therefore, the flux Φ, scales as the number n of the turns, all crossed by the same current, I. On the other hand, the voltage induced across the coil by the time-dependent B field is almost n times the voltage induced across a single turn. It follows that the inductance, which links the voltage induced across the coil to the derivative of the current, I, is expected to scale as n^2.

The same conclusion may be derived using the concepts of partial self-inductances and mutual inductances. The partial self-inductance of each side is given by (6.2): $L_{i,i} \cong (\mu/2\pi) \cdot l \cdot \ln[2l/(w+t)]$. Considering a single-turn square inductor, if the number of turns doubles, the number of partial self-inductance terms also doubles. In addition, there are now two positive mutual inductances per side; those between the adjacent conductors. Since the distance between the two segments is much smaller than their length, $d \ll l$, (6.3) leads to $L_{i,j} \cong (\mu/2\pi) \cdot l \cdot \ln(2l/d)$. The value can be very close to the self-inductance $L_{i,i}$, because the spacing between the conductors will be similar to the typical

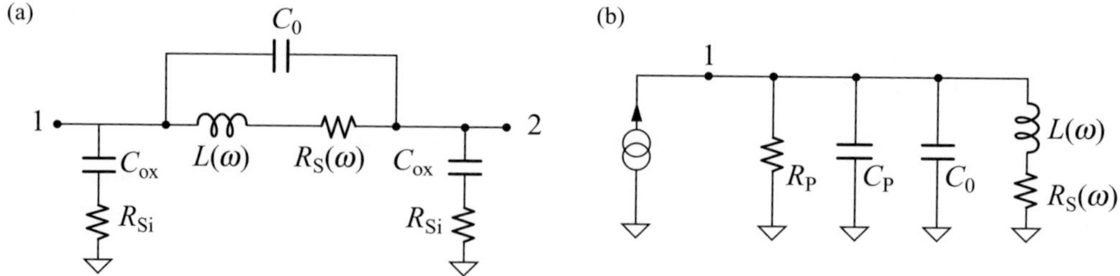

Figure 6.3 Two-port models of inductor: (a) physically based model and (b) equivalent circuit for the evaluation of the input impedance and Q factor

dimensions of the conductor cross-section. The total loop inductance is, therefore, increased by a factor four.

In general, if the inductor has n turns, there are n self-inductances $L_{i,i}$ and $n(n-1)$ mutual inductances $L_{i,j}$ per side. Each of the n conductors is influenced by the other $n-1$. If $L_{i,i} \cong L_{i,j}$, the total inductance is increased by a factor $n + n(n-1) = n^2$ with respect to the single-turn case. The argument is approximate, since the negative mutual inductances have been neglected. However, it is consistent with the classical result holding for a long solenoid, whose inductance per unit length is proportional to the square of the number of turns per unit length.

6.2.3 Inductor losses, models and parasitics

There is a wealth of literature dealing with loss mechanisms in integrated inductors and with the component models. References [4–16] are a very limited subset of the contributions on the topic.

Figure 6.3(a) shows a first-order model of the whole inductor. It is useful as a starting point to discuss the inductor losses, the dependence of the quality factor and of the component self-resonance on the device parameters.

The series resistance R_S accounts for the finite conductivity of the metal lines, and depends on the metal thickness but also on the coil geometry. Square inductors feature a series resistance larger than ideal components with circular loops. Octagonal inductors better approximate the circular shape and have a smaller series resistance for the same loop length.

The capacitance C_{ox} represents the parasitic capacitance between the coil and the substrate. The resistance R_{Si} accounts for the losses due to the displacement current injected into the substrate through C_{ox}. Often a capacitance C_{Si} is placed in parallel with R_{Si}, to represent the effect of the finite dielectric relaxation time. C_0 instead represents the capacitive effect between the conductors, in particular between the underpasses in Figure 6.2 and the upper conductors of the coils.

Note that both the resistance R_S and the inductance are dependent on the operating frequency. [12] One reason why R_S depends on frequency is the well-known *skin effect*. As the frequency rises, the current flows in the outer region of the metal layer. The series

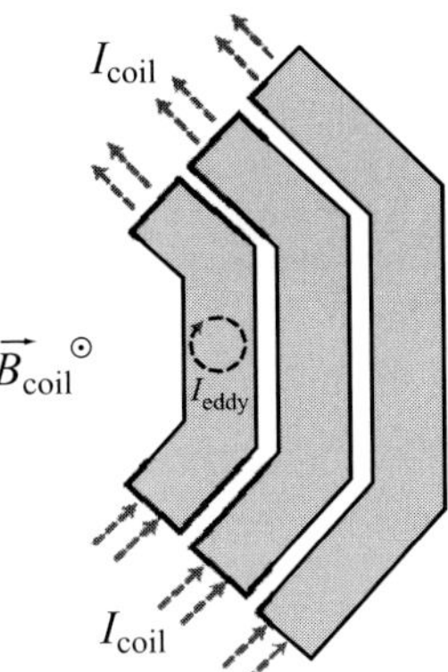

Figure 6.4 Proximity effect of eddy currents

resistance rises with frequency, following the dependence of the skin depth: $\delta = \sqrt{2\rho/\omega\mu}$, where ρ is the conductor resistivity. At 5 GHz the skin depth of copper is less than 1 μm.

A non-uniform distribution of the current density is also caused by the interaction between neighbouring segments. This effect is explained with the help of Figure 6.4, which shows a portion of a three-spiral inductor. The time-dependent flux generated by the outer loops of the coil induces the so-called eddy currents in the inner segments. The total current in the inner segments is not uniformly distributed, but is higher in the internal side of the turn, increasing the losses. [5, 13] The eddy currents decrease the magnetic flux through the inductor area, reducing the inductance value. This effect, which is referred to as the *proximity effect*, depends on frequency.

In spiral inductors the impact of the proximity effect can be more relevant than the skin effect on the single wire. Since the inner turns do not contribute so much to the total inductance, while adding more losses, they may be removed. This choice leads to designing so-called 'hollow' inductors. [5]

The variable magnetic field also causes eddy currents in the substrate. Assuming a B field orthogonal to the silicon surface, the current induced in the substrate is proportional to $B\omega/\rho_{Si}$, where ρ_{Si} is the substrate resistivity. The power dissipated in the substrate will be proportional to $B^2\omega^2$. Since the field B is proportional to the current flowing in the coil, these losses are taken into account by adding to R_S a contribution that increases proportionally to ω^2. These eddy current a also causes reduction in the B flux, and a reduction in the inductive impedance.

All these effects are described by properly setting the values and the frequency dependence of the lumped components in the model shown in Figure 6.3(a).

6.2.4 The inductor quality factor

The inductor quality factor is of paramount importance for our purposes. It has been derived in Chapter 4 as $\omega L/R_S$. Since the model in Figure 6.3(a) is much more complex than just an inductor with a series resistance, the result needs to be revised.

The Q factor of the inductor is often measured by shunting one port, let us say port 2 in Figure 6.3(a), and measuring the admittance of the resulting network. Figure 6.3(b) shows the equivalent circuit corresponding to the measurement condition, where the series C_{ox} and R_{Si} has been replaced by C_{P} and R_{P}. It can easily be shown that:

$$R_{\mathrm{P}} = \frac{1}{\omega^2 C_{\mathrm{ox}}^2 R_{\mathrm{Si}}} + R_{\mathrm{Si}}\ ;\ C_{\mathrm{P}} = C_{\mathrm{ox}} \cdot \frac{1}{1 + \omega^2 C_{\mathrm{ox}}^2 R_{\mathrm{Si}}^2}. \tag{6.4}$$

The quality factor of the component is then calculated as $Q = -\mathrm{Im}(y)/\mathrm{Re}(y)$. In the simple case of an inductance in series with a resistance, the expression reduces to the usual term $\omega L/R_{\mathrm{S}}$. For the circuit in Figure 6.3(b), it is, instead, given by:

$$Q = \frac{\omega L}{R_{\mathrm{S}}} \cdot \frac{R_{\mathrm{P}}}{R_{\mathrm{P}} + \left[(\omega L/R_{\mathrm{S}})^2 + 1\right] R_{\mathrm{S}}} \cdot \left[1 - \frac{R_{\mathrm{S}}^2\,(C_0 + C_{\mathrm{P}})}{L} - \omega^2 L\,(C_0 + C_{\mathrm{P}})\right]. \tag{6.5}$$

The first term is $\omega L/R_{\mathrm{S}}$. The second term accounts for the substrate losses. The third term is more difficult to justify. It is related to the electrical energy stored in the stray capacitors, which increases with frequency. As a matter of fact, at low frequencies the Q value given by (6.5) increases with ω. Then the quality factor peaks and starts to decrease, becoming zero for a value of ω that is referred to as the inductor self-resonance frequency.

The presence of a self-resonance in Figure 6.3(b) is not surprising. The inductor and its parasitic capacitors can resonate. At the inductor self-resonance frequency, the reactive admittance of the entire network is zero and the same happens to the quality factor, which is defined as $Q = -\mathrm{Im}(y)/\mathrm{Re}(y)$. Beyond this frequency, such a Q value even becomes negative.

Now a question arises. If a resonant tank is built by placing such an inductor in parallel with a capacitance C, can the values from (6.5) be used to compute the overall Q factor of the tank according to (4.6)?

The answer is certainly negative. The values derived as $-\mathrm{Im}(y)/\mathrm{Re}(y)$ should be handled with care. Whenever the ratio $-\mathrm{Im}(y)/\mathrm{Re}(y)$ is derived from a purely inductive or capacitive component, it gives the same values, which are obtained from the energetic Q definition given by (4.5). But in the more general case, as for the model network in Figure 6.3(b), this is not true any more. In this case, the ratio $-\mathrm{Im}(y)/\mathrm{Re}(y)$ is zero at the resonance and therefore cannot be coincident with the Q value which, at resonance, should account for the -3 dB bandwidth.

In conclusion, direct measurements of the admittance $y(\omega)$ have to be used to derive all the parameters of the inductor model in Figure 6.3(b), while the Q factor given by (6.5) cannot be used to compose the total Q factor of the resonant tank embodying that inductor. The procedure should be different. Once the model is set, the Q factor of the non-ideal inductor should be derived by separately computing the capacitive and inductive Q terms of the reactive elements according to their energetic definitions in (4.5):

$$Q_L = \frac{\omega L}{R_{\mathrm{S}}} \quad \text{and} \quad Q_{(C_{\mathrm{P}}+C_0)} = \omega_0 R_{\mathrm{P}}\,(C_{\mathrm{P}} + C_0)\,.$$

The Q factor of the overall resonant tank made by placing the non-ideal inductor in parallel with a capacitor C will then be obtained by first calculating the overall capacitive Q factor

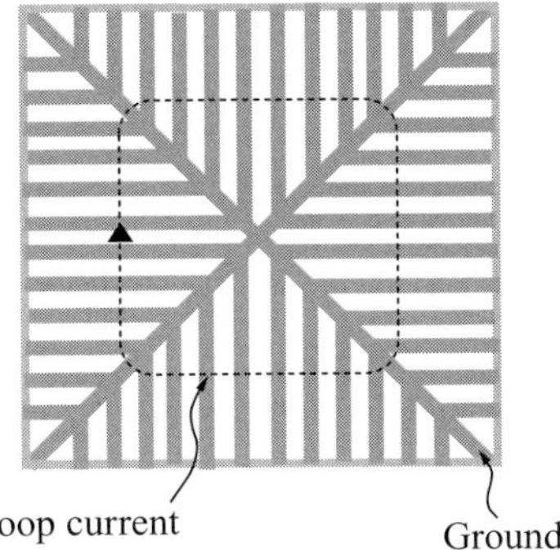

Figure 6.5 Top view of patterned ground shield

due to the parasitic capacitances and to the capacitor C, $Q_{(C_P+C_0)}$ and Q_C, then combining the inductive and the capacitive Q as in (4.6):

$$\frac{1}{Q} = \frac{1}{Q_L} + \frac{1}{\dfrac{C}{C_0 + C_P + C} Q_C + \dfrac{C_0 + C_P}{C_0 + C_P + C} Q_{(C_0+C_P)}}.$$

Another way is by resorting to circuit simulations: the tank capacitor C is placed in parallel with the inductor model; the shape of the resulting resonance is computed and the Q factor of the tank is derived from the ratio between the resonance frequency and the -3 dB bandwidth. [7]

6.3 Inductor topologies

6.3.1 Spiral inductors

The substrate losses caused by the resistance R_{Si} in Figure 6.3(a) are a major concern. In principle, these losses can be avoided if R_{Si} is made infinite. The substrate would act as an open circuit and no power would be dissipated by the currents flowing through it. This condition corresponds either to using a very high-resistivity substrate or to etching away the silicon below the inductors. [17] Both solutions are difficult and expensive in standard integrated technologies. The alternative is to reduce the substrate losses by shunting R_{Si} with a short. Also, in this case, no power will be dissipated in R_{Si}. In practice, the solution leads one to introduce a grounded conductor between the coil and the substrate. [6, 10] Incidentally, note that according to (6.4) R_P goes to infinity both when R_{Si} is infinite and when $R_{Si} = 0$. In turn, if R_P is infinite the second factor in (6.5) reduces to one.

The grounded conductor, which can be made of aluminium or doped polysilicon, shields the electric field and stops the electrically induced displacement currents in the substrate. Instead, the eddy currents, which are magnetically induced, will still circulate in the shield. Since the shield is closer to the metal lines, these currents can be even stronger than those induced in the substrate. Narrow slots cut in the conducting shield, like those shown in Figure 6.5, strongly reduce these losses, while still effectively shielding the electric field.

It is worthwhile to note that the inductance value and the losses depend on currents induced in the surrounding conductors. Hence, not only simulations, but even measurements on a stand-alone inductor may fail to give the precise values of the inductor parameters when the component is used in a silicon circuit. The use of a ground shield helps instead to improve the component reproducibility, by better isolating the inductor from the nearby environment, at least to some extent.

Whenever Q is limited by R_S, another option to improve the quality factor is to shunt several metal levels together in parallel, using vias. In this way, the series inductance L remains practically unchanged (it slightly decreases because the substrate is closer) while the series resistance R_S can decrease. In practice, the quality factor does not improve proportionally to the number of stacked metal levels. This is due mainly to the lower conductivity of the bottom metals. For this reason, the top metal always carries most of the current and thus the benefit of shunting the metal lines is reduced. The number of vias does not have a significant effect on Q and on the self-resonance frequency and leaves the inductance unaffected. This result can be exploited by halving the number of vias to reduce the simulation time during electromagnetic analyses. Since the substrate is closer, the self-resonance frequency decreases.

An alternative is to connect the metal levels in series, leading to the so-called 'stacked inductor'. Taking a two-level structure, with identical geometries on each level, a nearly four-fold increase of the inductance is expected. In fact, there are two coils and the magnetic flux is doubled. Since the series resistance is nearly doubled, the quality factor is potentially twice the value of the single coil. In general, multi-level series-connected inductors are useful for increasing the inductance per unit area. However, at high frequencies the quality factor gets quickly worse, owing to the closer substrate and, which is more important, the overlapping capacitance between stacked metals. In addition, the self-resonance frequency may be low. These inductors are used as RF chokes, when the high inductance-to-area ratio is a key parameter.

6.3.2 Microstrip and coplanar waveguide inductors

The increase of an integrated circuit's operating frequencies requires smaller reactive components. Spiral inductors are not suitable for synthesizing inductance values below 100 pH: narrow-diameter coils have a poor quality factor because of proximity effects. The design of inductors for frequencies higher than 10 GHz usually follows a different approach: the inductor is implemented with a single transmission line shorted at one end. Research on this subject is expected to be pushed forward as silicon integrated circuits are used for applications in the millimetre-wave region.

A transmission line, shorted at one end, shows an inductive impedance when its length l is smaller than $\lambda/4$. In the ideal lossless case, the magnitude of this impedance, ωL_S, is about $Z_0 \cdot \gamma l$, [4,18] where $\gamma = 2\pi/\lambda$ is the propagation constant and $Z_0 = \sqrt{L/C}$ is the line characteristic impedance, i.e. the square root of the ratio between the inductance and capacitance *per unit length*. Therefore, the inductance of the shorted line is $L_S = Z_0 \cdot l/v$, with v the wave propagation velocity. In practice, the design of integrated transmission lines

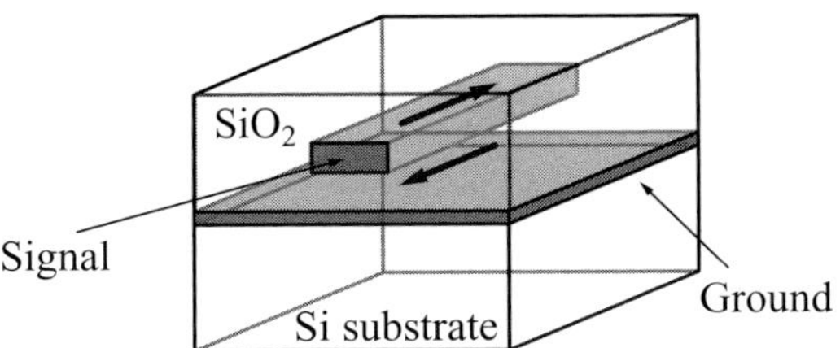

Figure 6.6 Microstrip line running on ground plane

relies on the use of electromagnetic simulations, the only way to account properly for the substrate response, skin and proximity effects.

Microstrip is the most common planar transmission line. Microstrip lines, integrated in silicon processes, are implemented using the top metal layer as the signal line and the bottom metal layer for the ground plane (Figure 6.6). The ground metal plane avoids, in part, electromagnetic field penetration into the substrate, thus reducing the induction of eddy currents in the substrate. Since the ground plane is the reference plane, it is crossed by a current equal to that flowing in the microstrip, but with the opposite direction. It follows that the series resistance is the sum of the resistance of the signal line and of the ground plane and grows as the square root of the frequency, because of the skin effect. The ground plane also avoids coupling between adjacent transmission lines.

A limitation of microstrips integrated in silicon processes is the low inductance per unit of length. Because of the short distance between the ground plane and the signal line (a few micrometres), the volume where the magnetic energy is stored is small and typical inductance values are around 0.1–0.5 pH/μm with an inductive quality factor that hardly exceeds 10 at 40 GHz. To increase Z_0 the designer can decrease the width of the line. In this way, however, the series resistance also increases.

An alternative to microstrips is the use of a coplanar waveguide (CPW), in which the signal line is surrounded by two coplanar ground planes; [19] CPW allows for a higher quality factor, up to 30 at 40 GHz, because of the storage of magnetic energy in a wider region of space. The designer has to control both the line width and the line spacing to modify the characteristic impedance: the trade-off between the quality factor and the parasitics is more relaxed than in microstrips. However, electric fields can penetrate the substrate, resulting in higher shunt losses, owing to displacement currents. Another drawback is that wide ground planes are required to avoid coupling between parallel CPWs, leading to a larger area occupation. A possible solution to these issues is the shielded CPW, [20] where, with respect to CPW, a ground plane is added beneath the signal line; the bottom ground plane is slotted to prevent inductance degradation.

Example 6.3 Substrate impact on partial inductance evaluation

The inductor design is usually performed by relying extensively on an electromagnetic simulator. However, the first-order estimate of the inductance value may be obtained using the expressions (6.2) and (6.3) for the partial self-inductances and mutual inductances. These results, obtained by Grover, [1] hold strictly for metal segments in air. The aim of the

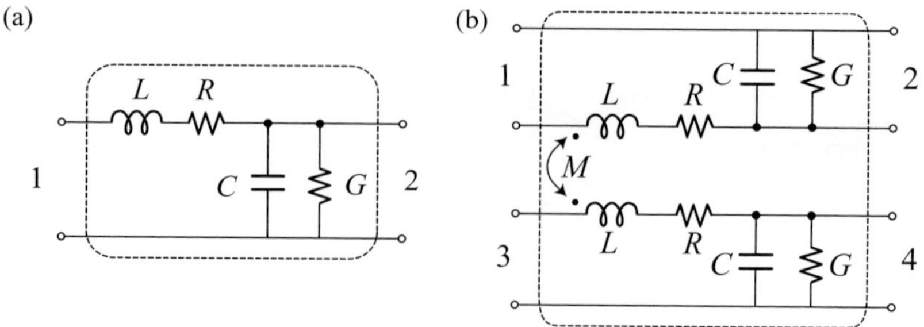

Figure 6.7 Lumped-element equivalent circuit of an incremental length of a transmission line: (a) isolated line, (b) two coupled lines

first example is, therefore, to quantify the impact of the substrate resistivity on the estimate. All the simulations refer to a 1 GHz frequency.

Let us consider a metal segment placed over a silicon substrate, with a rectangular cross section, 3 μm thick, and metal resistivity of 3×10^{-6} Ωcm. The metal segment is surrounded by a SiO_2 layer, 5.25 μm thick and runs at 0.95 μm from the silicon substrate. The relative dielectric constants are $\varepsilon_{Si} = 11.9$ and $\varepsilon_{ox} = 3.9$. The simulations have been performed using Agilent Momentum, imposing the open boundary condition (i.e., zero electric field at infinity). The substrate resistivity has been changed around the $0.1 - 10$ Ωcm range covered by RF technology processes. The S parameters of the metal segments were derived by connecting two generators with 50 Ω internal impedance at each port of the line. The S parameters were then converted into the equivalent model of a transmission line, whose single element is depicted in Figure 6.7(a). The numerical procedure is described in [21].

Since the only source of the magnetic potential is the current flowing in the metal segment, the L value derived from the simulations is equal to $L_{1,1} = (1/I) \int \vec{A}_{1,1} d\vec{\gamma}_1$, i.e., the partial self-inductance of the metal segment. The simulated inductance is shown versus the substrate resistivity for a 10 μm wide and 1 mm long metal wire in Figure 6.8(a). Even if the partial self-inductance in (6.2) is not strictly proportional to the length, the inductance per unit length, in pH/μm, was derived by dividing the total inductance by 1 mm $= 10^3$ μm. Note that, at 1 GHz, for a substrate resistivity above 5 Ωcm the inductance value no longer depends on the resistivity value and approaches the partial self-inductance of 1.1 pH/μm, which can be obtained using (6.2).

In contrast, when the substrate becomes very conductive, the system metal line–oxide–substrate approximates a line over a ground plane. The magnetic field no longer penetrates the substrate. It remains confined in the 0.95 μm thick oxide, and the inductance drops. In these cases, therefore, (6.2) provides an overestimate of the real inductance value.

Figure 6.8(b) shows the values obtained by taking the resistivity constant as 5 Ωcm, and by changing the metal width. The results derived from (6.2) are close to the simulation outcomes, which is not surprising. Figure 6.8(a) shows that 5 Ωcm is high enough not to affect the accuracy of (6.2).

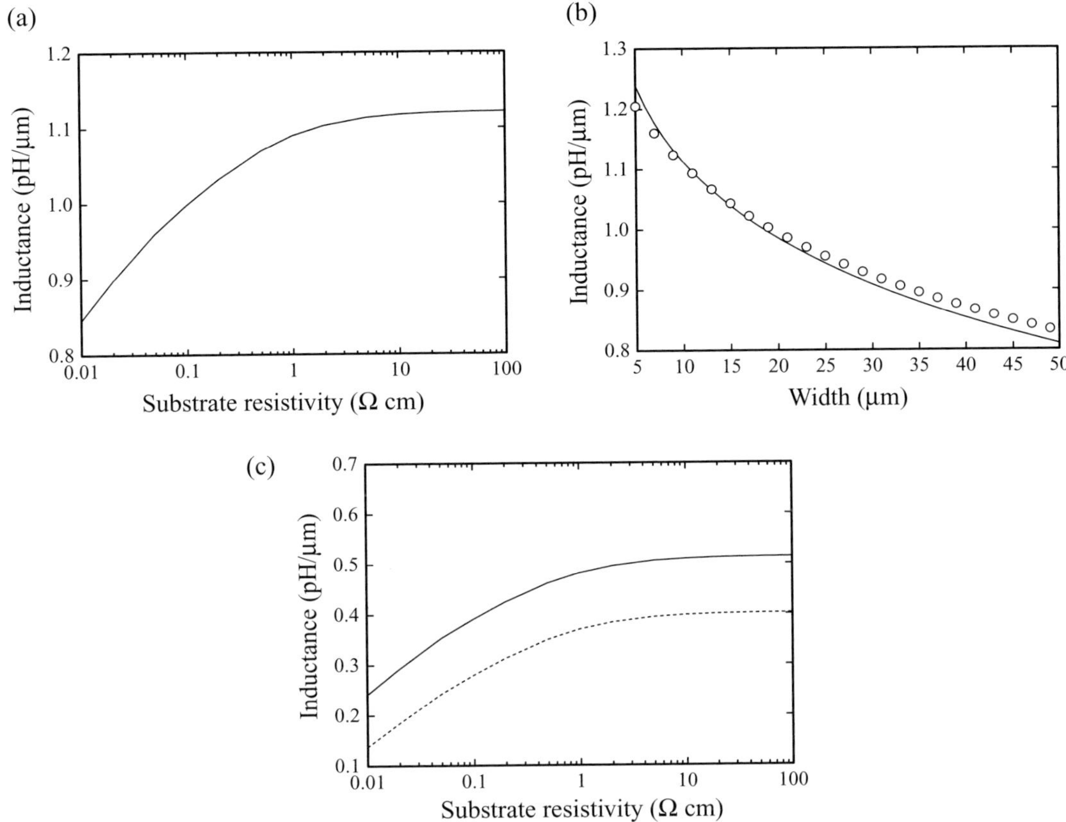

Figure 6.8 Partial inductances from electromagnetic simulations: (a) self-inductance of a metal segment versus substrate resistivity; (b) self-inductance versus metal width from Momentum (solid line) and theory (circles); (c) mutual inductance between two metal segments versus substrate resistivity, for spacings d of 50 μm (solid line) and 100 μm (dashed line)

The partial mutual inductance between two 10 μm wide and 1 mm long metal segments has also been simulated. The four-port S parameters have been used to derive the parameters of the equivalent circuit of the two coupled transmission lines depicted in Figure 6.7(b). The mutual inductance between the two inductors is the partial mutual inductance of one metal segment, which is shown versus the substrate resistivity in Figure 6.8(c). The solid and dashed lines are obtained by setting the distance between the centres of the metal segments as 50 μm and 100 μm, respectively.

Note again that, at 1 GHz, for a substrate resistivity above 5 Ωcm the mutual inductance values no longer depend on the resistivity and they approach the values obtained from (6.3), being 0.55 pH/μm and 0.42 pH/μm, respectively. The slight discrepancy between simulations and the values from (6.3) can be reduced by taking the d value as the geometric mean distance and not as the distance between the centres of the two wires.

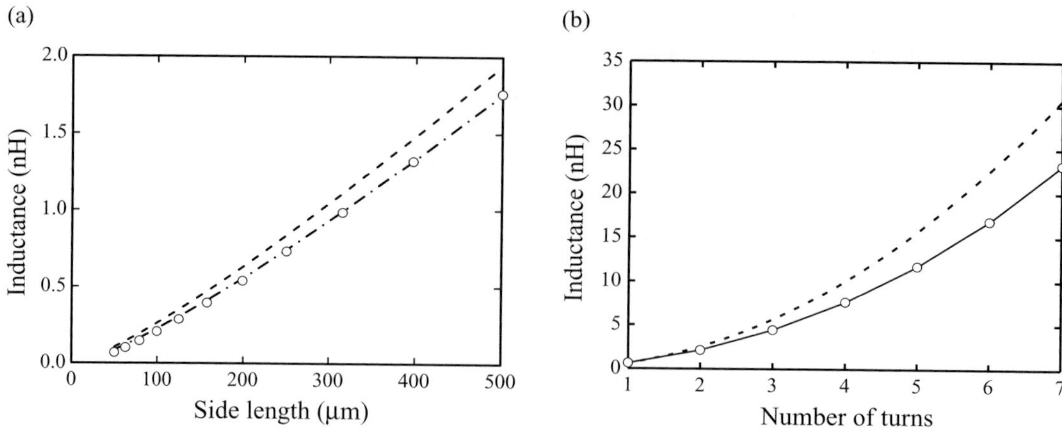

Figure 6.9 Squared spiral inductor: (a) inductance versus side length from simulations (circles) and from partial inductance calculations (dashed line from self-inductances only and dot–dashed line from self-inductances and mutual inductances); (b) inductance versus number of turns n from simulations (circles) and theoretical n^2 dependence (dashed line)

Example 6.4 Square inductor: inductance evaluation

This example deals with the design of a square inductor. The substrate resistivity is still 5 Ωcm and the line is 10 μm wide. Figure 6.9(a) shows the simulation results for a single-turn coil (circles) versus the side length. The dashed line is instead calculated as four times the partial self-inductance, $L_{i,i}$ from (6.2), of each side, thus neglecting the negative contributions of mutual inductances. The dash–dotted line has been obtained, instead, by taking account of the mutual self inductance, $L_{i,j}$ from (6.3), thus taking the total inductance as $4 \cdot \left(L_{i,i} - L_{i,j}\right)$.

Figure 6.9(b) shows the inductance dependence on the number of turns. The simulations have been performed for a hollow square inductor, taking a 10 μm line width, a 2 μm spacing between the lines and a 100 μm inner side length. When increasing the number of turns, from one to seven, the inductor expands externally, keeping the inner area constant at 100 μm × 100 μm. The dashed line shows, for comparison, the n^2 dependence. Because of the negative mutual inductances, the total inductance (circles) rises slower than n^2, the exponent being about 1.8.

6.4 Integrated varactors

6.4.1 Varactor key performance

Frequency tuning in integrated LC VCOs is accomplished using a variable capacitance, i.e., a varactor or varicap. In this application, the component will, therefore, be characterized by its quality factor and by the capacitance ratio factor.

The quality factor is usually limited by the unavoidable resistance in series with the variable capacitance. It is, therefore:

$$Q = \frac{1}{\omega_0 C R} = \sqrt{\frac{L}{C}} \cdot \frac{1}{R}.$$

Assuming a constant series resistance, which is usually an acceptable approximation, Q decreases as the capacitance increases.

The capacitance ratio factor k is instead:

$$k = \frac{C_{\text{max}}}{C_{\text{min}}},$$

which is the ratio between the maximum and the minimum capacitance values, and should not be confused with the varactor gain K_{VCO}.

To the first order, the frequency range $\Delta\omega_0$ covered by the VCO is related to the capacitance variation by $\Delta\omega_0/\omega_0 \cong -\Delta C/2C$. A more detailed result is:

$$\begin{aligned} \frac{\omega_{\text{H}} - \omega_{\text{L}}}{(\omega_{\text{H}} + \omega_{\text{L}})/2} &= 2 \cdot \frac{\sqrt{1/LC_{\text{min}}} - \sqrt{1/LC_{\text{max}}}}{\sqrt{1/LC_{\text{min}}} + \sqrt{1/LC_{\text{max}}}} \\ &= 2 \cdot \frac{\sqrt{k} - 1}{\sqrt{k} + 1}. \end{aligned} \tag{6.6}$$

It turns out that a capacitance ratio equal to two leads to a 34% tuning range; a value that may appear to be sufficient for most of the applications. For example, a 19% tuning range is able to cover the frequency interval from 5 to 6 GHz, which includes the bands of some WLAN standards, such as the IEEE 802.11a. In reality the tuning range must also cover the spread of the reactive components. In particular, the capacitance may have a value that is as much as 30% different from the nominal target. The tuning range must, therefore, be larger than just the signal bandwidth, to compensate for the process and sometimes also temperature variations. For an oscillator covering the WLAN standards, a 35% tuning range would be a reasonable request.

Last but not least, the parasitics should be taken into account. A parasitic capacitor, C_{P}, partly because of the varactor structure itself, is always in parallel with the varactor. It adds to both C_{max} and C_{min}. Taking $C_{\text{max}}/C_{\text{min}} = 2$ and $C_{\text{P}} \cong C_{\text{min}}$, (6.6) gives about a 20% tuning range, which can be too narrow.

6.4.2 Diode varactors

Varactors can be implemented using either junction diodes or MOS structures. In the first case the capacitance follows the dependence of the small-signal junction capacitance with the reverse voltage, V_{R}. [22] The capacitance is $C_{\text{V}} = C_0\,(1 + V_{\text{R}}/V_{\text{BI}})^{-m}$, where V_{BI} is the junction built-in potential, and m depends on the junction doping profile. For an ideally abrupt junction m is 0.5; in real cases it is less. The losses are essentially a result of the

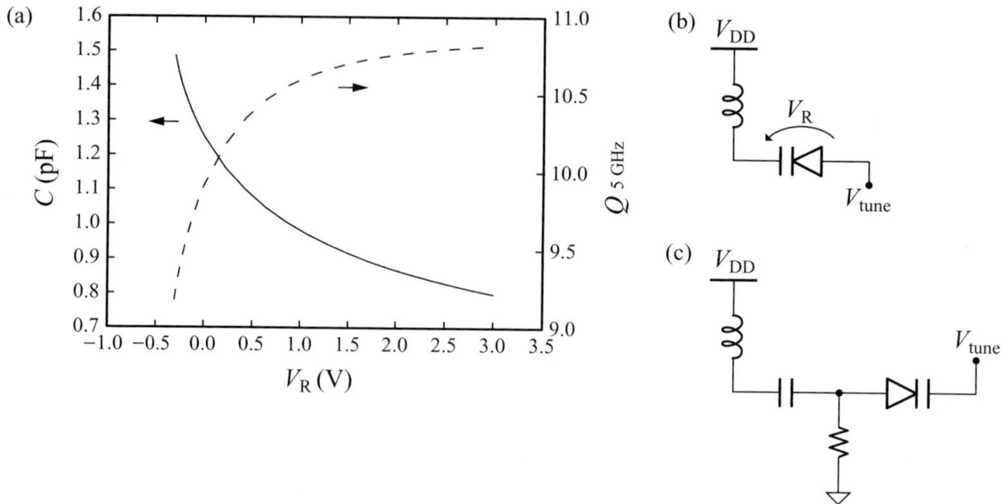

Figure 6.10 Varactors: (a) capacitance and Q factor of a diode, (b) LC tank with varactor, (c) varactor a.c. coupling

finite resistance of the carrier path in silicon. Components appearing in a technology library are optimized to reduce both losses and the parasitic capacitance. [23]

Bipolar or BiCMOS RF processes have process options suitable for varactor junction diodes: base-emitter or base-collector junctions. No specific options are, instead, available in standard CMOS technologies; therefore source-well or drain-well junctions may be used. Figure 6.10(a) shows the capacitance and the Q value of a diode varactor, as obtained from a simulation. The capacitance ratio k can be increased by forcing the diode close to forward bias, but the quality factor drops very quickly at 0.3–0.5 V, limiting the maximum k values to below two.

From the design perspective, there are two other important issues to cover: (i) the VCO is embedded within a PLL, therefore the unavoidable leakage current of the pn junction discharges the PLL loop filter capacitance, leading to spurious tones at the reference frequency; (ii) the varactor must be reverse-biased. That is quite obvious, but it has some consequences. Owing to the parasitic junction capacitance from the varactor n diffusion to the p substrate, the varactor cathode should be connected to the tuning voltage node. Otherwise, it would add a large parasitic capacitance to the tank. Therefore, the topology in Figure 6.10(b), which has been quite common in the oscillator topologies discussed so far, is not the best from this standpoint.

A different topology is shown in Figure 6.10(c), where a large capacitance a.c. couples the varactor to the inductor. The resistance forces the d.c. voltage of the varactor anode to ground. In both solutions, the tuning voltage can swing between ground and supply, while leaving the diode reverse-biased. The second case is, however, critical for the presence of the bias resistor, which is in parallel with the tank. If the resistance is too low, the Q is degraded; large R values affect the noise performance instead, as discussed in detail in Example 7.3.

Although a d.c. reverse bias is imposed, the peaks of the oscillation superimposed on the d.c. voltage drive the pn junction towards forward biasing. Because of the increasing

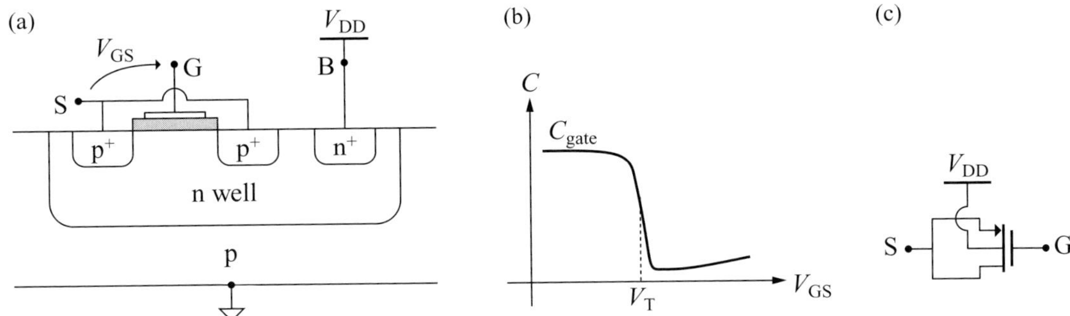

Figure 6.11 Inversion-mode pMOS varactor: (a) cross section, (b) *C–V* curve, (c) circuit symbol

junction conduction, the oscillation peaks are clamped. Therefore, the achievable oscillation amplitude may be limited, not by the saturation of the active element as shown in Chapter 4, but by the presence of the varactors.

6.4.3 MOS varactors

Some of the impairments of pn junctions can be avoided using MOS varactors. As soon as CMOS became a realistic option for RF circuits, a lot of attention was devoted to these solutions. References [24–27] are a limited selection of results on this topic.

MOS varactors exploit the capacitive response of the metal–oxide–semiconductor structures. [22] The device in Figure 6.11(a) is a pMOS in an n-well with the drain and the source tied together. Note that the n^+ diffusion is connected to the power supply V_{DD}. Figure 6.11(b) shows the small-signal capacitance, as a function of V_{GS}. For $V_{GS} < V_T$, the device is in inversion and the gate capacitance is, essentially, the oxide capacitance. When $V_{GS} > V_T$, the *C–V* response is given by the series of the oxide capacitance and the depletion capacitance, which is much smaller. In reality, as V_{GS} increases, the depleted regions shrink, thus causing the slight growth of the capacitance with V_{GS}. However, as far as the well is biased to V_{DD}, the device does not reach the accumulation mode for the V_{GS} range of interest. This varactor is an inversion mode pMOS (i-pMOS). Figure 6.11(c) shows the corresponding symbol. The complementary device, the inversion mode nMOS (i-nMOS) is made on a p substrate connected to ground. The n^+ drain and source diffusions are tied together. The cross-section, *C–V* characteristic and symbol can easily be derived, starting from those of the i-pMOS.

Another option, for a MOS varactor, is the one reported in Figure 6.12(a). The structure is different. The capacitance is measured between the bulk and the gate, while the p^+ diffusions are not present. When V_{GB} is below the flat-band voltage V_{FB}, the small-signal capacitance is the series of the oxide and the depletion capacitance (Figure 6.12(b)). Instead, when V_{GB} increases, beyond the flat-band voltage V_{FB}, an accumulation layer of electrons is created below the gate and, again, the capacitance is almost equal to the gate oxide contribution. This varactor is called accumulation-mode pMOS (a-pMOS). Its symbol, shown in Figure 6.12(c), highlights the absence of drain and source contacts. The transition between the

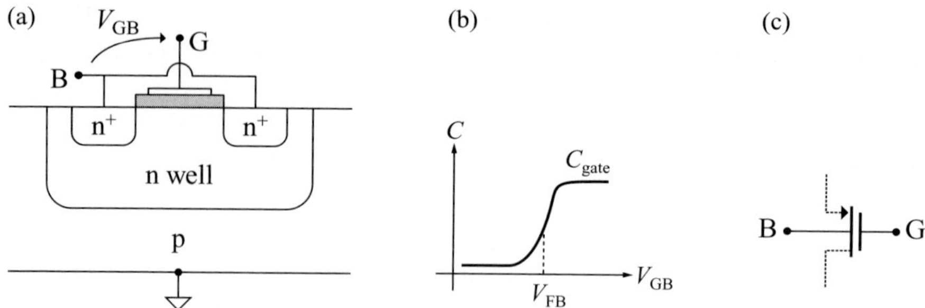

Figure 6.12 Accumulation-mode pMOS varactor: (a) cross-section, (b) C–V curve, (c) circuit symbol

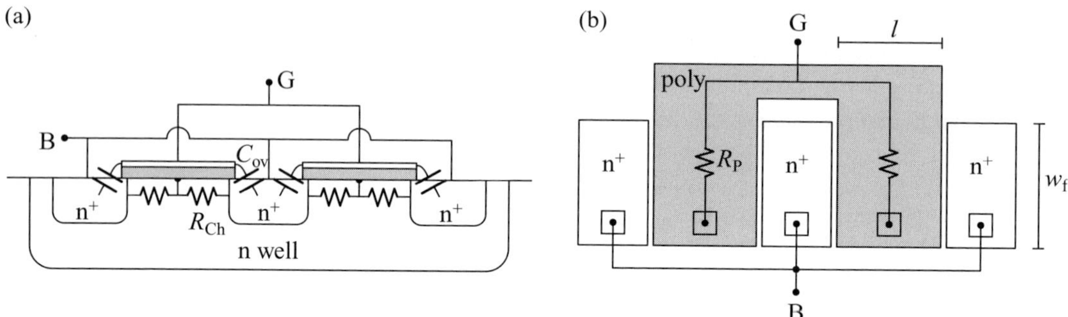

Figure 6.13 Loss resistances and parasitic capacitances in a MOS varactor with two fingers: (a) cross-sections, (b) top view

minimum and the maximum capacitance is less sharp than in inversion-mode components. Note that, unlike the other MOS components, an accumulation nMOS (a-nMOS) is feasible only if placed within an isolated p well, which is available only in triple well technologies.

MOS varactors do not suffer from leakage current or forward bias like diode components. For both inversion and accumulation types, the terminal featuring the lower parasitic capacitance is the gate, while the other terminal is loaded by diffusion capacitances. So the gate is usually linked to the inductor and the well contact is driven by the tuning voltage.

6.4.4 Tuning range and losses

Figure 6.13(a) shows the main parasitic capacitances and resistances of an a-pMOS. A similar discussion can be extended to the other MOS varactor types. As for standard MOS transistors, varactors are usually implemented with multi-fingered devices. For the sake of simplicity, only two fingers are represented in Figure 6.13(b). As mentioned, the minimum quality factor occurs when the capacitance is maximum, thus in accumulation (or in inversion, for an inversion-mode device).

The main contributions to losses are two resistances, both in series with the variable capacitance: R_{Ch} in Figure 6.13(a) is the channel resistance of the accumulation layer, while R_P in Figure 6.13(b) is the gate resistance due to the finite polysilicon conductivity.

Taking as l and w the length and the width of the gate (for the varactor in Figure 6.13(b) it is $w = 2w_f$, where w_f is the finger width), respectively, these resistances are proportional as:

$$R_{Ch} \propto \frac{l}{w} R_{Ch}^{Sq}, \qquad R_P \propto \frac{1}{N_f} \frac{w}{l} R_P^{Sq},$$

where R_{Ch}^{Sq} and R_P^{Sq} are the sheet resistances (*per square*) of the channel and of the polysilicon and N_f is, instead, the number of fingers. Usually, by increasing N_f, the second resistance becomes less important and the minimum quality factor becomes proportional to l^{-2}. In fact, by neglecting R_p and taking the gate capacitance as $C_{ox} \cdot wl$, with $C_{ox} = \varepsilon_{ox}/t_{ox}$, the quality factor is:

$$Q_{min} \propto \frac{1}{\omega_0 \cdot C_{ox} \cdot wl \cdot R_{Ch}} \propto \frac{t_{ox}}{\omega_0 \cdot \varepsilon_{ox} \cdot l^2 \cdot R_{Ch}^{Sq}}.$$

For a given technology, the minimum quality factor improves by reducing the channel length.

Unfortunately the opposite is true for the capacitance ratio k and, therefore, for the tuning range. Figure 6.13(a) highlights the overlap capacitances, C_{ov}, between gate and contact diffusions. This term is proportional to the gate width, w, and can be written as $C_{ov}^0 w$, where C_{ov}^0 is the overlap capacitance per unit of gate width. To emphasize the effect of the overlap capacitance on the k ratio, it is also useful to write the depletion capacitance of the varactor as $C_{de}^0 \cdot wl$. Clearly, C_{de}^0 is the depletion capacitance per unit of gate area. Owing to the larger C_{ox} value, the minimum varactor capacitance may be approximated with the depletion capacitance. It turns out that:

$$k = \frac{C_{ox} \cdot wl + C_{ov}^0 \cdot 2w}{C_{de}^0 \cdot wl + C_{ov}^0 \cdot 2w} \cong \frac{C_{ox} \cdot l}{C_{de}^0 \cdot l + 2 \cdot C_{ov}^0}.$$

In the second step, the overlap capacitance at the numerator has been neglected. Note that even if the stray capacitance reduces the tuning range, the capacitance ratio becomes closer to the ideal value, C_{ox}/C_{de}^0, as l increases. A trade-off, therefore, exists between the tuning range, which improves as l increases, and the quality factor, which, instead, calls for short l values.

A final remark about the impact of the scaling trend. [24] Since the oxide thickness t_{ox} is reduced proportionally to the channel minimum length, the quality factor should ideally scale as l^{-1}. However, this conclusion has to be taken with some care. Still, in some cases, the need for a smoother slope of the *C–V* characteristic suggests that the oxide thickness of the varactor component should not be scaled, as in standard transistors. Moreover, when the minimum length is used, R_P may become significant, and the preceding expression for the Q dependence does not actually hold any more.

As far as the tuning range is concerned, note that the maximum achievable ratio, C_{ox}/C_{de}^0, may increase as the technology scaling proceeds following the $C_{ox} = \varepsilon_{ox}/t_{ox}$ increase. This trend is, unfortunately, partially compensated for by a growth of C_{de}^0, because of the rising substrate doping.

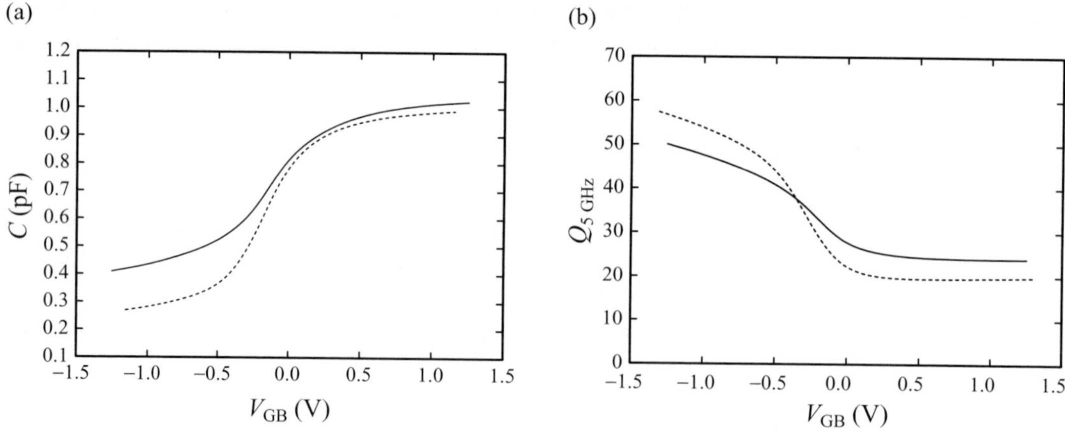

Figure 6.14 Accumulation pMOS varactors with 1 μm channel length in 0.25 μm (solid lines) and 0.13 μm CMOS (dashed lines): (a) *C–V* curves and (b) *Q* factor at 5 GHz from varactor model

Example 6.5 Varactor design and simulations

Let us consider the simulation results, at 5 GHz, for two accumulation pMOS varactors, in 0.25 μm and 0.13 μm CMOS technologies. To avoid having a too steep C–V characteristic, the varactors in the 0.13 μm technology feature the same oxide thickness as the 0.25 μm components. The C_{ox} value is, therefore, the same.

Figure 6.14 shows the capacitance and the quality factor as functions of the voltage across the two control terminals. The solid lines refer to the 0.25 μm technology, the dashed line to the 0.13 μm channel length. In both cases, taking $l = 1$ μm and 12 fingers with $w = 6.8$ μm each, the nominal maximum capacitance (the gate oxide capacitance) is set to 1 pF.

For the 0.13 μm technology, the capacitance ratio is higher, because of the reduced impact of parasitics on the overlap capacitances. The minimum quality factor is, instead, lower. This may be because of the larger substrate doping, which reduces the carrier mobility.

From the design perspective, the way in which the varactor is used is also important. As an example, let us consider the oscillator topology in Figure 4.16(a) and assume the gate terminal connected to the inductor, hence to the highest available voltage V_{DD}. With this choice, the bias across the varactor, V_{GB} in Figure 6.14(a), can only be positive, thus limiting the capacitance variation that can be exploited. The drawback may be circumvented by driving the tuning node to a voltage higher than V_{DD}. The reference may be generated by a charge pump, which will use thick-oxide transistors to hold the larger voltage properly. The situation is different for the complementary circuit in Figure 4.21, where the varactors can be biased with the gate terminal at about $V_{DD}/2$.

Figure 6.15 shows the dependence of the Q and k values as functions of the channel length l for the component in 0.25 μm technology. The total width is constant, $w = 6.8$ μm. Therefore as l increases the number of fingers is correspondingly reduced, to keep the overall gate capacitance constant at 1 pF. Figure 6.15(a) shows the dependence of the minimum

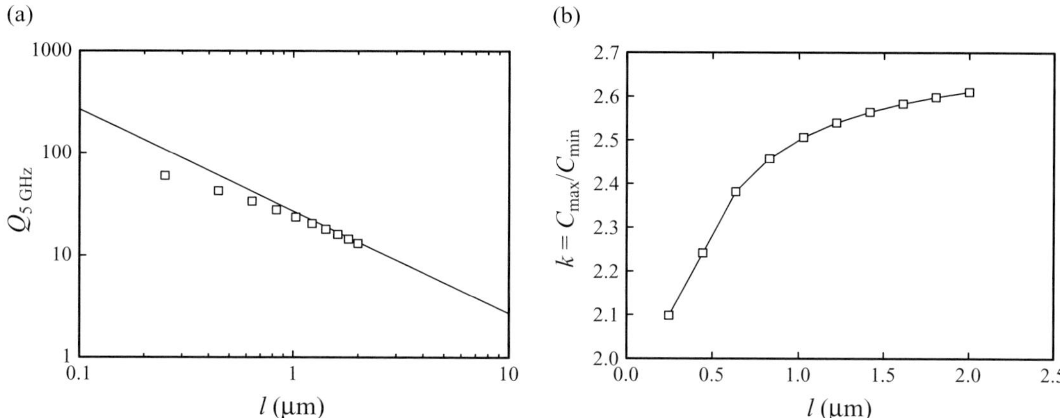

Figure 6.15 Accumulation pMOS in 0.25 μm CMOS: (a) Q factor versus channel length l from varactor model (squares) and theoretical l^{-1} dependence (solid line); (b) capacitance ratio. As l increases, the total capacitance is kept constant by reducing the number of fingers

quality factor $Q = 1/(\omega_0 CR)$ at 5 GHz. Since C is constant, and the channel resistance scales as l^{-1}, the quality factor improves as l^{-1} (solid line). When the channel becomes too short, the effect of the gate series resistance R_P becomes significant, and the Q improvement slows down.

Figure 6.15(b) shows the capacitance ratio k versus l. The ratio drops by reducing the channel length, because the capacitance of a single finger decreases while the overlap stray capacitance stays constant.

6.5 Switched tuning

It has already been pointed out that although a large varactor gain is necessary to cover a wide tuning range, this choice leads to some drawbacks. For instance, the level of the spurs discussed in Section 2.4 increases as the K_{VCO} value. The same happens to phase noise, because of the conversion effects discussed in Chapter 7.

A solution to this problem is given by the adoption of the switched (or digital) tuning technique. [28] In this scheme, the tank oscillation frequency may be changed in discrete steps by connecting some capacitors properly in parallel with the inductor or leaving them floating. This ‘coarse tuning’ makes it possible to split a large tuning range into smaller intervals. Within each interval the fine tuning is achieved using a varactor, driven, as usual, by V_{tune}. The K_{VCO} value is now the one corresponding to the fine tuning, and can be much smaller than the value that would be needed to cover the entire tuning range.

The switched tuning can be implemented with MOS varactors. Instead of biasing the device with an analogue voltage, a binary control signal can force the varactor into the depletion region (minimum capacitance) or into the accumulation/inversion region (maximum

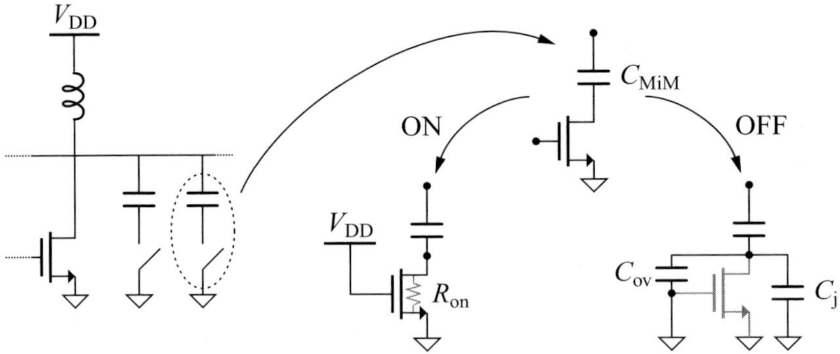

Figure 6.16 Switched capacitors for frequency tuning

capacitance). However, the simpler approach is to use a constant capacitor, usually a metal–insulator–metal (MiM), together with a series-connected MOS switch, as in Figure 6.16. When the MOS is on, the metal capacitance C_{MiM} loads the tank. In the off state, ideally, the capacitance is floating, and does not load the tank.

In reality, when the MOS is in the triode region, its resistance is approximately given by:

$$R_{on} = \frac{1}{(w/l)\mu C_{ox}(V_{DD} - V_T)}.$$

This resistance limits the capacitor quality factor to $Q = 1/(\omega_0 R_{on} C_{MiM})$. The resistance may be reduced by increasing the MOS width, w.

On the other hand, when the switch is off, the capacitance loading the tank is not zero. It is given by the series of the MiM capacitor and the parasitic capacitance C_P of the switch, which usually dominates, being much smaller than C_{MiM}. The capacitance C_P includes gate–source or gate–drain overlap capacitance $C_{ov}^0 w$ and drain-bulk capacitance C_j. The MiM bottom-plate parasitic capacitance is usually negligible. All these parasitic capacitances are proportional to the MOS width w. The off capacitance of the switch may, therefore, be written as $C_P^0 w$, and the capacitance ratio k becomes $k \cong 1 + (C_{MiM}/C_P^0 w)$. Again, a trade-off between capacitance ratio and quality factor exists. A wider MOS switch has a lower R_{on} and a higher quality factor, at the expense of a reduced capacitance ratio.

Technology scaling improves both the quality factor (R_{on} decreases) and, partially, the capacitance ratio. As the device scaling proceeds, the junction capacitance per unit area is expected to rise, because of the increase of the bulk doping concentration. On the other hand, the scaling of the source or drain contact area and the diffusion depth is limited by the reliability of the contact formation during manufacturing. This argument suggests that the capacitance ratio of a switched MiM should improve more slowly than that of MOS varactors. The application of the switched tuning technique in oscillators and the use of mixed analogue and digital tuned oscillators in PLLs is discussed in Section 7.4.1.

6.6 References

[1] F. W. Grover, *Inductance Calculation: Working Formulas and Tables*, Princeton, NJ: Van Nostrand, 1946.

[2] H. M. Greenhouse, Design of planar rectangular microelectronics inductors, *IEEE T. Parts, Hybrids Packaging*, **10**, Jun. 1974, 101–9.

[3] J. C. Slater and N. H. Frank, *Electromagnetism*, New York, NY: McGraw-Hill, 1947, reprinted New York, NY: Dover, 1969.

[4] J. R. Long and M. A. Copeland, The modeling, characterization and design of monolithic inductors for silicon RF IC's, *IEEE J. Solid-St. Circ.*, **32**, Mar. 1997, 357–69.

[5] J. Cranickx and M. Steyaert, A 1.8 GHz low-phase-noise CMOS VCO using optimized hollow spiral inductors, *IEEE J. Solid-St. Circ.*, **32**, May 1997, 736–44.

[6] C. P. Yue and S. S. Wong, On-chip spiral inductors with patterned ground shields for Si-based RF IC's, *IEEE J. Solid-St. Circ.*, **33**, May 1998, 743–52.

[7] K. O, Estimation methods or quality factor of inductors fabricated in silicon integrated circuit process technologies, *IEEE J. Solid-St. Circ.*, **33**, Aug. 1998, 1249–52.

[8] A. M. Niknejad and R. G. Meyer, Analysis, design and optimization of spiral inductors and transformers for Si RF IC's," *IEEE J. Solid-St. Circ.*, **33**, Oct. 1998, 1470–81.

[9] A. Zolfaghari, A. Chan and B. Razavi, Stacked inductors and transformers in CMOS technology, *IEEE J. Solid-St. Circ.*, **36**, Apr. 2001, 620–8.

[10] S. Yim and K. O, The effects of ground shield on the characteristics and performance of spiral inductors, *IEEE J. Solid-St. Circ.*, **37**, Feb. 2002, 237–44.

[11] Y. Cao *et al.*, Frequency-independent equivalent-circuit model for on-chip spiral inductors, *IEEE J. Solid-St. Circ.*, **38**, Mar. 2003, 419–26.

[12] N. A. Talwakar, C. P. Yue and S. S. Wong, Analysis and synthesis of on-chip spiral inductors, *IEEE T. Electron Dev.*, **52**, Feb. 2005, 176–82.

[13] K. Y. Tong and C. Tsui, A physical analytical model of multilayer on-chip inductors, *IEEE T. Microw. Theory*, **53**, Apr. 2005, 1143–9.

[14] O. Murphy *et al.*, Design of multiple-metal stacked inductors incorporating an extended physical model, *IEEE T. Microw. Theory*, **53**, Jun. 2005, 2063–72.

[15] J. Guo and T. Tan, A broadband and scalable model for on-chip inductors incorporating substrate and conductor loss effects, *IEEE T. Electron Dev.*, **53**, Mar. 2006, 413–21.

[16] W. Gao and Z. Yu, Scalable compact circuit model and synthesis for RF CMOS spiral inductors, *IEEE T. Microw. Theory*, **54**, Mar. 2006, 1055–64.

[17] J. Chang, A. Abidi and M. Gaitan, Large suspended inductors on silicon and their use in a 2 μm CMOS RF amplifier, *IEEE Electr. Device L.*, **14**, May 1993, 246–8.

[18] D. M. Pozar, *Microwave Engineering*, New York, NY: Wiley, 2nd edn, 1997.

[19] C. Doan, S. Emami, A. Niknejad and R. Brodersen, Millimeter-wave CMOS design, *IEEE J. Solid-St. Circ.*, **40**, Jan. 2005, 144–55.

[20] A. Natarajan, A. Komijani and A. Hajimiri, A fully integrated 24-GHz phased-array transmitter in CMOS, *IEEE J. Solid-St. Circ.*, **40**, Dec. 2005, 2502–14.

[21] W. R. Eisenstadt and Y. Eo, S-parameter-based IC interconnect transmission line characterization, *IEEE T. Compon. Hybr.*, **15**, Aug. 1992, 483–90.

[22] R. S. Muller and T. I. Kamins, *Device Electronics for Integrated Circuits*. New York, NY: J. Wiley & Sons, Inc., 1977, 2nd edn, 1986.

[23] T. Souyuer and R. G. Meyer, High-frequency phase-locked loops in monolithic bipolar technology, *IEEE J. Solid-St. Circ.*, **24**, Jun. 1989, 787–95.

[24] F. Svelto, P. Erratico, S. Manzini and R. Castello, A metal-oxide-semiconductor varactor, *IEEE Electr. Device L.*, **20**, Apr. 1999, 164–6.

[25] A. Porret, T. Melly, C. Enz and E. Vittoz, Design of high-Q varactor for low-power wireless applications using a standard CMOS process, *IEEE J. Solid-St. Circ.*, **35**, Mar. 2000, 337–45.

[26] P. Andreani and S. Mattisson, On the use of MOS varactors in RF VCOs, *IEEE J. Solid-St. Circ.*, **35**, Jun. 2000, 905–10.

[27] J. Maget, M. Tiebout and R. Kraus, MOS with n- and p-type gates and their influence on an LC-VCO in digital CMOS, *IEEE J. Solid-St. Circ.*, **38**, Jul. 2003, 1139–47.

[28] A. Kral, F. Behbahani and A. A. Abidi, RF-CMOS oscillators with switched tuning, *Proc. IEEE 1998 Custom Integrated Circuits Conf.*, May 1998, 555–8.

7 Noise up-conversion in VCOs

7.1 Introduction

The analytical framework for the estimation of the oscillator phase noise provided in Chapter 5 needs to be supplemented by other mechanisms for frequency conversion of the noise source spectra. In particular, the presence of voltage-dependent capacitances either placed intentionally within the tank to provide resonance tuning or because of junctions of active devices, cause loading of low-frequency noise sources into the oscillator bandwidth. This may seriously compromise the close-in phase noise, because of the presence of large flicker noise of scaled technologies.

Wide tuning ranges and low close-in phase noise are mutually exclusive requirements. The limitation in the tuning voltage range dictated by the low-voltage supply of modern silicon technologies and the wide tuning range required to cover the LO band and to compensate for process and temperature variations may require high VCO tuning sensitivity K_{VCO}. This problem is traditionally solved by employing a variable capacitance with an abrupt C–V characteristic. However, the oscillation frequency is no longer set by the tank resonant frequency, but becomes highly dependent on oscillation amplitude. Through this mechanism, all the noise sources present in the oscillator circuit that cause amplitude modulation noise produce, in turn, frequency and phase noise. [1]

Other mechanisms not related to voltage-dependent capacitances exist, and these may cause additional noise up-conversion. Some examples are the modulation of the active device delay and the effect of a low resonator quality factor.

7.2 Tuning curve and sensitivity coefficients

A cross-coupled oscillator employing a-pMOS as varactors is shown in Figure 7.1. Since this device has a variable capacitance centred around 0 V of gate-to-bulk voltage, the gates have to be biased to $V_{\mathrm{DD}}/2$, to span the whole C–V curve with a tuning voltage varying between 0 and V_{DD}. For this reason, the voltage supply of the oscillator is set to $V_{\mathrm{DD}}/2$.

In VCOs, the varactor does not work in the small-signal regime and, therefore, it is quite crude to assume that its capacitance is simply equal to the incremental value $C(V_{\mathrm{tune}})$ at the bias point and that the oscillation frequency is given by $\omega_0(V_{\mathrm{tune}}) = 1/\sqrt{LC(V_{\mathrm{tune}})}$.

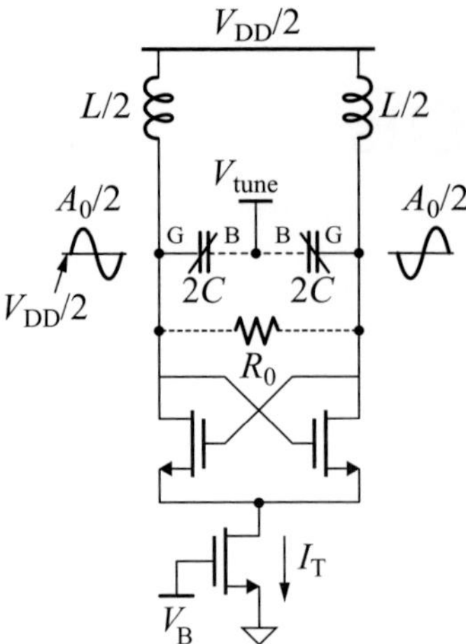

Figure 7.1 CMOS oscillator with a–pMOS varactors

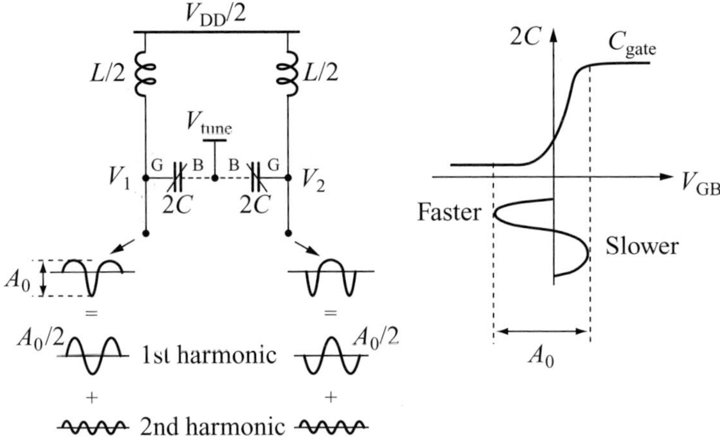

Figure 7.2 Resonator with non-linear capacitances: second-harmonic distortion due to variable capacitances

Figure 7.2 shows a non-linear differential tank with two MOS varactors. The differential equations describing its time evolution can easily be obtained by equating the current flowing into the inductor with the current in each of the two capacitors. The outcome is qualitatively obvious: the voltage waveform across the tank is not perfectly harmonic. Figure 7.3 shows the single-ended waveforms $V_1(t)$ and $V_2(t)$. Their shape can easily be explained. When the voltage signal drives the varactor to increase its capacitance, the oscillation slows down. When the varactor is pushed in the opposite direction to decrease its capacitance, the oscillation period shortens. The positive and negative peaks of the voltage waveforms have different magnitudes. At each peak, the voltage across one varactor reaches its maximum value, while the voltage across the other varactor is a minimum. Since charge conservation

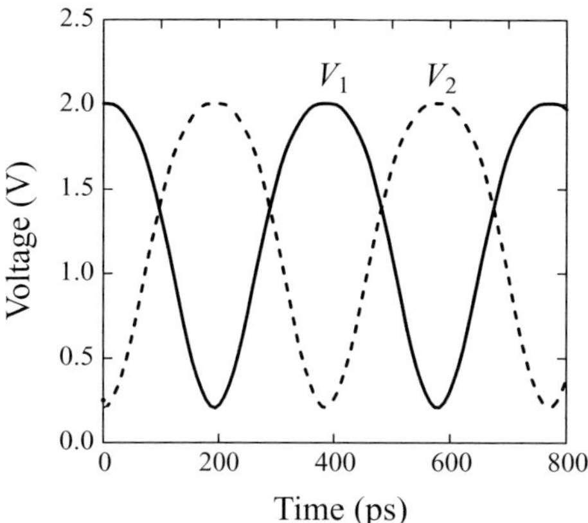

Figure 7.3 Simulated voltage waveforms at outputs of resonator with non-linear capacitances

must hold, the capacitor with the higher capacitance has lower peak voltage and vice versa (see Figure 7.2).

Most of the waveform power is concentrated in the first and second harmonics (Figure 7.2). The first harmonic of V_1 is 180° out of phase with respect to the first harmonic of V_2, while the second harmonics of the two waveforms are in phase. The output of the oscillator is the differential voltage and is symmetrical. The dominant harmonic is the fundamental at ω_0; however, residual odd harmonics can be present. The common mode, instead, varies at $2\omega_0$ with residual even harmonics.

By changing the d.c. bias V_{tune} the oscillation radial frequency ω_0 changes. This dependence is referred to as the tuning curve and its prediction for tanks with actual varactors is not trivial without relying on a complete circuit simulation. An analytical approximation has been given assuming a varactor with a step-like $C(V)$. [2] However, a simpler model may help to understand its dependencies better and to have first-order values useful for design purposes.

Let us consider the oscillator in Figure 7.1 using a-pMOS varactors. Their $C(V)$ curve may be approximated with the step-like $C(V)$ curve, depicted in Figure 7.4(a). In this approximation, an abrupt change occurs at $V_{\text{GB}} = V_C$. The bias voltage across the non-linear capacitor is $V_{\text{GB0}} = V_{\text{DD}}/2 - V_{\text{tune}}$. During each oscillation cycle, there is a change of the varactor capacitance. In Figure 7.4(a), the $C(V)$ curve is shown along with the oscillation waveform centred around the d.c. bias V_{GB0}. When V_{GB0} is lower than $(V_C - A_0/2)$, the tank capacitance equals C_{min} over the entire oscillation swing; hence, the output frequency is $\omega_{\text{H}} = 1/\sqrt{LC_{\text{min}}}$. When V_{GB0} is higher than $(V_C + A_0/2)$, it is given by $\omega_{\text{L}} = 1/\sqrt{LC_{\text{max}}}$ (Figure 7.4(b)). If V_{GB0} falls in the range $(V_C \pm A_0/2)$, the frequency can be linearly interpolated between the two limits, thus deriving the tuning curve shown in Figure 7.4(c). [3]

This simple argument highlights the fact that the oscillation frequency depends not only on the d.c. bias V_{GB0} but also on the oscillation amplitude. The oscillation amplitude A_0 sets the width of the transition window of the tuning curve. Therefore, if A_0 increases, the tuning

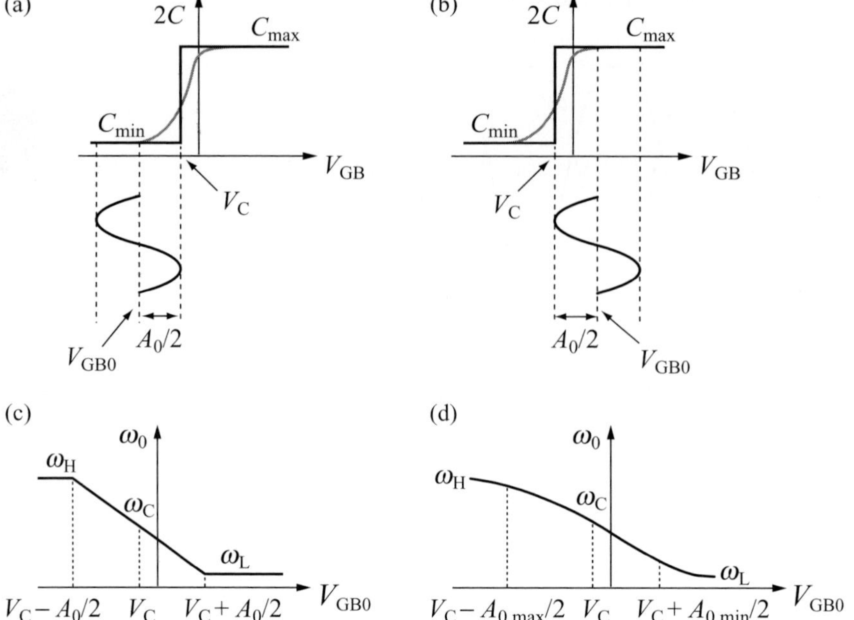

Figure 7.4 Oscillator with MOS varactor: step-like $C(V)$ curve with varactor bias V_{GB0} so that the voltage waveform spans only the curve region of (a) C_{min} and (b) C_{max}. Approximations of the tuning curve: (c) piecewise linear approximation making V_{GB0} span at constant A_0 and (d) actual dependence

characteristic steepness decreases. The larger A_0 is, the more the tuning curve departs from the steep small-signal curve.

Since the oscillation frequency depends on both the tuning voltage and the oscillation amplitude $\omega_0 = \omega_0(V_{tune}, A_0)$, its partial derivatives define two sensitivity coefficients. More precisely:

$$K_{VCO} = \left.\frac{\partial \omega_0}{\partial V_{tune}}\right|_{A_0}$$

is the VCO gain. It represents the sensitivity of the oscillation frequency to the d.c. bias across the varactor at constant oscillation amplitude. On the other hand:

$$K_{A_0} = \left.\frac{\partial \omega_0}{\partial A_0}\right|_{V_{tune}}$$

is the frequency sensitivity to the amplitude variation at constant d.c. bias across the varactor. This mechanism can be referred as AM-to-FM (amplitude to frequency modulation) conversion.

Both of these coefficients can be estimated by referring to the curve in Figure 7.4(c). The VCO gain K_{VCO} represents the slope of the tuning curve in the transition region. It can, therefore, be taken as:

$$K_{VCO} \approx \frac{\omega_H - \omega_L}{A_0}. \tag{7.1}$$

The tuning curve in the transition region can, therefore, be written as:

$$\omega_0(V_{\text{tune}}) \approx \omega_C + K_{\text{VCO}}(V_C - V_{\text{GB0}}) = \omega_C + \frac{\omega_H - \omega_L}{A_0}(V_C - V_{DD}/2 + V_{\text{tune}}). \tag{7.2}$$

In practical implementations the range ΔV_{tune} that V_{tune} can span is smaller than V_{DD}, whereas A_0 is comparable to V_{DD}. In this case, the linear approximation holds over the entire tuning range and the maximum tuning range is

$$T_{\text{range}} = K_{\text{VCO}} \cdot \Delta V_{\text{tune}} \approx (\omega_H - \omega_L)\frac{\Delta V_{\text{tune}}}{A_0}, \tag{7.3}$$

reduced by a factor $\Delta V_{\text{tune}}/A_0$ with respect to the maximum value $(\omega_H - \omega_L)$.

Moreover, by differentiating (7.2) with respect to A_0 and by using (7.1) we also get:

$$K_{A_0} = \left.\frac{\partial \omega_0}{\partial A_0}\right|_{V_{\text{tune}}} = -\frac{\omega_H - \omega_L}{A_0^2}(V_C - V_{DD}/2 + V_{\text{tune}}) = \frac{K_{\text{VCO}}}{A_0}\left(\underbrace{V_{DD}/2 - V_{\text{tune}}}_{V_{\text{GB0}}} - V_C\right). \tag{7.4}$$

In practice, the oscillation amplitude is not constant along the tuning range. As shown in Chapter 6, a maximum varactor quality factor is achieved when the capacitance is equal to $C_{\min}$. Therefore, the oscillation amplitude, which is proportional to the tank impedance $R_0 = \omega_0 L \cdot Q$, is a maximum or a minimum when the oscillation frequency is a maximum or a minimum. The resulting tuning curve will not be linear, as shown in Figure 7.4(d). A detailed derivation of the tuning curve and the AM-to-FM conversion factor in the case of pn junction varactors can be found in [4].

7.3 Noise up-conversion from varactors

Figure 7.5 shows the noise sources whose low-frequency components may be up-converted to the resonator bandwidth. Since ω_0 depends on both the tuning voltage V_{tune} and the oscillation amplitude A_0, any low frequency noise affecting these variables will cause frequency modulation at the oscillator output and, therefore, phase noise. For V_{tune} variations, we may write:

$$\Delta\omega_0 = K_{\text{VCO}} \cdot \Delta V_{\text{tune}}.$$

The resulting output phase is derived by integrating the instantaneous frequency. Assuming a frequency modulation at ω_m, it is

$$\phi(t) = \int_0^t [\omega_0 + \Delta\omega_0 \cdot \cos(\omega_m \cdot t')]\,dt' = \omega_0 t + \frac{\Delta\omega_0}{\omega_m} \cdot \sin(\omega_m \cdot t)$$

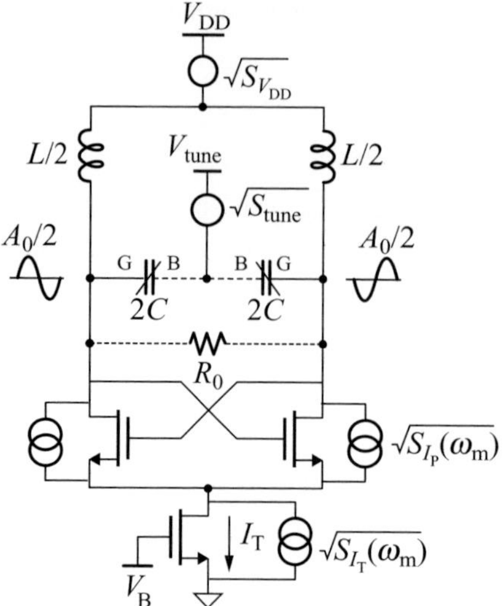

Figure 7.5 CMOS oscillator with low-frequency noise sources

and the resulting phase modulation is written as

$$\Delta\phi(\omega_m) = \frac{K_{VCO}\Delta V_{tune}}{\omega_m},$$

leading to a contribution to the single sideband-to-carrier ratio:

$$\mathcal{L}(\omega_m) = \frac{\Delta\phi^2}{4} = \frac{K_{VCO}^2}{2\omega_m{}^2} \cdot \frac{(\Delta V_{tune})^2}{2}.$$

In the most general case, denoting as $S_{tune}(\omega_m)$ the low-frequency noise at the tuning node, the resulting contribution to the single sideband-to-carrier ratio is given by:

$$\mathcal{L}(\omega_m) = K_{VCO}^2 \frac{S_{tune}(\omega_m)}{2\omega_m^2}. \tag{7.5}$$

The varactor d.c. bias is also set by the voltage supply V_{DD} in this circuit topology. Thus, a low-frequency supply disturbance has a similar impact, leading to another contribution as (7.5).

In a similar way, the frequency modulation from a low-frequency–amplitude noise can be quantified using K_{A_0},

$$\Delta\omega_0 = K_{A_0} \cdot \Delta A_0,$$

and the contribution to $\mathcal{L}(\omega_m)$ is given by:

$$\mathcal{L}(\omega_m) = K_{A_0}^2 \cdot \frac{S_{AM}(\omega_m)}{2\omega_m^2}, \tag{7.6}$$

where $S_{AM}(\omega_m)$ is the low-frequency noise spectrum affecting the oscillation amplitude.

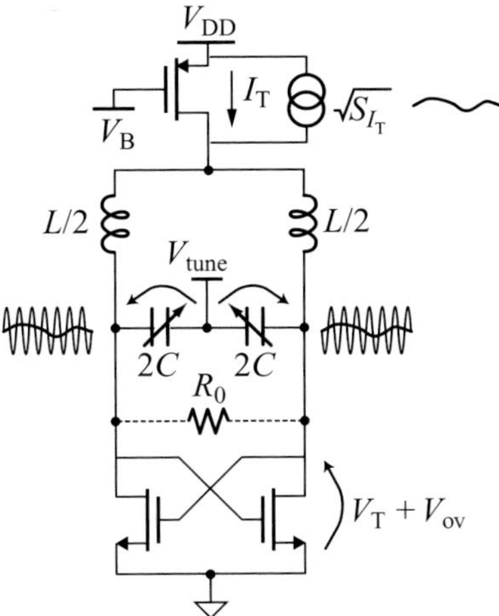

Figure 7.6 Top-biased CMOS cross-coupled oscillator

In practice, in Figure 7.1, the amplitude noise may result from low-frequency noise of the bias source I_T. In the current-limited regime, this current is related to the oscillation amplitude by the expression $A_0 = (2/\pi)\, I_T R_0$ obtained in Section 4.4. Therefore, the amplitude modulation noise spectrum is given by $S_{AM}(\omega_m) = (\partial A_0/\partial I_T)^2 \cdot S_{I_T}(\omega_m) = (2R_0/\pi)^2 \cdot S_{I_T}(\omega_m)$. If the oscillator is operated in the region where the amplitude starts to saturate, the conversion from current-modulation to amplitude-modulation noise is lower.

If alternative topologies of oscillators are employed where the varactor bias depends also on the oscillator bias current, the current-generator noise produces an additional contribution to phase noise. This phenomenon can be referred to as conversion from common mode modulation (CMM) to frequency modulation. [5] Referring, for instance, to the oscillator in Figure 7.6, the average output voltage is given by the FET gate–source voltage or $V_T + I_T/2g_m$, if the FETs operate under carrier-velocity saturation. Therefore, tail current noise produces varactor bias noise. The resulting phase noise can be estimated by applying (7.5) again.

The larger the varactor non-linearities, the higher the potential contribution of low-frequency noise up conversion to the VCO phase noise. This is another reason leading to a trade-off between tuning range and noise performance. It adds to the other source of degradation of the noise performance at wide tuning range already encountered in Chapter 6, where the varactor's quality factor has been shown to degrade at large capacitance ratios C_{max}/C_{min}.

Example 7.1 Flicker noise up-conversion in a MOS varactor-tuned oscillator

This example shows the derivation of the tuning curve and flicker-induced noise in the MOSFET cross-coupled oscillator shown in Figure 7.5, whose oscillation amplitude has been derived in Example 4.2 and phase noise has been derived in Example 5.3.

To study the effect of variable capacitances on the flicker-induced noise, the two linear capacitors of 6 pF in the tank have been replaced by two a-pMOS varactors, whose C–V characteristic and quality factor are depicted in Figure 7.7(a). These varactors are available

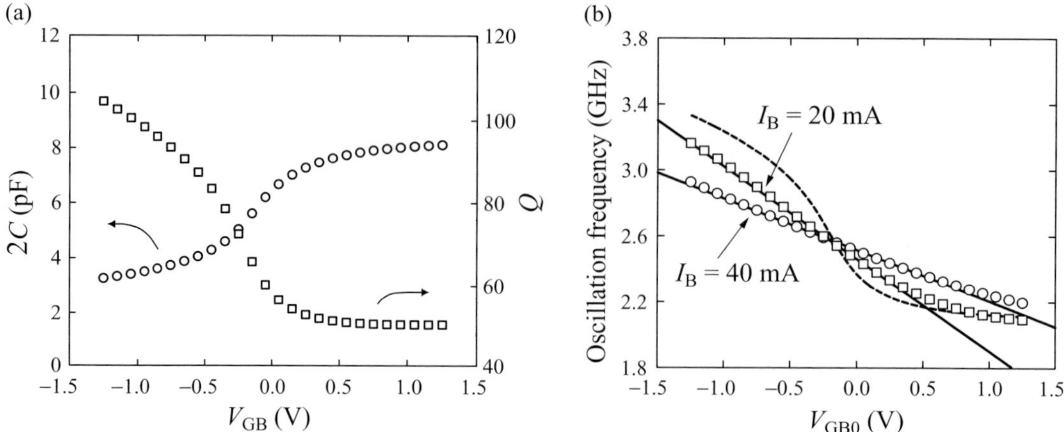

Figure 7.7 Tuning curve estimation of the CMOS cross-coupled oscillator of Example 7.1: (a) $C(V)$ curve (circles) and quality factor Q (squares) of the varactor; (b) resulting tuning curve from simulations (circles and squares), from theory proposed in Section 7.2 (solid lines) and from erroneous estimation $1/\sqrt{LC}$ (dashed line)

in a 0.25 μm CMOS process. Each varactor has a minimum capacitance $C_{min} = 3.3$ pF and a capacitance ratio $C_{max}/C_{min} = 2.45$. The inductor is still an integrated coil with $L/2 = 0.7$ nH per tank side.

Given these capacitance values, the two boundary frequencies are $f_L = 2.1$ GHz and $f_H = 3.3$ GHz. In practice, large tuning ranges in oscillators cause wide variations of the equivalent parallel loss resistance $R_0 = \omega_0 L Q$, because this resistance depends on ω_0 both directly and via the quality factor. In this example, for a fair comparison with the results obtained with constant tank capacitances, a resistor of 170 Ω has been placed in parallel with the tank, so that the resulting R_0 has a weak dependence on ω_0 and ranges between 170 Ω and 190 Ω. This value of R_0 is very close to the one used in the examples in Chapters 4 and 5.

With a tail current I_T of 20 mA, the oscillation amplitude varies between 2.2 and 2.4 V over the tuning range. From (7.1), the tuning sensitivity $K_{VCO}/2\pi$ is given by the ratio between $f_H - f_L$ and the oscillation amplitude. Therefore, it is expected to be about (1.2 GHz)/(2.2V) ≅ 550 MHz/V. By increasing I_T to 40 mA, the oscillation amplitude is almost doubled and, consequently, the estimated $K_{VCO}/2\pi$ is halved.

The tuning curve has been achieved from circuit simulations and the results are shown in Figure 7.7(b). The solid lines, which are straight lines with slopes equal to the estimated values of $K_{VCO}/2\pi$, agree very well with the simulated dots. However, the dashed curve, which is obtained as $1/\sqrt{LC(V_{GB0})}$, provides an incorrect estimation of the tuning curve.

As the oscillation amplitude increases, the tuning sensitivity lowers. As a result, the tuning range for a given tuning voltage range is reduced. For $I_T = 40$ mA and $\Delta V_{tune} = 2.5$V, the boundary frequencies cannot be reached and the maximum tuning range as given by (7.3) and confirmed from the circuit simulations is limited to about 700 MHz.

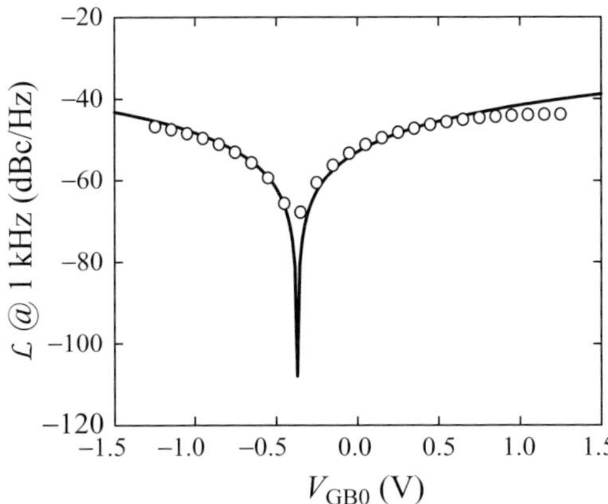

Figure 7.8 Simulated phase noise at 1 kHz offset from carrier of oscillator of Example 7.1 (circles) and theoretical estimation of noise induced by tail flicker (solid line)

The flicker noise originating from the tail generator causes amplitude noise, which is converted into frequency noise by the presence of the varactors. The FET employed as a current generator exhibits a flicker noise of about S_{I_T} (1 kHz) $= 4\,(\mathrm{nA})^2/\mathrm{Hz}$. If the oscillator operates in the current-limited regime, there is a linear relationship between amplitude and bias current. Therefore, the resulting amplitude modulation noise is:

$$S_{AM}(\omega_m) = \frac{4R_0^2}{\pi^2} \cdot S_{I_T}(\omega_m).$$

Otherwise, the factor $2R_0/\pi$ should be replaced by the actual slope of the $A_0(I_T)$ curve.

From (7.4), the frequency sensitivity K_{A_0} to amplitude-modulation noise can be estimated as $K_{VCO}(V_{GB0} - V_C)/A_0$. Therefore, substituting the expressions for K_{A_0} and S_{AM} into (7.6), the phase noise can be written as

$$\begin{aligned} \mathcal{L}(\omega_m) &= K_{A_0}^2 \cdot \frac{S_{AM}(\omega_m)}{2\omega_m^2} \\ &\approx \frac{2}{\pi^2} \cdot \frac{R_0^2 S_{I_T}(\omega_m)}{A_0^2} \cdot \frac{K_{VCO}^2 \cdot (V_{GB0} - V_C)^2}{\omega_m^2} \end{aligned} \tag{7.7}$$

The phase noise has a quadratic dependence on the varactor bias voltage. Figure 7.8 shows the theoretical phase noise at 1 kHz offset from the carrier (solid line, given by (7.7)) compared with SpectreRF simulation results (circles). The theoretical expressions, although very approximate, agree very well with the simulated points and demonstrate that the noise coming from the tail is the limiting noise contribution at low offsets, because of the AM-to-FM conversion. The solid line approaches zero for a varactor bias of about −0.35 V, since, at that bias, the sensitivity K_{A_0} is zero. Nevertheless, the simulated noise is not zero, as the output noise at that bias is set by the flicker coming from the switching pair. The latter can be minimized either by proper choice of the transistor size and type [6, 7] or by employing the tail resonator at twice the oscillation frequency. [8]

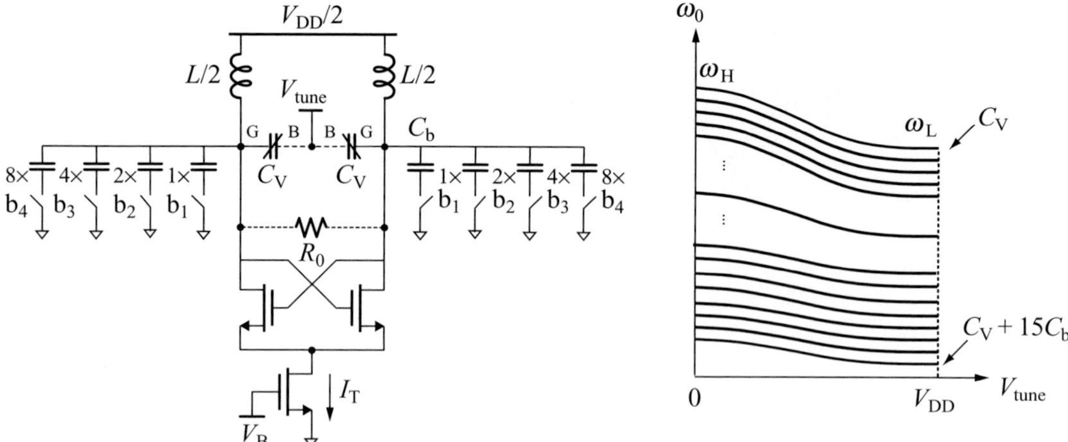

Figure 7.9 (Left) CMOS oscillator employing mixed analogue and digital frequency control. (Right) family of tuning characteristics

7.4 Topologies and methods to minimize for up-conversion

7.4.1 Low tuning sensitivity and automatic frequency control

To mitigate these effects drastically, a viable method is to reduce the VCO gain K_{VCO}. This requirement, however, conflicts with the need to design VCO cells capable of spanning a wide tuning range. Since the voltage tuning range is limited by device reliability issues, the larger the tuning range the larger the required gain.

Most of the time, a reasonable compromise is reached by using the mixed analogue and digital frequency control shown in Figure 7.9. The choice is to cover the tuning range by using many overlapping tuning curves with moderate slopes, instead of a single tuning curve with a high K_{VCO}. As discussed in Section 6.5, the bank of capacitors can alternatively be implemented by using accumulation-mode or inversion-mode varactors, which are switched from their maximum to their minimum capacitance, or by using MiM capacitors in series with switches (as in Figure 7.9). In the first option, even if the varactors are biased in the almost flat regions of their C–V characteristic, the oscillating voltage still spans the C–V curve. Therefore, some residual AM-to-FM conversion from the switched varactors still exists. In this respect, MiM capacitors are preferable.

An automatic calibration algorithm chooses the proper VCO tuning characteristic by switching the bank of capacitors. Various approaches exist to the realization of an automatic frequency control (AFC) circuit. Two main options are shown in Figure 7.10. In the arrangement in Figure 7.10(a), the calibration starts by opening the phase-locked loop and connecting the VCO analogue tuning (TUNE) to a reference voltage, typically in the middle of the tuning voltage range. After that, the logic applies a first-guess digital word to the digital tuning input of the VCO. A frequency discriminator senses the frequency difference between the REF and DIV signals and increases or decreases the digital tuning word, to

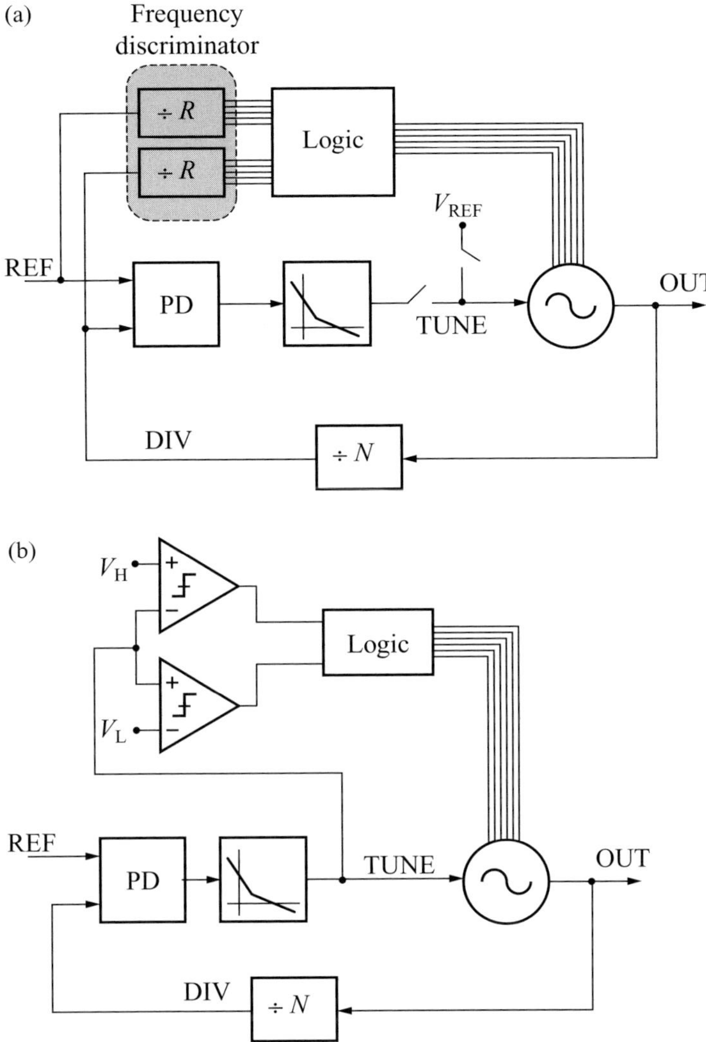

Figure 7.10 PLL with automatic frequency control (AFC): (a) based on frequency discrimination; (b) based on tuning voltage monitoring

compensate the frequency difference. A binary search algorithm reduces the number of steps required to find the correct VCO characteristic. Examples of this realization are discussed in [9, 10].

Typical constraints in the design of AFC loops are represented by the specified channel switching time of the frequency synthesizer. If N is the number of bits of the VCO digital tuning input and T is the time required by the frequency discriminator to resolve the frequency difference with a certain accuracy, a binary-search algorithm requires, at most, $(1 + N)\,T$ seconds. There may not be enough time for calibration when changing channels. Therefore, calibration is only possible at the synthesizer power-up.

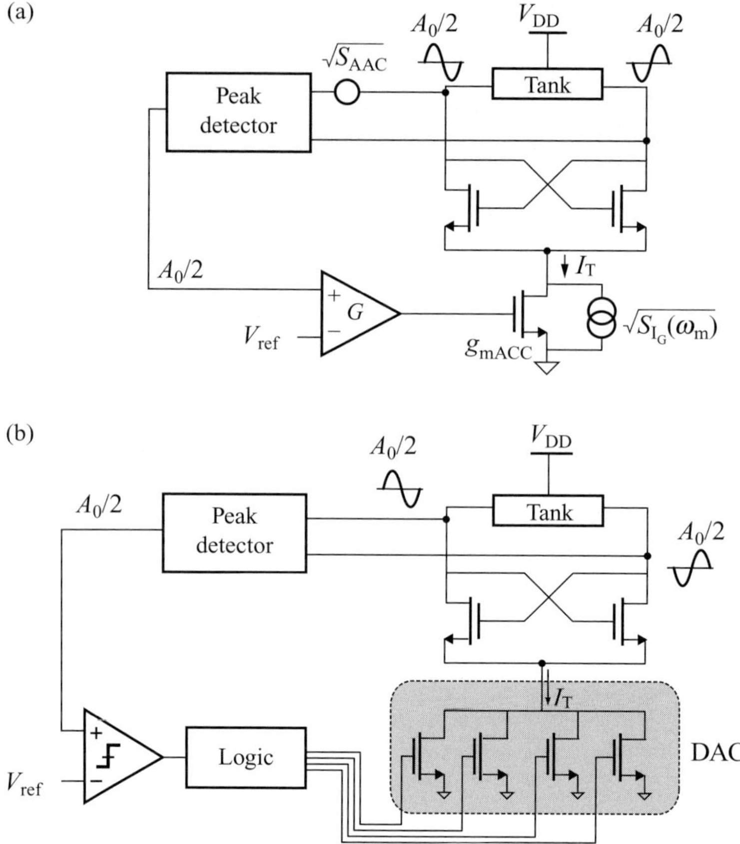

Figure 7.11 Oscillators with automatic amplitude control (AAC): (a) analogue and (b) digital implementations

A typical alternative approach is shown in Figure 7.10(b). The loop is always closed, while two comparators detect if the analogue tuning voltage exceeds the allowed range between V_H and V_L. In such a case, the finite-state logic modifies the digital tuning word coherently until the analogue tuning is brought back to inside the allowed voltage range.

7.4.2 Low-noise automatic amplitude control (AAC)

In principle, amplitude modulation caused by the noise source within the oscillator circuit could be attenuated by adding a control loop, which detects the oscillation amplitude and compensates for its variations. An analogue automatic amplitude control loop (AAC) (shown in Figure 7.11(a)) senses the oscillation amplitude with a peak detector and controls the bias current to keep A_0 to a constant level V_{ref}. Examples of AAC circuit implementations can be found, for instance, in [11 and 12].

Automatic amplitude control is commonly employed for purposes other than lowering the amplitude modulation noise. To optimize the oscillator's FoM, the oscillation amplitude has to be kept at its maximum along the whole tuning range. However, even at a constant resonator quality factor Q, the equivalent resistance $R_0 = \omega_0 L Q$ (and in turn A_0) depends on the oscillation frequency. The AAC also ensures (i) a safe and fast oscillation start-up, which is particularly required in systems adopting multiple VCOs, which are turned on and off to cover different frequency ranges; (ii) a well-defined level of output power, which is important for properly driving the prescaler following the VCO in the PLL; (iii) the possibility of operating the oscillator transconductor in a linear regime. Since the AAC is a mechanism for oscillation amplitude stabilization, the oscillator active stage does not serve as an amplitude limiter. This choice may give benefits on noise performance. [11]

The AAC is a non-linear circuit, owing to the joint presence of the oscillator and of the peak detector, and the noise analysis deserves special care. The amplitude A_0 is the controlled variable; the peak detector is modelled as a linear block with a gain of about 1/2 (since it comes from single-ended waveform rectification) followed by an amplifier, which drives the tail current generator of the VCO. The link between A_0 and I_T is given, as usual, by $A_0 = (2/\pi) I_T R_0$. Figure 7.11(a) also shows the equivalent voltage noise source of the AAC peak detector and gain stage, represented by the spectrum S_{AAC}, and the current noise source due to the current generator, represented by the spectrum S_{I_G}. The AAC open-loop gain has a bandwidth limited by the dominant pole ω_{AAC}, which is introduced to guarantee loop stability. Therefore, the open-loop gain can be written as

$$G_{\mathrm{loop}}(\omega) = \frac{\frac{1}{2} G \cdot g_{m_{\mathrm{AAC}}}}{1 + \mathrm{j}\omega/\omega_{\mathrm{AAC}}} \cdot \frac{2}{\pi} R_0$$

and the current power spectral density associated to the VCO tail current is given by:

$$S_{I_T} = S_{I_G} \left| \frac{1}{1 + G_{\mathrm{loop}}} \right|^2 + S_{\mathrm{AAC}} \left| \frac{1}{\frac{2}{\pi} R_0} \cdot \frac{G_{\mathrm{loop}}}{1 + G_{\mathrm{loop}}} \right|^2 .$$

It follows that only the current generator noise S_{I_G} is filtered out from d.c. up to the AAC GBWP. The resulting oscillator-amplitude noise within the AAC bandwidth is approximately given by the input-referred AAC voltage noise:

$$S_{\mathrm{AM}}|_{\text{in-band}} = \frac{S_{I_G}}{\frac{1}{4} G^2 \cdot g^2_{m_{\mathrm{AAC}}}} + S_{\mathrm{AAC}} \approx S_{\mathrm{AAC}}.$$

In other terms, the loop is able to detect the amplitude modulation produced by the tail noise and to compensate for it. But it is not able to compensate for the S_{AAC} voltage noise source, which is equivalent to noise on V_{ref} and which produces amplitude-modulation noise. This noise is ultimately converted into frequency-modulation noise, as discussed in the previous section.

Since it may be hard to maintain the peak detector noise to negligible levels, AACs are scarcely used to lower the amplitude-modulation noise. If an amplitude calibration method is necessary for the other reasons listed above, a digital AAC loop may be used (Figure 7.11(b)). Although less accurate than analogue loops, the digital solution does not add noise.

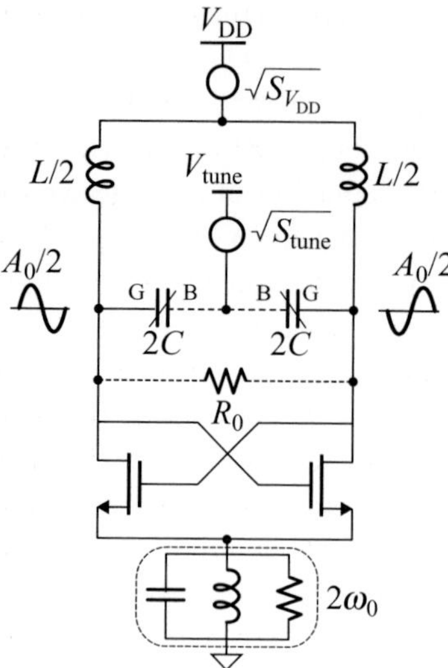

Figure 7.12 CMOS oscillator without tail-current generator

The bias current is digitally controlled by a current DAC replacing the current generator and the control word is set by comparing the oscillation amplitude with the reference level V_{ref}. In this case, the only contribution to the bias noise comes from the current generators of the DAC, as in the original oscillator without AAC. Therefore, the digital AAC regulates the amplitude without increasing the amplitude modulation noise. A current DAC with few bits (three or four) is sufficient, since the required amplitude accuracy and the maximum current variations are limited.

7.4.3 Cross-coupled oscillator without tail-current generator

One may wonder whether the tail-current generator is necessary in differential oscillators or if it is even detrimental for performance, since it has been shown in the previous section to provide dominant flicker noise. Both single-type and complementary cross-coupled oscillators with no bias current generator have been designed. [3, 13]

In complementary cross-coupled oscillators with no tail transistor, the transistors are sized to have the same over-drive in the balanced condition. In this way, the balanced output voltage is $V_{DD}/2$ and each output node can have a rail-to-rail swing, leading to a differential amplitude of $A_0 = V_{DD}$. This topology intrinsically regulates the bias current, to produce full-swing oscillation. Therefore, even in the presence of spread in the resonator peak impedance, the oscillation amplitude is always equal to the available voltage headroom. The current drained from the power supply has a large component at $2\omega_0$ and its average value depends on the supply voltage and on the parallel resonator resistance R_0.

Figure 7.12 shows a single-type cross-coupled oscillator employing a network resonant at $2\omega_0$ in place of the tail-current generator. This network biases the FET sources at the lower rail, while it acts as a high impedance for the FET switching behaviour. Hence, it increases the achievable oscillation amplitude and reduces the common-mode current component at $2\omega_0$.

The $1/f^3$ noise of the oscillator can now only originate from flicker of the MOS transistor, acting as switches. The flicker noise of these switches can be modelled by a voltage generator in series with the gate of each device. This noise is, in principle, sampled by a pulse train at $2\omega_0$, at the voltage zero crossings. As a result, the noise component at ω_m produces drain current tones at $2n\omega_0 \pm \omega_m$, flowing in a non-linear varactor oscillating at ω_0. The mixing of these harmonics produces frequency modulation of the carrier at ω_0.

However, the switching pair is expected to feature lower flicker noise than the current generator for two main reasons. (i) The FETs operate in the triode region for large portions of the oscillation period; hence, they exhibit lower current flicker noise than the tail transistor that operates steadily in saturation. (ii) Switched MOSFETs are known to have lower flicker noise than transistors biased in the steady-state condition. [14]

The voltage-supply lowering in scaled CMOS technologies has raised interest in oscillator cross-coupled topologies with no current generator, even if some concerns are related to their poor power supply rejection. The power supply affects the amplitude of oscillation. Therefore, it has to be generated by a low-noise low-drop-out regulator. However, a similar problem also arises in the topology with the current generator: the amplitude is set by the bias current and by the tank parallel resistance. Even if the former is independent of supply variations, the latter suffers from variations over the tuning range and process spreads. A low-noise amplitude control loop is therefore necessary, in that case.

The power supply also affects the frequency of oscillation, since it sets the varactor bias voltage. Again, this problem is common to both topologies, whether or not they employ a tail-current generator. Immunity to power supply variations can be achieved by employing either differential varactors [15, 16] or a network that feeds the supply variations back to the tuning voltage.

Example 7.2 MOSFET oscillator design with low flicker-induced noise

Figure 7.13 shows a double cross-coupled oscillator. This oscillator, in 0.25 μm CMOS technology, with $V_{DD} = 2.5$V and biased with I_T of about 2 mA, oscillates around 5 GHz. The resonator employs two spiral inductors, each with inductance $L/2 = 1$ nH, and two a-pMOS varactors. In this example, practical losses of inductor and varactors are taken into account.

Figure 7.14(a) shows the $C(V)$ curve of the half-tank capacitance plotted versus the varactor bias voltage V_{GB0} in the $[-V_{DD}/2, +V_{DD}/2]$ range. The capacitances $2C$ of the tank are the varactor capacitance plus a parasitic of about 550 fF, due to the inductor, the transistors' gates and the drain-to-substrate junctions. The effective maximum-to-minimum capacitance ratio C_{max}/C_{min} is about two. The curve in Figure 7.14(a) is a step-like approximation of the $C(V)$ curve, where $C_{max} = 1.2$ pF, $C_{min} = 0.7$pF and $V_C = 0.2$ V. The differential oscillation amplitude A_0 ranges between 1.6 and 2.1 V.

In the double cross-coupled topology, the output average voltage, which biases the varactors, is not set directly by the voltage supply as in the single cross-coupled oscillator. If the current flows symmetrically in the two sides of the circuit and the transistors operate in velocity saturation, the output average voltage is about $V_{DD} - |V_{TP}| - I_T/2g_{mP}$, where V_{TP}

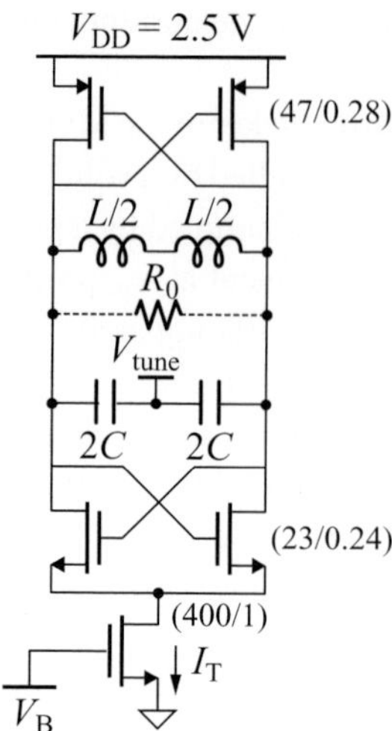

Figure 7.13 CMOS double cross-coupled oscillator

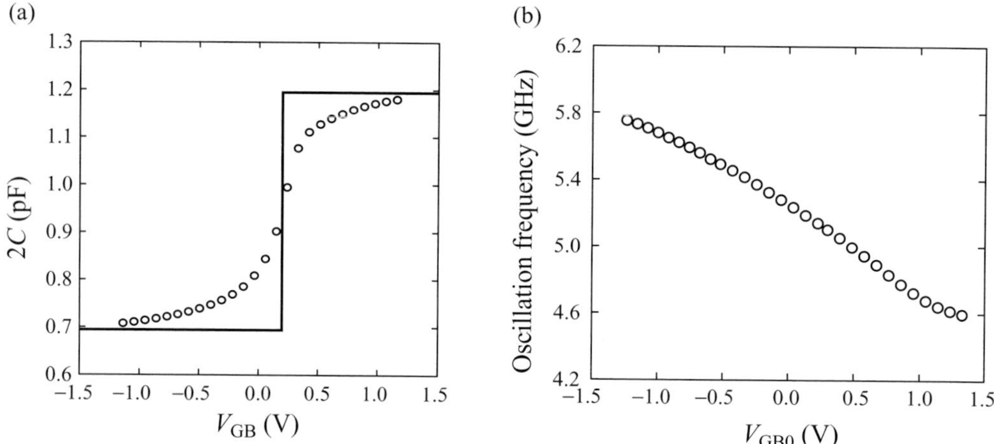

Figure 7.14 Oscillator in Example 7.2: (a) tank capacitance C–V characteristic; (b) tuning curve

and g_{mP} are the pFET threshold voltage and transconductance. This is set to about $V_{DD}/2$. The limits of the frequency range given by C_{max} and C_{min} are 4.6 GHz and 6.0 GHz. The tuning curve from circuit simulations is shown in Figure 7.14(b). The tuning range within the tuning voltage range is 21% (between 4.6 and 5.7 GHz).

Let us consider the impact of the low-frequency noise from the tail-current generator on the output phase noise. A tone at ω_m injected from the oscillator tail gives rise to both common-mode and amplitude modulation, as depicted in Figure 7.15. A transistor-level simulation would directly provide the sensitivity $(\mathrm{d}\omega_0/\mathrm{d}I_T)$. The fastest procedure is to run a transient simulation of the circuit biased at current I_T, then to repeat the simulation at $(I_T + \Delta I_T)$ and to detect the frequency shift $\Delta\omega_0$. The sensitivity is then given approximately

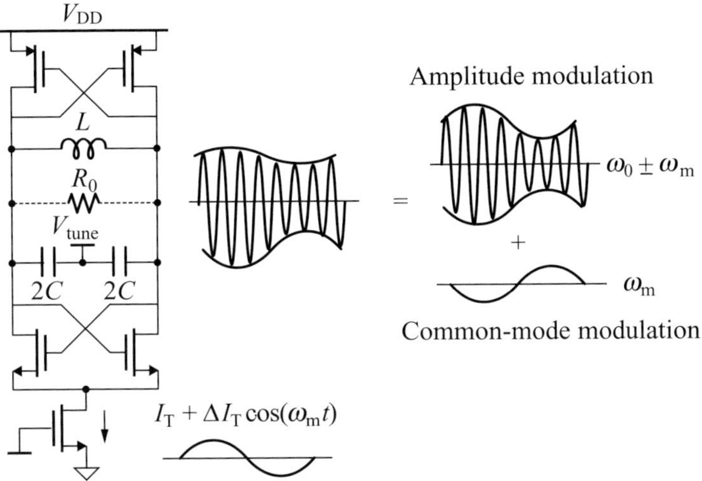

Figure 7.15 Tail flicker causes both amplitude modulation and common-mode modulation, which both modulate the oscillation frequency

by the ratio between $\Delta\omega_0$ and ΔI_T. Once the sensitivity is known, the contribution to $\mathcal{L}$ can be estimated as:

$$\mathcal{L}(\omega_m) = \left(\frac{d\omega_0}{dI_T}\right)^2 \cdot \frac{S_{I_T}(\omega_m)}{2\omega_m^2}.$$

However, for first-order evaluations, the designer can rely on sensitivity estimations from the linear model. The sensitivity of the frequency to the tail current can be written as the sum of the sensitivity due to the common-mode voltage variation and that due to the amplitude variation.

The linear approximation discussed in Section 7.2 suggests that the common-mode sensitivity is constant in the transition region

$$(V_C - A_0/2) \leq V_{GB0} \leq (V_C + A_0/2)$$

and it is equal to

$$K_{VCO} \approx (\omega_H - \omega_L)/A_0 = 2\pi \cdot 660\,\text{Mrad/(Vs)}.$$

It falls abruptly to zero outside this range (see dashed line in Figure 7.16). The amplitude modulation sensitivity K_{A_0} is instead zero at $V_{GB0} = V_C$ and rises linearly as the d.c. bias moves towards the extremes of the transition region. In reality, the coefficients will have smoother dependences, like those obtained from circuit simulations and shown as circles and squares in Figure 7.16. They both fall at the extreme of the tuning range, where the described conversion mechanisms become less effective.

Since the common-mode voltage is approximately

$$V_{CM} = V_{DD} - |V_{TP}| - I_T/2g_{mP},$$

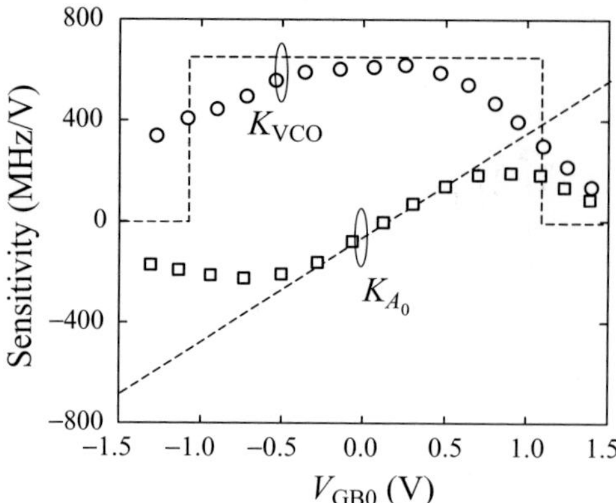

Figure 7.16 Sensitivity coefficients: frequency sensitivity to common-mode modulation K_{VCO} and to amplitude modulation K_{A_0} from circuit simulations (circles and squares, respectively) and from first-order approximations (dashed lines)

a bias current variation ΔI_T causes a common-mode change $\Delta V_{CM} = -\Delta I_T/2g_{mP}$ and the resulting frequency shift is:

$$\left.\frac{\partial\omega_0}{\partial I_T}\right|_{CM} = \frac{\partial\omega_0}{\partial V_{CM}} \cdot \frac{\partial V_{CM}}{\partial I_T} = -K_{VCO} \cdot \left(-\frac{1}{2g_{mP}}\right).$$

The contribution to the frequency shift due to the modulation of the output amplitude follows instead from (7.4):

$$\left.\frac{\partial\omega_0}{\partial I_T}\right|_{AM} = \frac{\partial\omega_0}{\partial A_0} \cdot \frac{\partial A_0}{\partial I_T} \approx K_{A_0} \cdot \frac{A_0}{I_T} \approx K_{VCO} \cdot \frac{(V_{GB0} - V_C)}{I_T},$$

where a linear dependence between amplitude and current has been assumed. In practice, approaching the voltage-limiting regime, the derivative $\partial A_0/\partial I_T$ is lower than A_0/I_T. Therefore, the actual sensitivity will be lower than the estimated value.

The resulting total sensitivity is given by:

$$\frac{d\omega_0}{dI_T} = \left.\frac{\partial\omega_0}{\partial I_T}\right|_{CM} + \left.\frac{\partial\omega_0}{\partial I_T}\right|_{AM} \approx K_{VCO}\left[\frac{1}{2g_{mP}} + \frac{(V_{GB0} - V_C)}{I_T}\right]. \tag{7.8}$$

Using (7.8) and $1/g_{mP} = 100\ \Omega$, the frequency sensitivity $d\omega_0/dI_T$ is estimated to be about $2\pi \cdot 40$ Mrad/(smA) at $V_{GB0} = V_C$. Note that at this bias the second term in (7.8) is zero. The tail current flicker noise of $4(\text{nA})^2/\text{Hz}$ at 1 kHz limits the phase noise $\mathcal{L}$ to −25 dBc/Hz at 1 kHz or, equivalently, to –85 dBc/Hz at 100 kHz.

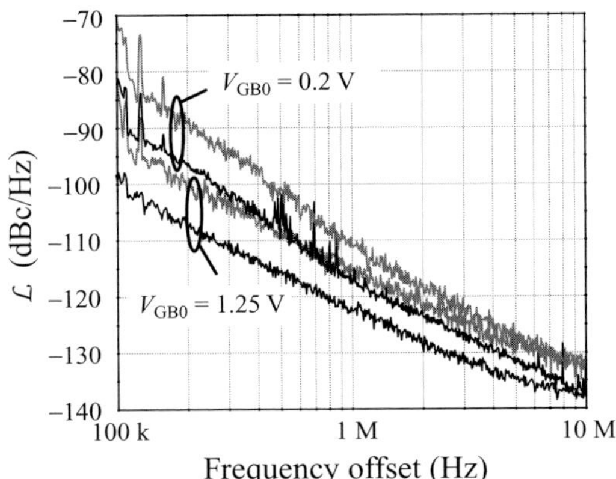

Figure 7.17 Measured spectra of oscillator described in Example Example 7.3

Figure 7.17 (grey lines) shows the phase noise spectra, measured with varactor biasing $V_{GB0} = 0.2V$ and 1.25 V, where the total sensitivity is maximum and minimum, respectively. The extrapolated $1/f^3$ noise at 100 kHz is −83 and −94 dBc/Hz, respectively. The noise at $V_{GB0} = 0.2$V agrees closely with the first-order estimation. The variation of the close-in phase noise over the tuning range is significant and the corner frequency moves by a decade, from 200 kHz to 2 MHz, seriously limiting the performance of the VCO along the tuning range, even when locked in a wideband PLL. At high K_{VCO}, the double cross-coupled topology has, in general, higher flicker-induced noise than the bottom-biased cross-coupled oscillator topology shown in Figure 7.1, since the conversion from common-mode to frequency modulation is dominant over the conversion from amplitude to frequency modulation.

As proposed in this section, lower up-converted noise can be achieved by removing the bias generator. The spectra measured at $V_{GB0} = 0.2$ V and 1.25 V are shown as the black lines in Figure 7.17. For both tuning voltages, the $1/f^3$ noise is reduced by about 10 dB when the tail-current generator is taken away. The $1/f^2$ region of the phase spectrum is also improved, since the tail-current generator was also responsible for the dominant white-noise component in the oscillator shown in Figure 7.13.

7.5 Other mechanisms of noise up-conversion

Even in the absence of varactors, oscillators exhibit up-conversion of flicker noise. In Section 5.4.1, a noise source with non-zero average ISF has been shown to be responsible for flicker noise up-conversion, or, equivalently, a noise source with non-real c_1 coefficient of the HTF. In practice, this happens with asymmetric HTF functions, see for instance [8].

In bipolar stages, where the transistor delay depends linearly on the bias current, an additional up-conversion effect may come into play. [17] It has already been shown in Section 4.2.5 that the phase shift caused by the active element makes the oscillation frequency shift from the resonance $1/\sqrt{LC}$ value. This delay may be ascribed, not only to the transistor transit time, but also to the input pole caused by the presence of the base resistance $r_{\mathrm{bb}'}$.

A change in the bias current affects the loop delay introduced by the active transconductor, causing a modulation of the oscillation frequency:

$$\Delta\omega_0 = -\frac{\omega_0}{2Q}\Delta\theta \approx -\frac{\omega_0}{2Q}\cdot\frac{\partial\theta}{\partial I_{\mathrm{B}}}\cdot\Delta I_{\mathrm{B}},$$

where $\Delta\theta$ and ΔI_{B} are the loop-delay variation and the bias-current variation. The additional θ shift may be ascribed to a pole at ω_{P}. Its variation with the bias current may be written as:

$$\begin{aligned}\frac{\partial\theta}{\partial I_{\mathrm{B}}} &= -\frac{\partial}{\partial I_{\mathrm{B}}}\arctan\left(\frac{\omega}{\omega_{\mathrm{P}}}\right)\\ &\approx -\omega\frac{\partial}{\partial I_{\mathrm{B}}}\left(\frac{1}{\omega_{\mathrm{P}}}\right) = -\frac{\omega}{\omega_{\mathrm{P}}^2}\cdot\frac{\partial\omega_{\mathrm{P}}}{\partial I_{\mathrm{B}}}.\end{aligned}$$

Therefore, the peak frequency deviation can be written as:

$$\Delta\omega_0 \approx \frac{1}{2Q}\cdot\left(\frac{\omega_0}{\omega_{\mathrm{P}}}\right)^2\cdot\frac{\partial\omega_{\mathrm{P}}}{\partial I_{\mathrm{B}}}\cdot\Delta I_{\mathrm{B}}.$$

Non-linear capacitances or device delays are not the only mechanisms responsible for low-frequency noise up-conversion. Another mechanism can be ascribed to the low tank Q factor and to non-linearity of the oscillator active device. If the Q factor is so low that the higher-order harmonics of the transconductor cannot be neglected, the output voltage is no longer sinusoidal. The main consequence is that its frequency is not simply set by the Barkhausen criterion applied to the fundamental harmonic. The criterion must also be fulfilled for the higher-order harmonics. Groszkowski shows that the oscillation frequency normalized to the tank resonance frequency ω_0 is given approximately by:

$$\frac{\omega}{\omega_0} \simeq 1 - \frac{1}{2Q^2}\sum_{n=2}^{\infty}\left[\frac{n^2}{n^2-1}\left(\frac{I_n}{I_1}\right)^2\right], \tag{7.9}$$

where I_n is the nth Fourier coefficient of the transconductor output current and Q is the tank quality factor. [18]

In cross-coupled oscillators operated at high amplitudes, the current waveform generated by the transconductor approaches a square wave and the Fourier coefficients of the current are $I_n = I_1/n$ (with odd n). Thus, from (7.9), the oscillation approaches the limit $\omega_l = \omega_0 \cdot [1 - 1/(8Q^2)]$, which is offset from the tank resonance frequency. Considering a bipolar differential pair, we should also expect that as the amplitude is lower than twice the thermal voltage V_{T}, the oscillation frequency is approximately ω_0. As the amplitude rises and exceeds $2V_T$ it should approach ω_l. Therefore, some AM-to-FM conversion exists only at low amplitudes. This can be estimated as:

$$K_{A_0} = \frac{\partial\omega}{\partial A_0} \approx \frac{\omega_0}{8Q^2}\cdot\frac{1}{2V_{\mathrm{T}}}$$

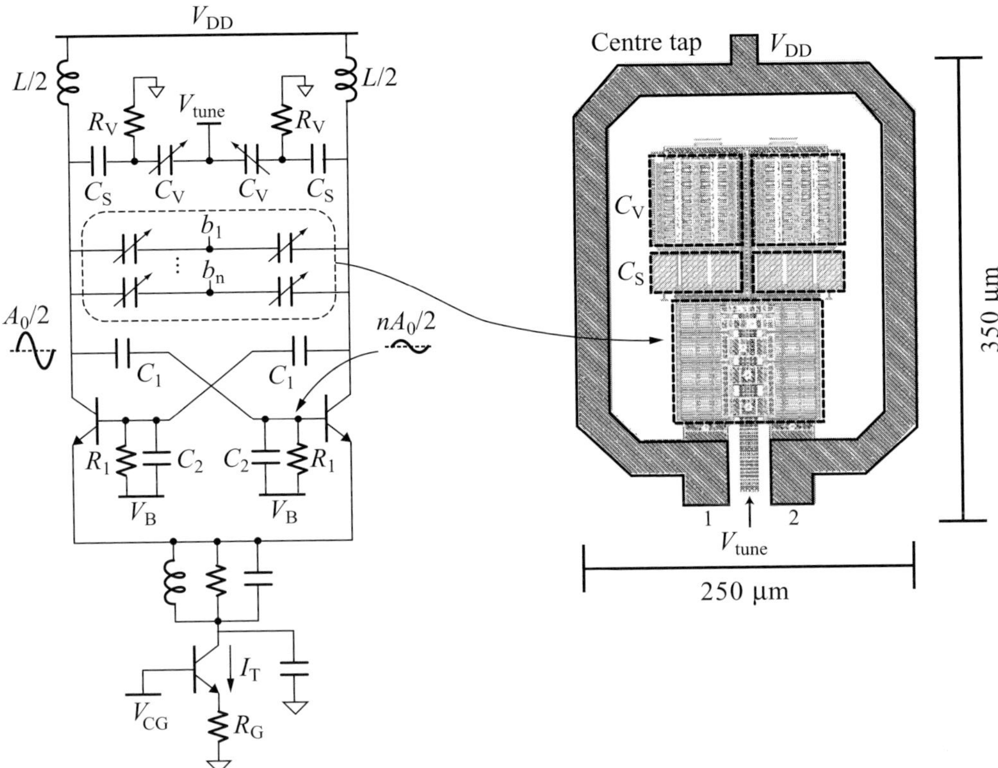

Figure 7.18 Bipolar cross-coupled oscillator and layout of resonator

Example 7.3 Bipolar 5 GHz oscillator design

The examples deals with the design of a 5 GHz oscillator for a 256 QAM radio system. The integral phase noise requirement obtained from system-level simulations limits the spot noise of the 5 GHz carrier to about −100 dBc/Hz at 100 kHz and the corner frequency to less than 10 kHz. If the transmit and receive bands have to be covered by a single local oscillator in the presence of process spreads, a tuning range of about 40% around 5 GHz is required.

The bipolar cross-coupled oscillator with capacitive voltage divider in Figure 7.18 has already been discussed in Chapters 4 and 5. The voltage divider extends the oscillation amplitude range so that the transistor pair stays in the forward active region. The saturation of the bipolar must be avoided to attain higher amplitudes and to maintain the contribution of the transistors' base shot noise to negligible values.

Another reason is that saturation exacerbates the low-frequency noise up-conversion. The transistor delay increases and the parasitic C–B junction capacitances contributing to tank capacitance are driven toward forward bias. This not only reduces the maximum achievable oscillation frequency and narrows the tuning range, but also increases the frequency sensitivity to low-frequency noise from the tail generator. The *C–V* of a pn-junction becomes steeper in forward bias and the steeper the $C(V)$ characteristic, the stronger the AM-to-FM conversion.

In Section 4.4.4, the maximum oscillation amplitude to avoid saturation has been written as:

$$\begin{aligned} A_{0,\max} &= \frac{2}{1+n} \cdot (V_{\mathrm{DD}} - V_{\mathrm{Gsat}} - V_{\mathrm{BEon}} + V_{\mathrm{BCsat}}) \\ &\approx \frac{2}{1+n} \cdot (V_{\mathrm{DD}} - V_{\mathrm{Gsat}}), \end{aligned}$$

where n is the partition factor.

The drawback of the voltage divider is an increased weight of the collector shot noise and the base resistance thermal noise. Their contributions increase as $1/n$ (see Example 5.2). The oscillator noise factor is:

$$F \approx \frac{1}{2n} + \frac{g_{\mathrm{m}} r_{\mathrm{bb'}}}{n}.$$

Hence, the $1/f^2$ phase noise can be approximated by

$$\mathcal{L}(\omega_{\mathrm{m}}) = \frac{kT}{C} \cdot \left(\frac{\omega_0}{Q}\right) \cdot \frac{1}{\omega_{\mathrm{m}}^2} \cdot \frac{(1+n)^2}{4\,(V_{\mathrm{DD}} - V_{\mathrm{Gsat}})^2} \cdot \left(1 + \frac{0.5 + g_{\mathrm{m}} r_{\mathrm{bb'}}}{n}\right) \tag{7.10}$$

and an optimum value of n exists, which minimizes the overall noise. In this example, it is $n_{\mathrm{opt}} = 1/3$ and the phase noise at 100 kHz estimated from (7.10) is –110 dBc/Hz.

To fulfil the phase noise specification at 100 kHz, the up-converted low-frequency noise also has to be limited.

First, the K_{VCO} is greatly reduced by employing mixed analogue and digital tuning. Referring to Figure 7.18, the total tank capacitance of about 3.75 pF (in the centre of the tuning range) is divided into three contributions: (i) the varactors C_{V} in series with the capacitances C_{S}, (ii) the bank of i-nMOS varactors switched from maximum to minimum capacitance, (iii) the parasitic capacitances of the inductor coil, interconnections and transistors.

The varactor cathodes are d.c. biased to ground. Two MiM series capacitances C_{S} a.c. couple the varactors to the inductor and the transconductor and lower the K_{VCO}. The value of the resistor R_{B} has to be chosen properly. It has to be high enough not to load the LC tank:

$$R_{\mathrm{B}} \gg \frac{Q}{\omega_0 C_{\mathrm{V}}\,(1 + C_{\mathrm{V}}/C_{\mathrm{S}})},$$

where Q is the overall tank quality factor and C_{V} is the varactor capacitance. It has to be low enough so that its thermal noise contribution to phase noise is negligible. The thermal noise $4kTR_{\mathrm{B}}$ modulates the varactor bias and up-converts to phase noise through the VCO gain. Therefore, on the basis of (7.5), it follows that:

$$R_{\mathrm{B}} \ll \frac{2\omega_{\mathrm{m}}^2 \cdot \mathcal{L}(\omega_{\mathrm{m}})}{4kT \cdot K_{\mathrm{VCO}}^2}.$$

The tank inductance of 270 pH can be synthesized by a single-turn inductor. This structure features a higher quality factor, because of the absence of current-crowding effects typical

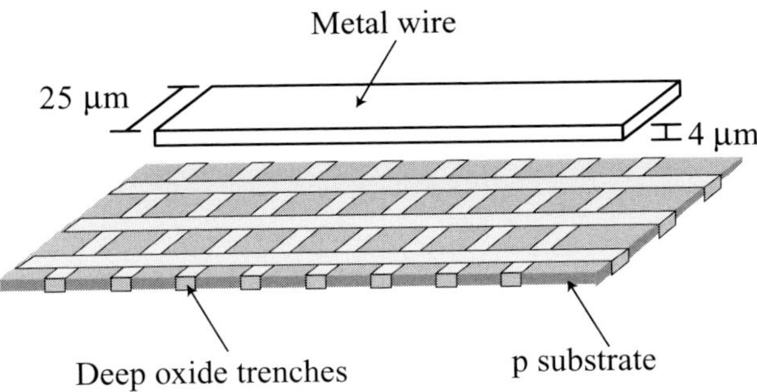

Figure 7.19 Inductor with deep oxide trenches

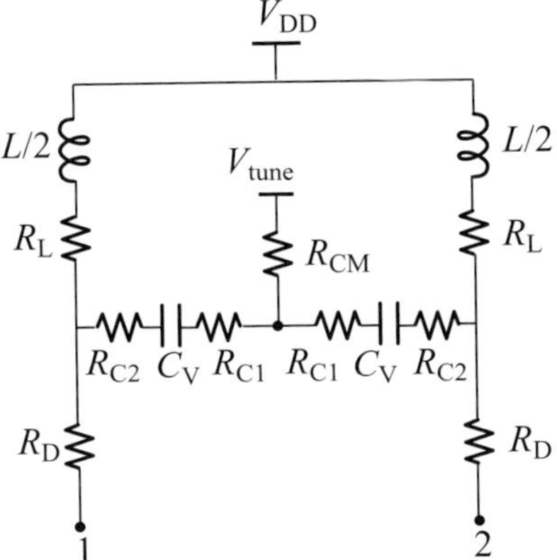

Figure 7.20 Losses at interconnections

of multi-turn inductors and a much lower parasitic capacitance thanks to the absence of inter-winding capacitances. Using the thick top metal layer maximizes the inductor quality factor. The deep oxide trenches underneath the inductor coil (see Figure 7.19) reduce the capacitance of the coil to the substrate, thus increasing its quality factor and self-resonance frequency. The simulated quality factor is about 23–28 in the 4–6 GHz range, where inductor losses are dominated by the metal series resistance.

The adoption of low L values increases the sensitivity of the quality factor to parasitic resistances, thus requiring a careful layout. Figure 7.20 highlights the LC tank parasitic resistances of interconnections. The drain resistances R_D and the tune resistance R_{CM} have negligible impact on the overall quality factor. The former are in series with the high-impedance output of the differential active transconductor and the latter is connected to a balanced node. However, the series resistances R_{C1}, R_{C2} and R_L are especially harmful, as they add directly to the intrinsic series loss resistance of the reactive components. To minimize these parasitic resistances, the tank capacitance has been laid out inside the inductor coil (see Figure 7.18). The electromagnetic coupling between the inductor and the inner metal areas has negligible impact on the tank quality factor, as confirmed by electromagnetic simulations.

7.6 References

[1] C. Samori, A. L. Lacaita, A. Zanchi, S. Levantino and F. Torrisi, Impact of indirect stability on phase noise performance of fully-integrated LC tuned VCOs, *Proc. European Solid-State Circuits Conference*, Duisburg, Sep. 1999, 202–5.

[2] E. Hegazi and A. A. Abidi, Varactor characteristics, oscillator tuning curves, and AM-FM conversion, *IEEE J. Solid-St. Circ.*, **38**, Jun. 2003, 1033–39.

[3] S. Levantino, C. Samori, A. Bonfanti *et al.* Frequency dependence on bias current in 5 GHz CMOS VCOs: impact on tuning range and flicker noise upconversion, *IEEE J. Solid-St. Circ.*, **37**, Aug. 2002, 1003–11.

[4] S. Levantino, C. Samori, A. Zanchi and A. L. Lacaita, AM-to-PM conversion in varactor-tuned oscillators, *IEEE T. Circuits-II*, **49**, Jul. 2002, 509–13.

[5] C. Samori, S. Levantino and V. Boccuzzi, A −94dBc/Hz@100kHz, fully-integrated, 5-GHz, CMOS VCO with 18% tuning range for bluetooth applications, *2001 IEEE Custom Integrated Circuits Conference*, San Diego, May 6–9, 2001.

[6] B. De Muer, M. Borremans, M. Steyaert and G. Li Puma, A 2-GHz low-phase-noise integrated LC-VCO set with flicker-noise upconversion minimization, *IEEE J. Solid-St. Circ.*, **35**, Jul. 2000, 1034–8.

[7] A. Jerng and C. G. Sodini, The impact of device type and sizing on phase noise mechanisms, *IEEE J. Solid-St. Circ.*, **40**, Feb. 2005, 360–9.

[8] K. Hoshino, E. Hegazi, J. J. Rael and A. A. Abidi, A 1.5V, 1.7mA 700 MHz CMOS LC oscillator with no upconverted flicker noise, *Proc. European Solid-State Circuits Conference*, Sep. 2001, 337–40.

[9] H.-I. Lee, J.-K. Cho, K.-S. Lee *et al.* A $\Sigma\Delta$ fractional-N frequency synthesizer using a wide-band integrated VCO and a fast AFC technique for GSM/GPRS/WCDMA applications, *IEEE J. Solid-St. Circ.*, **39**, Jul. 2004, 1164–9.

[10] A. Ravi, G. Banerjee, R. E. Bishop *et al.* 10 GHz, 20 mW, fast locking, adaptive gain PLLs with on-chip frequency calibration for agile frequency synthesis in a 0.18 μm digital CMOS process, *Digest of Technical Papers of Symposium on VLSI Circuits*, Jun. 2003, 181–4.

[11] M. A. Margarit, J. L. Tham, R. G. Meyer and M. J. Deen, A low-noise, low-power VCO with automatic amplitude control for wireless applications, *IEEE J. Solid-St. Circ.*, **34**, Jun. 1999, 761–71.

[12] A. Zanchi, C. Samori, S. Levantino and A. L. Lacaita, A 2-V 2.5-GHz −104-dBc/Hz at 100 kHz fully integrated VCO with wide-band low-noise automatic amplitude control loop, *IEEE J. Solid-St. Circ.*, **36**, Apr. 2001, 611–19.

[13] M. Tiebout, Low-power low-phase-noise differentially tuned quadrature VCO design in standard CMOS, *IEEE J. Solid-St. Circ.*, **36**, Jul. 2001, 1018–24.

[14] E. A. M. Klumperink, S. L. J. Gierkink, A. P. van der Wel and B. Nauta, Reducing MOSFET 1/*f* noise and power consumption by switched biasing, *IEEE J. Solid-St. Circ.*, **35**, Jul. 2000, 994–1001.

[15] S. L. J. Gierkink, R. C. Frye and V. Boccuzzi, Differentially bathtub-tuned CMOS VCO using inductively coupled varactors, *Proc. European Solid-State Circuits Conference*, Estoril, Sep. 2003, 501–4.

[16] A. Bonfanti, S. Levantino, C. Samori and A. L. Lacaita, A varactor configuration minimizing the amplitude-to-phase noise conversion in VCOs, *IEEE T. Circuits-I*, **53**, Mar. 2006, 481–8.

[17] C. Samori, A. L. Lacaita, A. Zanchi, S. Levantino and G. Calì, Phase noise degradation at high oscillation amplitudes in LC-tank VCO's, *IEEE J. Solid-St. Circ.*, **35**, Jan. 2000, 96–9.

[18] K. K. Clarke and D. T. Hess, *Communication Circuits: Analysis and Design*, Reading, MA: Addison-Wesley, 1971.

8 Frequency division

8.1 Introduction

PLLs designed for frequency synthesis employ a variable-frequency divider, which provides programmability and enables the use of reference crystal oscillators. The two main issues related to the design of the frequency divider are the high input frequency and the programmability of the division factor. As will be made clearer in this chapter, wide variations of the division factor and high input frequency are opposing requests. To some extent, higher operation frequency can be traded against power consumption. However, as we will show, smart architectural choices can allow for much higher frequency of operation with the same power consumption and technology process. The input sensitivity that is the minimum required amplitude of the input signal is also of great concern in high-speed dividers.

Digital dividers are based on either flip-flops or latches, using either static or dynamic logics. Analogue implementations of divide-by-two circuits relying on injection locking or parametric amplification also exist. Depending on their implementation, dividers have different noise generation mechanisms. However, a lower phase noise generally demands higher power consumption.

This chapter discusses typical techniques for realizing programmable dividers, with particular emphasis on such topologies as allow for maximum speed in a given semiconductor process. The circuit implementations of the building blocks will be reviewed. Finally, methods for predicting the noise of digital frequency dividers will be examined.

8.2 Digital frequency dividers

Digital frequency dividers by M are modulo-M counters. The most significant bit (MSB) of a modulo-M counter provides a signal whose frequency is M times lower than the input clock frequency.

8.2.1 Binary and ring counters

Counters can be categorized into binary and ring counters, depending on their logic implementation. Binary counters are based on memory elements. They provide a binary number,

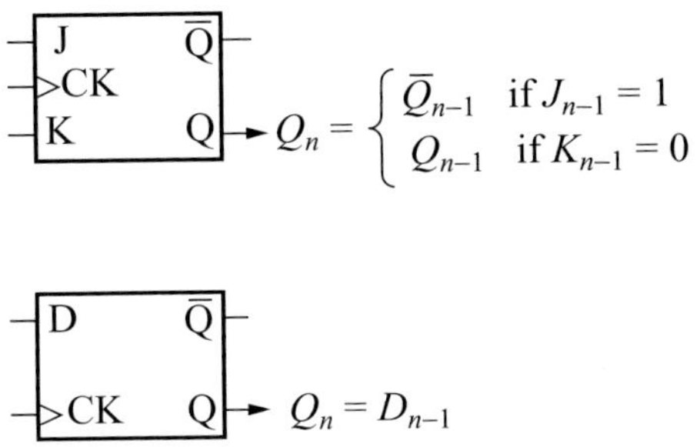

$$Q_n = \begin{cases} \bar{Q}_{n-1} & \text{if } J_{n-1} = 1 \\ Q_{n-1} & \text{if } K_{n-1} = 0 \end{cases}$$

$$Q_n = D_{n-1}$$

Figure 8.1 Building blocks of counters: edge-triggered J–K flip-flop and edge-triggered D flip-flop

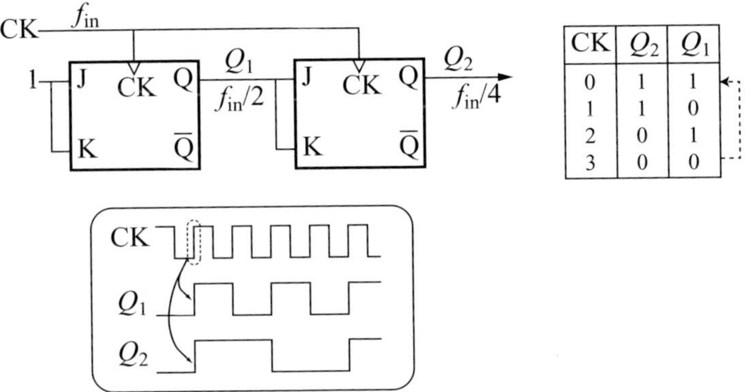

CK	Q_2	Q_1
0	1	1
1	1	0
2	0	1
3	0	0

Figure 8.2 Modulo-4 binary counter based on J–K flip-flops: schematic, state sequence, and output waveforms

which is incremented or decremented at each clock cycle. Their most typical building block is an edge-triggered J–K flip-flop,[1] whose function is shown in Figure 8.1. It is, essentially, a binary memory element. Ring counters are shift registers whose output is connected in feedback to their input. Their building block is the D flip-flop, also shown in Figure 8.1, which is a shift register. They provide individual digit outputs rather than a binary output.

An example of a modulo-4 counter based on J–K flip-flops is shown in Figure 8.2, along with the state sequence and the output waveforms.[2] The first flip-flop that has both inputs J and K connected to 1 toggles its state at each positive clock edge, i.e., it is a toggle flip-flop. Therefore, it divides the input frequency by 2. The second flip-flop toggles its state whenever the first flip-flop output is 1; therefore, its output frequency is equal to the input frequency divided by four ($f_{\text{in}}/4$). This counter is typically referred to as a binary counter, since its count Q_2Q_1 is an iterated binary countdown from 11 to 00. In general, a modulo-M binary counter requires $\log_2 M$ flip-flops.

[1] Alternatively, an S–R flip-flop can be used. However, the J–K flip-flop has the advantage of preventing a race condition occurring when both the S and R inputs are at logic 1 and the CK input changes from 0 to 1.

[2] In the time diagrams shown in Figures 8.2 and 8.5, the gate delays are neglected for the sake of clarity. In reality, in the counter shown in Figure 8.2, the presence of delays would shift the time diagram of Q_2 by one CK period.

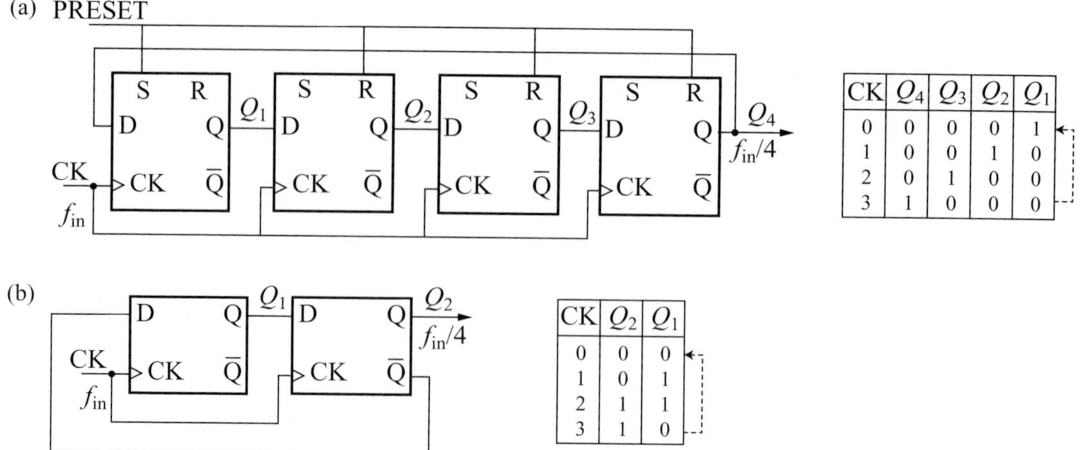

CK	Q_4	Q_3	Q_2	Q_1
0	0	0	0	1
1	0	0	1	0
2	0	1	0	0
3	1	0	0	0

CK	Q_2	Q_1
0	0	0
1	0	1
2	1	1
3	1	0

Figure 8.3 Modulo-4 ring counters: (a) straightring and (b) twisted-ring counter or Johnson counter

Another modulo-4 counter is shown in Figure 8.3(a). It is a four-bit shift register closed in feedback. At each positive clock edge, the 1 bit shifts to the following flip-flop, as shown in the counting sequence in the figure. Each one of the four outputs, for instance Q_4, provides a square wave at one quarter of the clock frequency with $1/4$ duty cycle. In the binary counter, the duty cycle of the output was, instead, $1/2$. Moreover, ring counters operate in a subset of the available number of states. The modulo-4 ring counter shown in Figure 8.3(a) uses four flip-flops and has 16 possible states, but only four states are used. For this reason, it must be forced initially into a valid state. Each D flip-flop has two additional inputs, S and R, which allow for setting and resetting it. The PRESET signal sets the initial count to $Q_4Q_3Q_2Q_1 = 0001$. If the D flip-flops have no set input (S), additional control logic can preset the divider correctly.

To reduce the number of flip-flops to $M/2$ and obtain an output signal with 50% duty cycle, the standard ring topology can be modified into the twisted ring topology shown in Figure 8.3(b). The inverted output of the last stage is fed back to the input of the first stage and the resulting state sequence is shown in the figure. This counter is often referred to as a Johnson counter. The resulting Q_1 is in quadrature with Q_2. In this case, flip-flop presetting can be avoided, since all the states are valid. This helps in simplifying the flip-flop circuit and increasing the maximum operating speed.

While the implementation of straight ring counters is straightforward for any integer modulo M, Johnson counters require additional control logic to override unwanted states. For instance, the Johnson counter in Figure 8.3(b) can be modified to achieve a frequency division by three, as shown in Figure 8.4. The additional AND gate makes the counter override the state $Q_2Q_1 = 10$. Both outputs Q_1 and Q_2 provide square waves running at $f_{in}/3$ with duty cycles equal to $2/3$ and $1/3$, respectively. The relative phase delay is equal to 120 degrees. Again, no presetting of the flip-flops has been used. Even if the counter starts in the invalid state $Q_2Q_1 = 10$, the following state is valid and the counter never comes back to the invalid state. The presence of an additional gate in the critical path makes divide-by-three circuits slower than their divide-by-two counterparts. In practice, they operate at roughly half the maximum frequency of divide-by-two circuits. [1]

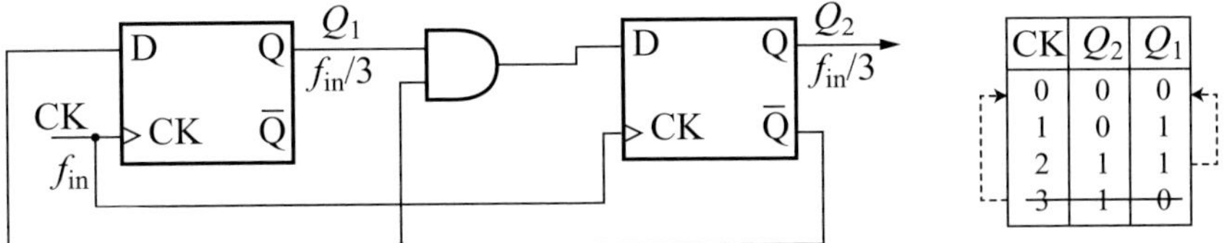

Figure 8.4 Modulo-3 Johnson counter

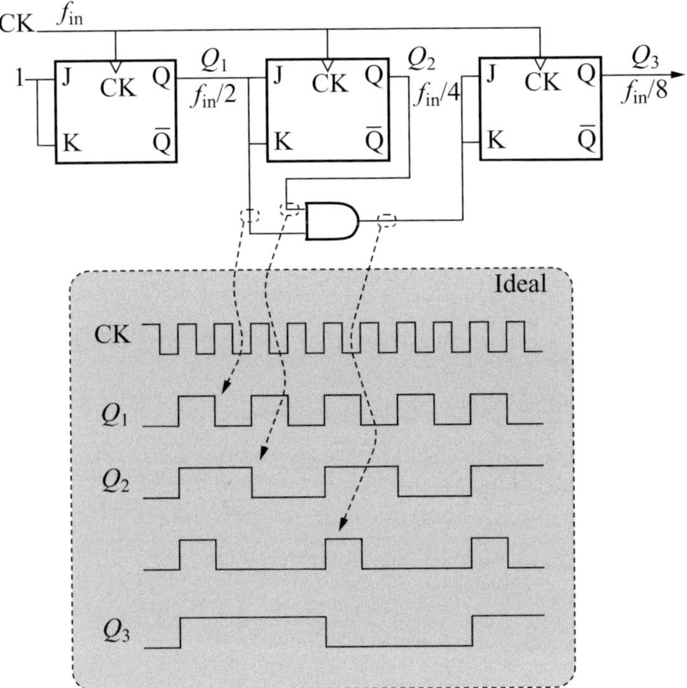

Figure 8.5 Modulo-8 binary counter

Binary counters require additional controlling gates to determine when each flip-flop is allowed to change state and when it is not. This is necessary even if the modulus is a power of two. Referring, for instance, to the modulo-8 counter shown in Figure 8.5, an additional AND gate is required to produce the correct J–K signal for the third flip-flop. In general, the required control logic can be obtained by looking at the state sequence of each flip-flop. In long dividers, the enable logic becomes intricate. Full-parallel or serial enable implementations exist, which may optimize either the speed or the capacitive load of each flip-flop. Typically, combinations of the two approaches are employed. [2]

8.2.2 Flip-flop-operated and latch-operated counters

Counters, like any sequential circuit, are generally built with edge-triggered flip-flops. Even if other possible implementations exist, edge-triggered flip-flops are commonly built as two level-triggered latches in master–slave configurations (Figure 8.6). A level-triggered latch

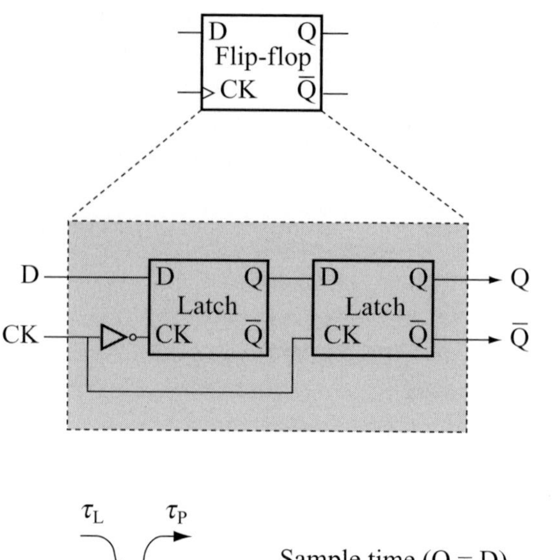

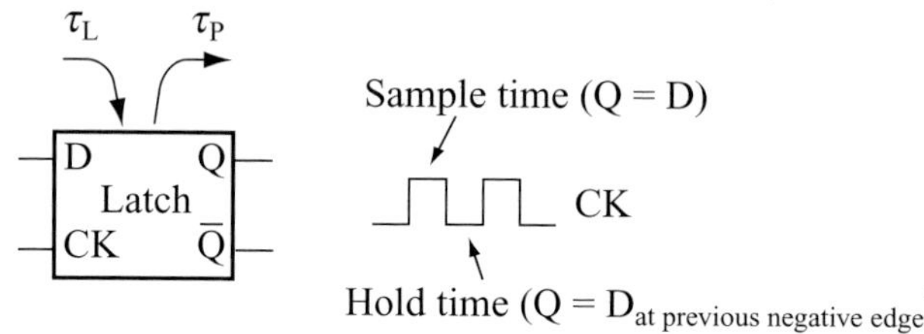

Figure 8.6 Master–slave implementation of positive-edge-triggered flip-flop and latch function

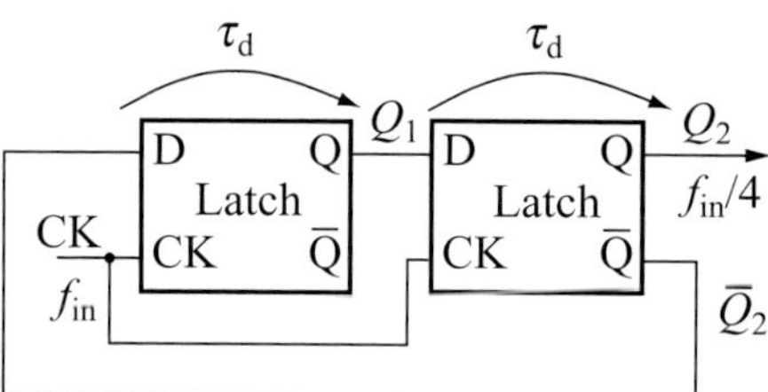

Figure 8.7 Latch-operated modulo-4 Johnson counter

is, essentially, a sampler circuit, as shown in the same figure. When the input clock CK is high, the output Q follows the data input D, that is, the latch is in transparent mode. When CK goes low, Q is held and D does not affect it any more. In the last case, the latch is in opaque mode. Then, the master–slave cell in Figure 8.6 is a positive-edge-triggered flip-flop.

Alternatively, a sequential circuit can be based on level-triggered latches rather than edge-triggered flip-flops. This approach results in higher speed and lower power consumption. The drawback is the limited range of the input frequencies. Figure 8.7 shows the modulo-4 Johnson counter, already shown in Figure 8.3(b), employing *level*-triggered latches instead of *edge*-triggered ones.

The propagation of data through a latch takes a delay time $\tau_d = \tau_L + \tau_P$, where τ_L and τ_P are the latching and propagation periods, respectively. The latching period (or hold time) is the time required by a latch to accept D. It starts either when CK goes high or when CK is already high and D changes; it ends when CK can go low without harming the new latched data. The D input data must be prevented from changing during the latching period. The propagation time is the additional time required by the latch to propagate the new data to the output. For simplicity, we are assuming that these delays are the same for transitions from 1 to 0 and from 0 to 1.

To guarantee proper operation, the latch must be transparent for a time shorter than the delay time τ_d. Otherwise, the following latch would start to latch the new data that has just passed through the cell. Therefore, if the clock is assumed to have a 50% duty cycle, the input frequency f_{in} must be higher than $(2\tau_d)^{-1}$. However, the clock period must be longer than τ_d, so that the data can reach the following latch before the next rising edge of CK. Thus, f_{in} must be higher than τ_d^{-1}. In practice, this Johnson counter divides the input frequency f_{in} by four if:

$$\frac{1}{2\tau_d} < f_{in} < \frac{1}{\tau_d}.$$

As expected, counters based on level-triggered latches can reach higher frequencies of operation, since latches have lower delay times than flip-flops; but their allowed input frequency has an inferior bound.

This circuit also has its own frequency of oscillation in the absence of an input clock. This frequency is $(4\tau_d)^{-1}$, as it acts like a two-stage ring oscillator with a sign inversion within the ring. For this reason, such frequency dividers can also be regarded as injection-locked oscillators. As known from the theory of injection locking by Adler, [3] an oscillator can be locked to the frequency and the phase of the injected signal, if this signal frequency falls inside the so-called locking range. This concept leads, again, to the existence of a minimum frequency of operation.

An example of a frequency divider designed with level-triggered latches can be found in [4].

8.2.3 Synchronous and asynchronous counters

The preceding examples have shown two different types of counters, both belonging to the category of synchronous counters. All memory elements are simultaneously triggered by the same clock. However, in high-speed frequency synthesizers, synchronous counters can consume large amounts of power and apply a significant capacitive load to the oscillator output. Another issue in synchronous dividers is related to race problems. Since all stages can change their state almost simultaneously, it may happen that the stages going to 0 are slower than the stages going to 1 and all stages are momentarily at 1. This condition may cause a glitch in the output waveform. Referring to Figure 8.5, if Q_2 is slower to go to 0 than Q_1 is to go to 1, the output of the AND gate experiences glitches. [2]

For these reasons, frequency dividers are often based on asynchronous counters. A toggle flip-flop, or a J–K flip-flop with J = K = 1, is a modulo-2 divider. Modulo-4 division can be performed by connecting the output of this flip-flop to the CK input of another flip-flop. Figure 8.8(a) shows a modulo-8 asynchronous counter based on toggle flip-flops. If the

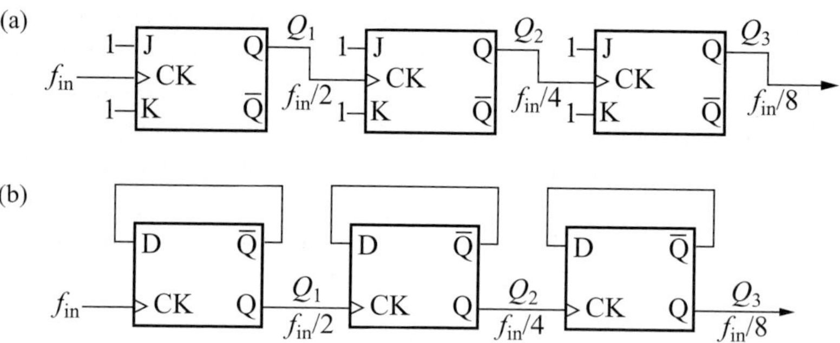

Figure 8.8 Asynchronous counters: (a) cascade of toggle flip-flops and (b) cascade of modulo-2 Johnson counters

flip-flops are triggered on the positive edge of the clock CK, the state following 000 is 111. In fact, when the first flip-flop toggles from logic 0 to 1, all the following flip-flops toggle. Therefore, this arrangement operates backward counting. Forward counting can be obtained either by connecting the $\overline{Q}$ output of each flip-flop to the following CK input or by using negative-edge-triggered flip-flops. The power consumption of asynchronous counters can be reduced with respect to synchronous counters, since each stage operates at half the frequency of the previous stage.

A D-type flip-flop with its D input fed from its own inverted output also makes a modulo-2 counter. Cascading another identical stage increases the division factor by a factor of two. Figure 8.8(b) shows a modulo-8 asynchronous counter based on modulo-2 Johnson counters. The output overflow of each stage ripples from stage to stage. For this reason, these counters are also called ripple counters. The consequence of the overflow ripple is instability of the output count. Referring to the circuit in Figure 8.8(b), the transition from $Q_3Q_2Q_1 = 000$ to 111 is not smooth. On the clock edge, the output Q_1 goes first to 1, changing the apparent count to 001. This triggers Q_2 to rise, changing the apparent count to 011. Finally, this triggers Q_3 to rise and leads the count to 111. This is not a problem when the counter is used as a frequency divider, since the instantaneous count is not important. However, as we will show, the delay caused by the overflow ripple limits the operating frequency of programmable dividers.

In practice, frequency dividers based on asynchronous counters consume less power than synchronous implementations, but add delay between the input clock and the output at the divided frequency.

8.3 Programmable dividers

The various implementations of counters so far presented allow frequency division by a fixed integer. However, phase-locked loops designed for frequency synthesis normally require variable frequency dividers. The modulus of the divider has to be varied over a certain range of integer numbers.

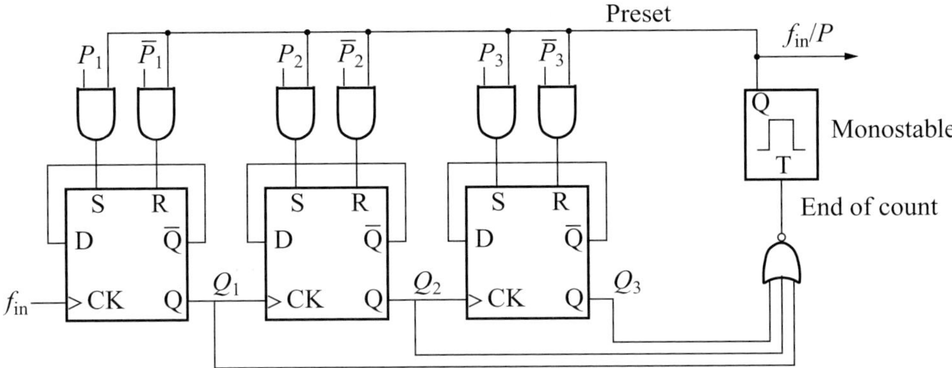

Figure 8.9 Presettable modulo-P asynchronous counter

A programmable divider could be realized in principle using a synchronous finite-state machine. The binary counters presented in the previous section can be modified by adding a more intricate combinational logic circuit, which generates the next state from the present state and provides programmability of the division factor. These other stages increase the capacitive load of each flip-flop, raising the total power dissipation. Thus, this implementation is not practical for high-speed frequency dividers.

On the other hand, the speed of a fully asynchronous divider with programmable modulus is limited by the delay of many flip-flops and intricate logic circuitry. Different approaches exist to allow programmable division for fast signals and they will be presented later. They are typically based on a combination of synchronous and asynchronous dividers. For this reason, the design of presettable asynchronous dividers will be reviewed first.

8.3.1 Presettable asynchronous dividers

The simplest way of designing a counter with a programmable modulus is by modifying the asynchronous Johnson counter shown in Figure 8.8(b). That modulo-8 counter counts backward from $Q_3Q_2Q_1 = 111$ to 000. The idea is to preset that counter to a different initial state P and to detect the final state F by means of an 'end-of-count' logic. Doing so, the counter counts between P and F. This concept is implemented in the schematic in Figure 8.9. The counter starts from the state defined by the preset number $P_3P_2P_1$ and it overflows when it reaches the state 000. In practice, the modulus is equal to P and it can be varied by varying P. A more complicated end-of-count logic can also make the final state variable with an input word F.

The main limitation of a presettable asynchronous divider is the low maximum operating frequency. Proper operation is guaranteed if the end-of-count signal presets the counter before the next clock edge arrives. In the example shown in Figure 8.9, the last state transition is between 001 and 000. This means that after the clock edge, the first flip-flop has to toggle. Therefore, the sum of the delays of one D flip-flop, the NOR gate, the T flip-flop, the AND gate and the set–reset time of the D flip-flop itself has to be less than one clock period. This condition sets a maximum operating clock frequency.

In this respect, choosing 111 as the end-of-count state would have been extremely disadvantageous. In that case, the last state transition is between 000 and 111, which requires the overflow to ripple over the whole flip-flop chain. The maximum allowed clock frequency would have been much lower than in the previous case.

8.3.2 Pulse swallowing

High-frequency operation is attained when the logic function is kept simple. As has been shown previously, the simplest dividers divide by fixed numbers. Therefore, the programmable divider could have a fixed-modulus high-speed divider as its first stage. In this context, the fixed-modulus divider is commonly referred to as a prescaler. Scaling down the signal frequency makes it possible to have a presettable asynchronous divider after the prescaler. This approach limits the clock loading and the overall divider power consumption, while allowing for high-speed operation. The obvious drawback, however, is the necessity of lowering the reference frequency to get the same frequency resolution. If a presettable modulo-P divider follows a modulo-M prescaler, the overall frequency division ratio is MP. Thus, the reference frequency has to be lowered exactly by the modulus of the prescaler M. This would imply narrowing the loop bandwidth, which may be undesirable.

A typical solution is the pulse-swallowing technique. [1, 2] If S input pulses are swallowed in the preceding arrangement, the output period becomes longer by S reference periods. Therefore, the overall frequency division ratio is $N = (MP + S)$, which can be varied in unity steps by changing S. A pulse-swallower circuit is shown in Figure 8.10. The block inside the dashed box generates the output signal, which is high for S edges of the input clock A and low for $(P - S)$ edges of the same input, assuming $P > S$. This output signal acts as a 'swallow enable'. The first stage is a dual-modulus prescaler, i.e., a high-speed counter whose modulus can be alternated between two numbers. The modulus control is set either to $(M + 1)$ or to M, depending on the output signal. Hence, the output signal has a period equal to:

$$S(M + 1)T_{\text{in}} + (P - S)MT_{\text{in}} = (MP + S)T_{\text{in}}$$

where T_{in} is the input clock period. The modulo-P counter is commonly referred as a program counter, while the modulo-S counter is called the swallow counter.

The inequality $P > S$ and the required division ratios identify a set of possible values of M, P and S. Ease-of-design and power-dissipation considerations allow identification of the most suitable triplet. A design example of the pulse-swallower circuit is proposed in Example 8.1.

Practical implementation requires synchronization, to compensate for the delay introduced by the asynchronous presettable counters. In particular, the modulus control (MC) should be synchronized to the dual-modulus prescaler output A. Moreover, prescalers and presettable counters are very often implemented in different fashions or logic families, owing to their different speed specification. Therefore, a level shifter is required after the prescaler, to compensate for the different voltage rails. Additional details can be found, for instance, in [5].

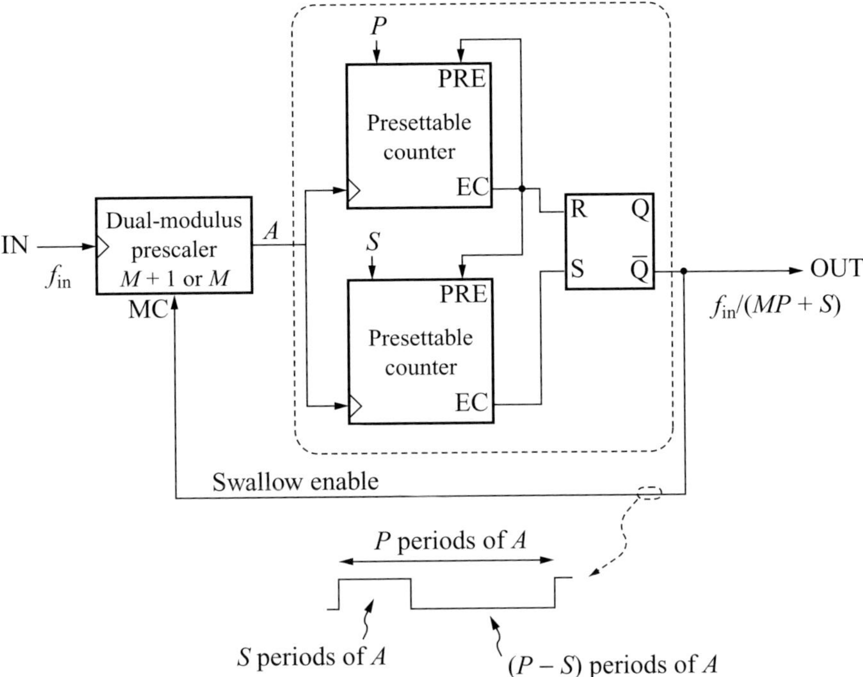

Figure 8.10 Programmable divider based on pulse swallowing. Each presettable divider can be implemented as shown in Figure 8.9. PRE represents the preset input, while EC is the end-of-count output

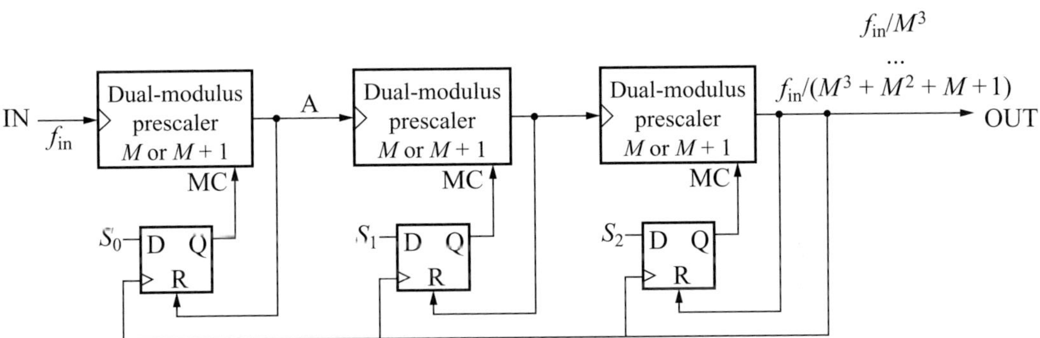

Figure 8.11 Programmable divider based on series of dual-modulus prescalers

8.3.3 Series of dual-modulus prescalers

Another architecture of programmable divider commonly employed in frequency synthesizers is shown in Figure 8.11. [2, 6] This topology is also based on pulse swallowing. If the input word $S_2S_1S_0$ is equal to 000, all the modulus control inputs MC are low and each divider overflows after M input pulses. Therefore, the overall division ratio is M^3. If only S_0 is 1, then the MC of the first prescaler is 1 and this prescaler divides by $M + 1$ for only one cycle of A. On the following edge of A, the D flip-flop is reset and the MC returns to logic 0. Thus, one input clock pulse has been swallowed and the overall division ratio becomes $M^3 + 1$. If only S_1 is 1, one pulse of A is swallowed. This corresponds to M pulses of the input clock. Therefore, the overall division is $M^3 + M$. Therefore, the division factor depends on the input word as follows: $N = M^3 + S_2M^2 + S_1M + S_0$. In general, a string of K dual-modulus prescalers provides a division by $M^K + S_{K-1}M^{K-1} + \cdots + S_1M + S_0$, which can be programmed by the input word $S_{K-1} \ldots S_1S_0$.

Compared with the pulse swallower, which requires three counters with different characteristics, this architecture is modular. It consists of a repetition of a dual-modulus prescaler. Most typically, modulo-2 or modulo-3 prescalers are used in the string. Therefore, if it is $M = 2$, the overall modulus can be varied between $N = 2^K$ and $2^K + 2^{K-1} + \cdots + 2 + 1 = 2^{K+1} - 1$. This is almost a variation of a factor of two in the division factor N, or a relative variation of N over its mean value equal to 66%. This is also the maximum output frequency range of the synthesizer over the mean frequency. When this range is insufficient, the modulus of the single prescalers can be increased or the string of prescalers can be used in combination with a set–reset counter. In this second option, the resulting architecture is, however, no longer modular.

An evolution of this architecture can be found in [7] and [8] based on the back propagation of the swallow-enable signal; [8] also provides an alternative method for extending the division range.

Example 8.1 Pulse-swallower divider

A frequency synthesizer suitable for a zero-IF receiver in the 2400–2480 MHz ISM band with 1 MHz channel spacing and employing an integer divider requires a reference of 1 MHz and a loop frequency division factor that is variable between 2400 and 2480. In the design of the pulse swallower in Figure 8.10, the first step is the choice of the moduli M, P and S of the dual-modulus prescaler, program and swallow counter, respectively. The total division factor $N = MP + S$ has to vary between 2400 and 2480. Some degrees of freedom exist, but also some constraints have to be considered.

To simplify the channel-select logic, we assume that only S can vary. Another simplification in the design of the dual-modulus prescaler comes if either M or $M + 1$ is a power of two. For proper operation of the pulse-swallower circuit, the modulus P has to be greater than S. One possible strategy is to choose S to be as low as possible so that the modulus of the presettable divider $P > S$ is a minimum. If S was varied between 1 and 81 to cover the 81 possible division ratios, then P should be higher than 81, that is

$$M < \frac{N - S}{P} = \frac{2399}{81} = 29.6. \tag{8.1}$$

Choosing $M = 16$, the required modulus for the program counter is:

$$P = \frac{N - S}{M} = \frac{2399}{16} = 149.$$

Therefore, N can be varied between 2400 and 2480 if S is changed between 16 and 96. The maximum value of S is lower than the chosen value of P. Therefore the initial assumption is satisfied. The chosen moduli are summarised in the first two rows of Table 8.1. In the

Table 8.1 *Design values for pulse swallower in Example 8.1*

M	P	S	N	$P > S$
16	149	16	2400	TRUE
16	149	96	2480	TRUE
32	75	0	2400	TRUE
32	75	80	2480	**FALSE**
32	76	48	2480	TRUE
22	109	2	2400	TRUE
22	109	82	2480	TRUE
9	256	96	2400	TRUE
9	256	176	2480	TRUE

same table, other possible design values are shown. Note that for $M = 32$, the inequality $P > S$ cannot be satisfied over the whole divider ratio range as predicted by (8.1). Such a limitation can be overcome by increasing P by one at the far end of the ratio range.

If the constraint of choosing M as a power of two is removed, values of M higher than 16 can be used with constant P over the whole range. An example with $M = 22$ is shown in the table. Another strategy in the choice of M could be to make the modulus of the program counter equal to a power of two, to simplify its implementation. If the program counter modulus is set to $P = 256$, then the choice of $M = 9$ covers all the desired division ratios (see the last two rows of Table 8.1).

8.4 Dual-modulus prescalers

8.4.1 Synchronous prescalers

Dual-modulus prescalers are typically implemented as synchronous Johnson counters. The synchronous approach and the use of D-type flip-flops over the more complicated J–K type allow for a higher frequency of operation. By modifying the modulo-3 counter in Figure 8.4, it is possible to derive the modulo-2/3 prescaler. This circuit is shown in Figure 8.12(a). The critical signal path now includes two controlling gates, making the circuit slower than a modulo-2 prescaler.

Using prescalers with higher moduli would relax the speed performance and the power consumption of the following presettable dividers in the pulse-swallower topology. A possible implementation of a modulo-4/5 prescaler is shown in Figure 8.12(b). If the input MC is 0, then Q_3 is always high and the subsequent AND gate operates as an inverter on Q_2. The resulting circuit is the modulo-4 Johnson counter shown in Figure 8.3(b). If MC is 1, the counter modulus becomes 5, as illustrated by the state sequence in Figure 8.12(b). Invalid states are always followed by valid states. The power consumption of this modulo-4/5 divider is higher than that of the modulo-2/3 divider, since it includes one more flip-flop running at high frequency. Equivalent implementations of prescalers can be obtained by applying De Morgan's laws. For instance, the NAND gates can be replaced

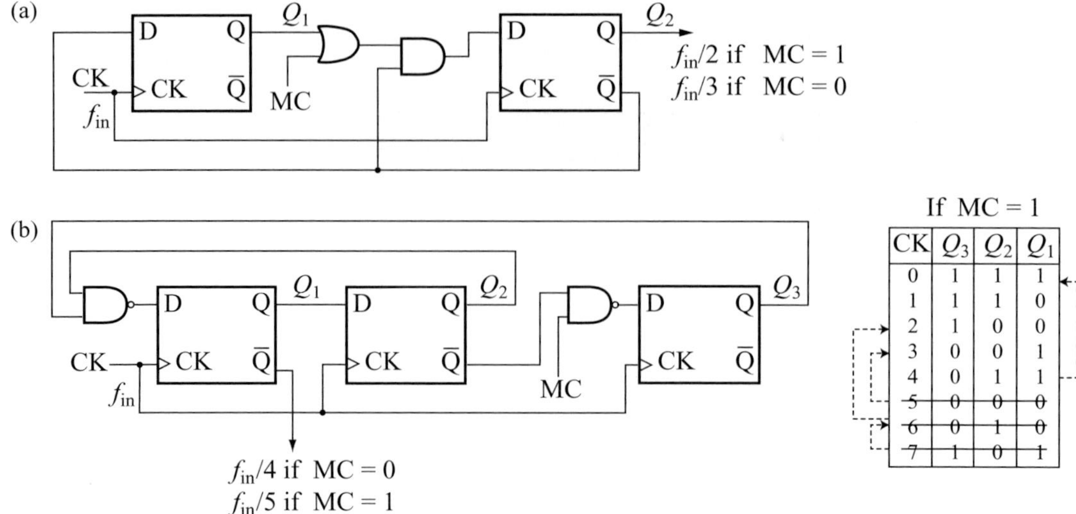

CK	Q_3	Q_2	Q_1
0	1	1	1
1	1	1	0
2	1	0	0
3	0	0	1
4	0	1	1
~~5~~	~~0~~	~~0~~	~~0~~
~~6~~	~~0~~	~~1~~	~~0~~
~~7~~	~~1~~	~~0~~	~~1~~

Figure 8.12 Synchronous dual-modulus prescalers: (a) modulo-2/3, (b) modulo-4/5

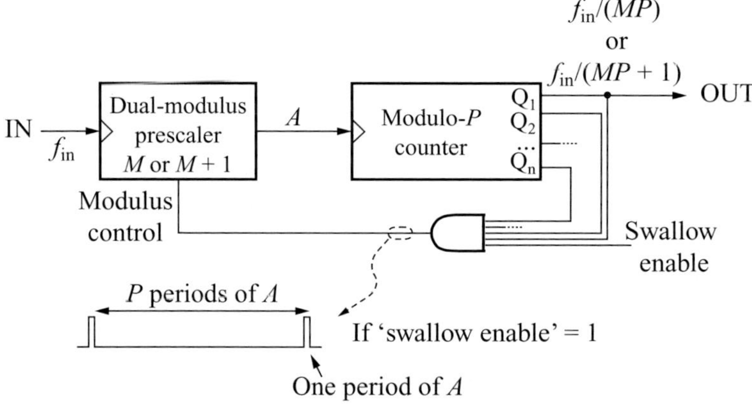

Figure 8.13 Mixed synchronous and asynchronous implementation of dual-modulus prescaler

with OR gates, since $\overline{A \otimes B} = \overline{A} \oplus \overline{B}$. This may have some advantages in reducing the number of transistors connected in series and improving the divider speed. [5]

8.4.2 Synchronous and asynchronous implementation

A method of increasing the modulus of the synchronous dual-modulus prescaler without raising the power consumption too much is based on the concept of pulse swallowing, as shown in Figure 8.13. The $M/(M+1)$ prescaler is typically a 2/3, 3/4 or 4/5 divider, while the modulo-P divider is an asynchronous Johnson counter, as shown in Figure 8.3(b). If the 'swallow enable' is high, an input pulse is swallowed and the frequency division

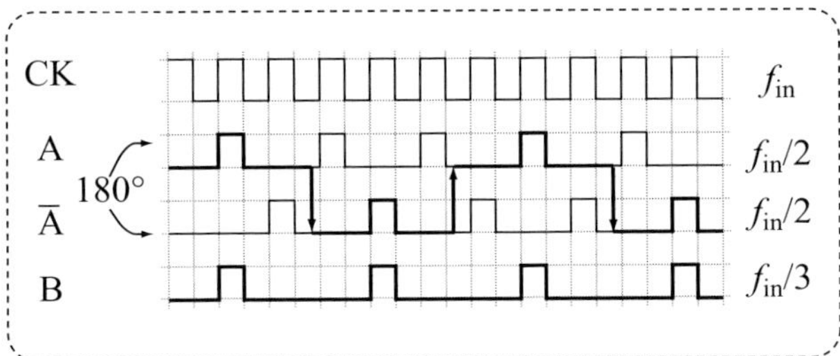

Figure 8.14 Concept of phase switching for modulo-3 division

ratio becomes $(MP + 1)$. Therefore, this arrangement attains $MP/(MP + 1)$ dual-modulus division by relying on an $M/(M + 1)$ prescaler.

Combining synchronous and asynchronous sections introduces potential race conditions. For instance, if the asynchronous section employs negative-edge-triggered flip-flops, little-time is left for the delay of the flip-flops and the controlling gate to feed back the modulus change to the dual-modulus synchronous divider. A better implementation uses positive-edge-triggered flip-flops in the asynchronous divider. [1] A practical example of a race condition and a method to avoid it are shown in Example 8.2.

8.4.3 Phase-switching prescalers

Other implementations of the dual-modulus prescaler have been investigated, to improve its maximum operating frequency in a given semiconductor process.

The key problem of prescalers employing the Johnson topology is the presence of control logic gates, allowing for division ratios that are not powers of two. As previously shown, frequency division by three requires an additional gate. The result is that dividers with other division factors achieve lower speeds than divide-by-two circuits.

Division by three could, in principle, be performed by combining two out-of-phase signals A and $\overline{\text{A}}$ at $f_{in}/2$, as shown in Figure 8.14. The signal B is given by A during the three CK pulses and by $\overline{\text{A}}$ during the following three CK pulses, and has frequency $f_{in}/3$. Since the rate of switching is also $f_{in}/3$, the signal controlling the phase switching can be derived by feeding back the signal B. The operating frequency can be increased with respect to a Johnson divide-by-three circuit, since the only element running at the input clock frequency is a divide-by-two circuit.

To use 50%-duty-cycle signals and to lower the operating frequency of the phase select block, common implementations of the phase-switching technique are modulo-4/5 dividers, as shown in Figure 8.15(a) and (b). [9] The second modulo-2 Johnson counter uses a master–slave flip-flop, which provides quadrature phases at the outputs of the two latches. Alternatively, a modulo-4 counter can be used. When MC is 1, a signal of period $5T_{in}$ is obtained by combining two signals with period $4T_{in}$ and a phase delay of 90 degrees, that is, a time delay of T_{in}. The circuit acts as a modulo-5 divider. The control signals CTRL driving the multiplexer MUX are obtained by the output signal, after a frequency division by four. When MC is 0, the OUT is always connected to the input of the MUX and the division

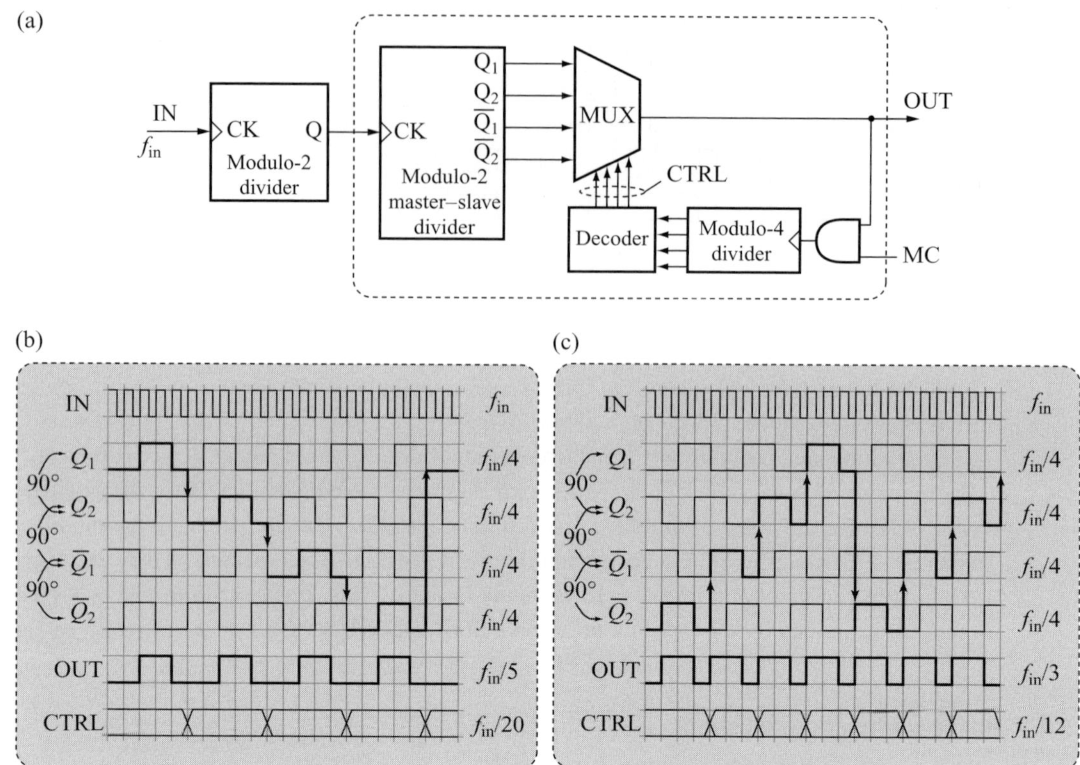

Figure 8.15 Phase-switching divider: (a) block schematic, (b) forward phase-switching scheme for modulo-4/5 division, (c) backward phase-switching scheme for modulo-3/4 division

ratio is 4. In practice, the block inside the dashed box in Figure 8.15(a) is a modulo-2/2.5 divider.

Compared with the modulo-4/5 synchronous Johnson counter shown in Figure 8.12(b), where three flip-flops were running at the maximum frequency, this phase-switching prescaler requires four modulo-2 dividers and a multiplexer acting as phase selector. However, only one of the modulo-2 dividers runs at the full clock speed.

The main limitation of phase-switching prescalers is the chance of generating glitches. Since the delays of the logic gates can vary considerably over process and temperature, it may happen that the output is switched from Q_1 to Q_2 before Q_2 has gone to 0 (see Figure 8.15(b)). As a result, a spike appears in the OUT signal.

Various solutions exist to get robust glitch-free phase switching. [9–12] One interesting method proposed in [13] uses the backward phase-switching scheme illustrated in Figure 8.15(c). Instead of switching from one signal to another, whose phase lags the previous one, the subsequent signal leads the previous one. In this way, the achieved output frequency is $f_{in}/3$. The interesting result is that if the switching instant is anticipated or delayed, no spike takes place. This concept can be extended to higher moduli by increasing the number of switched phases. For instance, a modulo-7.5/8 divider, after the modulo-2 division, requires

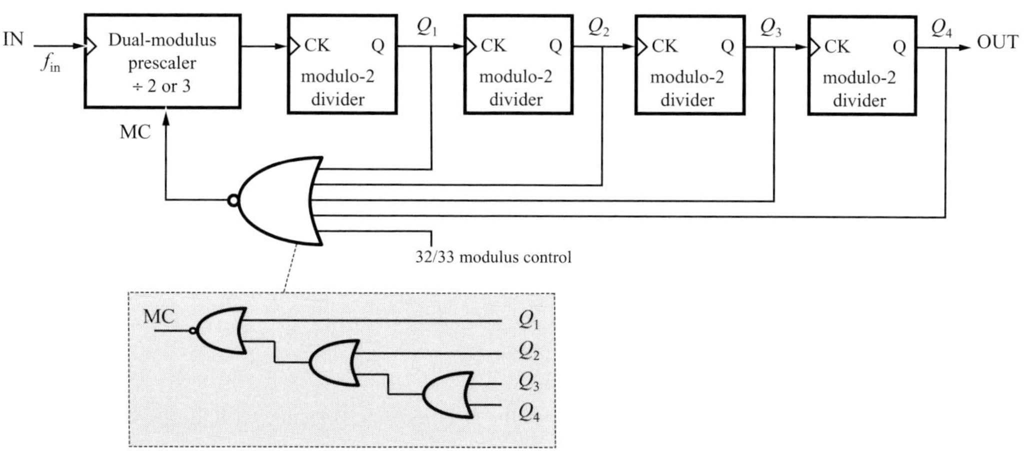

Figure 8.16 Functional schematic of modulo-32/33 divider of Example 8.2

a modulo-8 division, which provides eight phases separated by 45 degrees. [13, 14] Good matching between phases is required at the multiplexer; suitable circuit implementations can be found for instance in [15], [14] and [9].

The dual-modulus prescaler based on phase switching can be employed either in the pulse-swallower topology or in the series of dual-modulus prescalers to obtain a programmable divider.

Example 8.2 Design of a 32/33 dual-modulus prescaler

A block schematic of a modulo-32/33 prescaler is shown in Figure 8.16. The speed limitation of this prescaler is set by the feedback of the modulus control (MC). When the circuit divides by 32, the feedback control is masked. The most critical condition occurs when the prescaler is supposed to divide by 33. The division by 33 is obtained using an input 2/3-divider, which is forced to divide by 3 once every 32 input transitions by the modulus control. If the delay between the output of the 2/3-divider and the modulus control is higher than one input clock period, the circuit cannot perform the correct division.

To minimize this delay and to increase the maximum operating frequency, the modulus control MC is provided by a NOR gate and not by an AND gate. Since this prescaler performs backward counting, the AND gate would raise the modulus control when its inputs switch from $Q_4Q_3Q_2Q_1 = 0000$ to 1111. The modulus control delay would be the sum of four 2-divider delays plus the AND delay. Choosing instead a NOR gate, the modulus control is high when its inputs switch from $Q_4Q_3Q_2Q_1 = 0001$ to 0000. Thus, the modulus control delay is only a result of a single divider delay plus the NOR delay. The same result can be obtained by changing from a backward to a forward counting sequence. This can be achieved by using negative-edge-triggered instead of positive-edge-triggered flip-flops or by connecting $\overline{Q}$ instead of Q to the next stage in the asynchronous chain.

Circuit simulations show that this architectural choice improves the maximum operating frequency from 1 GHz to 3 GHz, employing the current-steering logic in a 0.35 μm

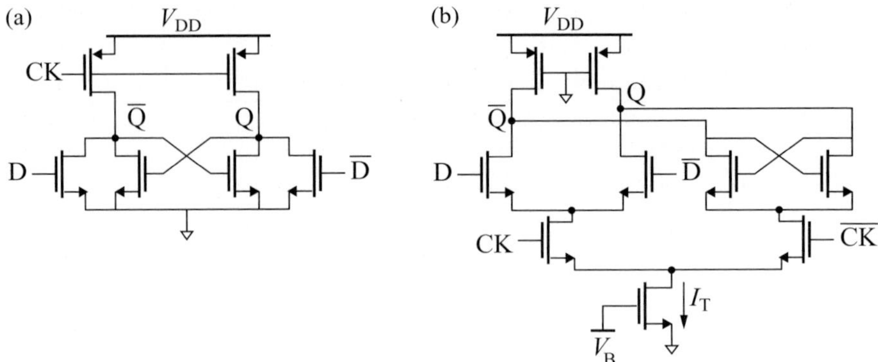

Figure 8.17 Some common implementations of the D-type latch: (a) static CMOS, (b) static CML latches

CMOS process. Following the same principle, the implementation of the NOR gate can be optimized. Realizing the 4-input gate as shown in the inset in Figure 8.16, the delay between Q_1 and the modulus control (which represents the critical path in the divider) is minimized.

8.5 Circuit implementation

Frequency dividers, even if they are based on digital counters, cannot be considered to be digital circuits. Their high frequency of operation requires accurate design, optimization and verification. Therefore, they cannot be synthesized automatically from a high-level register-transfer description relying on standard cells, as is the case for other state machines.

When the divider has to operate at high frequency with respect to the transistor cut-off frequency, circuit topologies have to be optimized for speed. Robust operation has to be guaranteed over process, voltage supply and temperature variations.

8.5.1 CMOS and current-mode logic

The main building block of the above-described counters is the D-type level-triggered latch. There are many possible implementations of this cell. A typical CMOS latch is shown in Figure 8.17(a). The CMOS latch has a rail-to-rail output swing, but also requires full-swing input signals. The CK input swing has to be wide enough to turn on the pMOS transistors, i.e., at least the transistor threshold plus the over-drive voltage $V_T + V_{OV}$. Since the CK input signal has a finite slope, this corresponds to a certain delay time before the cell would be able to sense the CK transition.

The most typical logic in high-frequency dividers is emitter-coupled or source-coupled logic (ECL or SCL), as well as current-steering or current-mode logic (CML). An example of a CML latch is shown in Figure 8.17(b). This logic is based on the use of differential stages. The current of a tail-current generator is switched between two branches by the CK input signal. A regenerative pair holds the data when the CK signal is low. In this example, the loads are pMOS transistors in the triode region. However, diode-connected transistors are also employed as loads. The amplitude of the differential output is set by the bias current I_T and by the load resistance R, and is $I_T R$. Since the Q outputs can be connected to either the D or the CK inputs of the following stage, the output swing has to be able to switch completely both the differential stages connected to the D and CK inputs. A differential stage with FET with a square law I–V switches completely when the differential input is higher than V_{ov} (where V_{ov} is the over-drive voltage of the FET biased at the whole I_T current). Therefore, with the swing equal to $I_T R$, the transistor pairs of the latch are switched completely if $V_{ov} \leq I_T R$. This result holds true so long as the conducting FET of the pair remains in the saturation region. In the case of the D-input pair, the gate–drain voltage of the conducting FET is $I_T R$ and must be lower than the threshold voltage V_T.

The common arguments justifying the high-speed capability of CML over CMOS are its reduced voltage swing and its current-steering operation. Since the swing needed to switch the CML stage is the V_{ov} against the $V_{ov} + V_T$ of the CMOS stages, less delay occurs before input sensing.

Other advantages of the CML over the CMOS logic are related to the use of differential circuits, which makes the current drained by the supply less variable. This is advisable to lower the switching noise applied at the power supply, which may be responsible for spurs and noise in the synthesizer output spectrum. Moreover, differential circuits are immune to coupled disturbances; they reject disturbances coming from the substrate and power supply due to other blocks.

In the implementation of the modulo-3 divider shown in Figure 8.4, an AND gate is required between the Q output of the first flip-flop and the D input of the second one. As has already been discussed, the delay time introduced by this additional gate reduces the maximum clock frequency of the modulo-3 prescaler with respect to a simpler modulo-2 divider. Merging this AND gate into the first latch of the second flip-flop reduces this delay time. Figure 8.18 shows a cell that merges the AND and the LATCH functions. The additional advantage of this arrangement is the re-use of the latch current I_T in the AND function. In practice, the function merging saves power and increases the operating speed.

In current-steering topologies, the tail-current generator can be eliminated. Figure 8.19 shows a pseudodifferential latch. The transistors connected to CK and $\overline{\text{CK}}$ in the original current-steering latch are split into two transistors respectively. The main advantage of this modification is the lower voltage headroom required for the cell. However, the drain currents of the clocked transistors are not well controlled, resulting in less predictable output swings.

8.5.2 Static and dynamic latches

Current steering topologies are fast, but also very power hungry. This is not only because of the static bias current of the differential stages, but also because of the fact that the typical

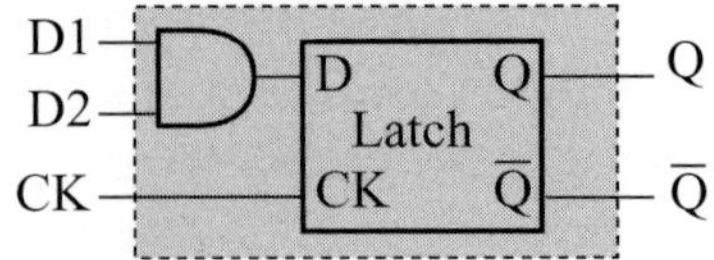

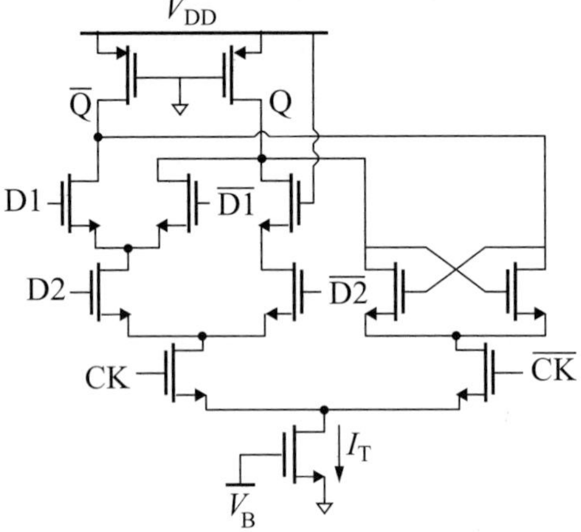

Figure 8.18 Merged AND gate and D-type latch in CML logic

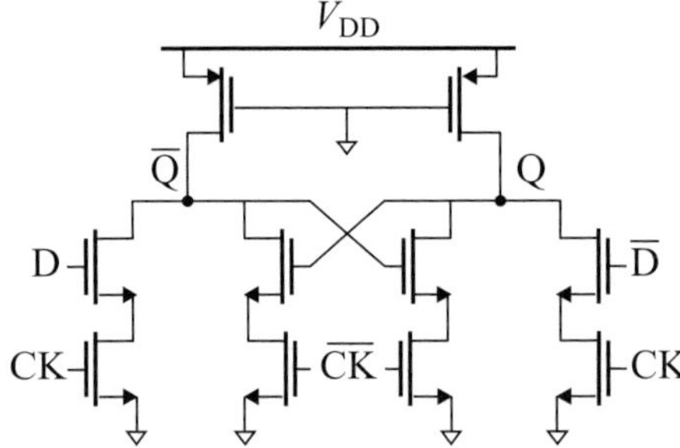

Figure 8.19 Pseudo-differential implementation of D-type latch

implementations of CML modulo-2-dividers require several transistors with tangled routing. This gives rise to a large interconnection stray capacitance, which has to be compensated for by increasing the bias current. The frequency divider can dissipate most of the power budget in a PLL; therefore, minimization of the overall synthesizer power consumption should start from the optimization of the frequency divider.

CMOS latches eliminate the static current consumption and reduce the number of transistors, but they are often too slow. This is mainly a result of higher voltage swing and the presence of the transistors in positive feedback. The positive feedback, which is the basis of any static memory element, introduces an additional delay time to the switching time of the cell.

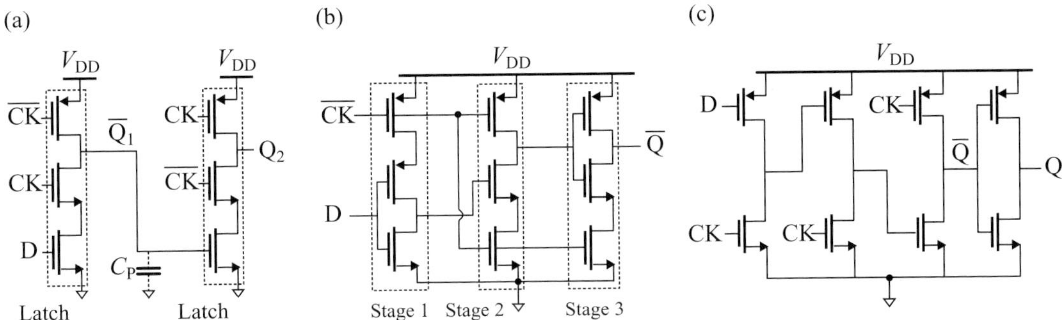

Figure 8.20 Dynamic D-type flip-flops: (a) dual-clock-phase pseudo-nMOS, (b) simple TSPC (with generic domino-logic cell), (c) extended TSPC

The alternative method to positive feedback in building a memory element is relying on the charge stored in a capacitance. This concept is implemented in dynamic memories. A dynamic pseudo-nMOS flip-flop is shown in Figure 8.20(a). When CK is high, the first latch acts as a pseudo-nMOS inverter. When the CK is low, the first latch becomes opaque and the charge is maintained in the parasitic capacitance C_P. Examples of frequency dividers based on this concept can be found, for instance, in [16]. The drawback of such logic is the lack of complementary outputs and the need of complementary CK inputs. In an asynchronous chain, the output of one modulo-2 counter is the input of the following one. Thus, an inverter has to be used, which produces skew between the two clock phases.

For this reason, flip-flops requiring only a single clock phase have been developed. Domino logic is a dynamic logic where one of the clock phases is used to pre-charge the output node to a wanted state. When CK is low, the stage is in a pre-charge period and the output is pre-charged to V_{DD}. When CK is high, the evaluation period starts and the output stays high or goes low depending on the implemented logic function. Conventional domino logic still requires a dual-phase clock, but single-phase implementations exist, like the one shown in Figure 8.20(b). This scheme is typically referred to as true single-phase clocking (TSPC). It is a modified version of the original TSPC flip-flop stage of Yuan and Svensson. [17–19] The transistors driven by the input clock are placed nearest to the power rails, to avoid the body effect and to lower the internal node capacitance. This results in higher speeds and a higher input sensitivity. A further extension of TSPC, which increases the achievable speed thanks to a reduction of the piled transistors, is shown in Figure 8.20(c). [20] However, this has all the drawbacks of pseudo-nMOS logic. It has static current, lower swings and noise immunity, and it requires additional care in design. [21]

The advantage of dynamic over static flip-flops in terms of power consumption is, mainly, due to the reduced capacitive loads. To compare the TSPC approach against the static CML design, both the size of the devices and the interconnection lengths have to be taken into account. Conventional CML and TSPC ÷2-dividers are shown for comparison in Figure 8.21. The CML divider in Figure 8.21(a) embodies 18 transistors, considering also the tail generators and the pMOS resistive loads. In this topology, the stacking of several transistors limits the achievable output swing. This limited swing has to switch the source-coupled

(a)

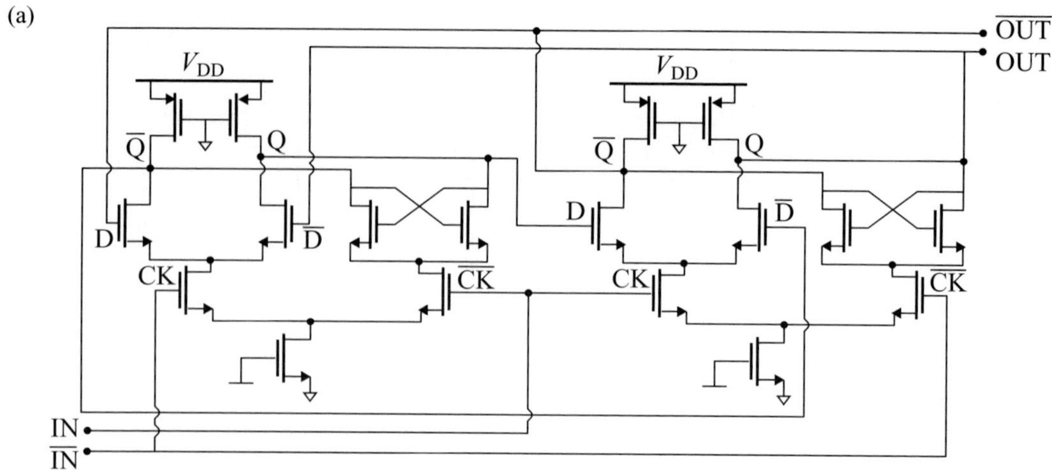

(b)

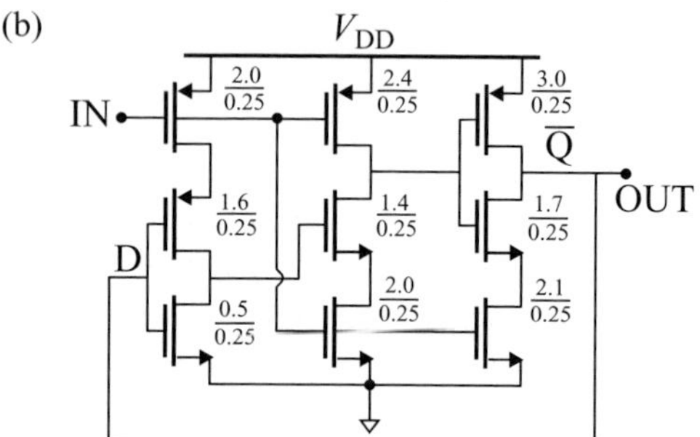

Figure 8.21 Circuit schematic of modulo-2 dividers: (a) CML, (b) TSPC

transistors of the following SCL stage completely; thus their aspect ratio W/L cannot be minimum. Furthermore, the transistors connected in positive feedback to hold the data need to have a voltage gain greater than one. Again, the aspect ratio of these transistors cannot be the minimum. Finally, the aspect ratio of the tail transistors is large because of the limited voltage headroom. Because of the non-minimum device sizes, the CML divider has high input capacitance and long metal interconnections between its flip-flops. Even with a careful layout, the interconnection parasitic capacitance can be larger than the transistors' input capacitance itself. Therefore, a buffer is often needed between the VCO and the CML divider, further increasing the power dissipation.

Although TPSC prescalers have a lower input sensitivity than SCL dividers, their lower input load allows the VCO to be connected directly to the divider. The small capacitance added by the prescaler to the LC resonator of the VCO does not limit the VCO tuning range significantly. Therefore, the prescaler can be driven by the large swing of the VCO output.

The drawbacks of dynamic logic are the lack of precise quadrature outputs, which may be required, for instance, in building phase-switching prescalers, and the existence of a minimum frequency of operation due to limited charge retention.

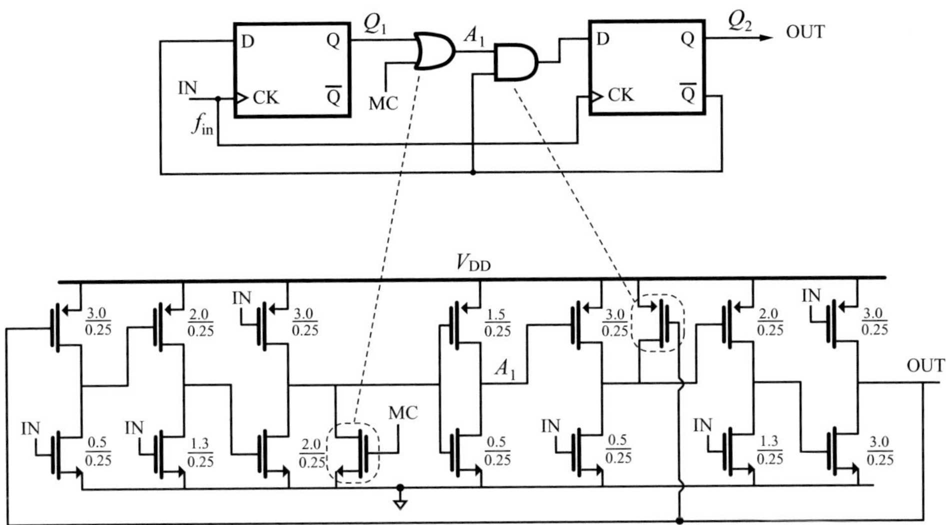

Figure 8.22 Modulo-2/3 divider in E-TSPC logic

The susceptibility to disturbances due to single-ended topology is typically negligible. However, the lack of a constant bias current causes the generation of switching noise, which is applied at the power supply and the local bulk. It is, therefore, advisable to use a power supply isolated by the analogue power supply of the PLL, such as the charge pump or, especially, the VCO supply.

An example of a modulo-2/3 divider in extended TSPC is shown in Figure 8.22. Again speed can be enhanced by merging the logic gates within the flip-flop circuits. The FET aspect ratio shown in the figure is obtained by employing the classical domino logic scaling. [21]

8.5.3 Practical design issues

Because the first stage of the divider circuit is biased by the d.c. level of the VCO output signal, care has to be taken to ensure proper operation over all process variations. This needs to be investigated with Monte Carlo simulations. The operating temperature has to be varied over a wide range. In MOS technology, a higher temperature causes lower carrier mobility and thus lower current capability. The circuit does not have to be sized to obtain a maximum operation speed with typical model parameters, but it must be robust for process variations and temperature changes. The presence of interconnection capacitance and temperatures of up to 100 °C may require overdesigning the bias current of the fast CML flip-flops by a factor of three, with respect to the initial design without interconnection capacitances and 25 °C temperature.

The floorplan of prescalers should be carefully optimized to minimize the length of the interconnections. Every effort should be made to compact the layout and keep parasitic capacitances and resistances as small as possible. Therefore, the typical layout of digital circuits in long rows should be avoided. Owing to high resistance of the polysilicon and high capacitance between poly and substrate, interconnections have to be made with metal, and the use of poly has to be limited to gates. Where necessary, 45 degree metal lines have to be used so that they would be shorter and their parasitics minimized. Wide metal tracks are used for power lines and on-chip decoupling capacitors should be placed physically close to the divider. The voltage supply and the ground have to be distributed through the top metal layer, so that their series resistance can be minimized. This supply, which is subject to fast current requests, will be corrupted by switching noise. Therefore, the p bulk and n-type wells should be decoupled by the rails. Guard rings of substrate contacts or oxide trenches, when available, help in coupling less switching noise from the dividers to the analogue blocks, such as the VCO.

8.6 Noise in digital dividers

The phase noise generated by the divider is typically neglected by PLL designers. However, in some cases, it can affect the synthesizer noise performance within the PLL bandwidth. If the division factor used within the loop is high, the divider noise power transferred to the PLL output is multiplied by the square of the division factor. The PLL in-band phase noise can represent the dominant term of the integral value, which may heavily degrade the signal-to-noise ratio in wideband communication systems based on QAM modulations. The divider noise power is multiplied by the square of the division factor, when it is transferred to the PLL output. This is also true for other stages processing either the PLL reference signal or the feedback signal from the divider. Those stages may be, for instance, the reference buffer, the inverters at the output of the frequency divider driving the phase detector or the phase detector itself.

Empirical models are typically employed to describe the phase noise of digital dividers based on different logic families, such as [22] and [23]. However, the existence of a fundamental trade-off between noise and power dissipation in digital frequency dividers has been demonstrated. [24]

The estimation of the divider noise is not straightforward. The various noise sources in the circuit affect the zero-crossing instants of the output signal and the resultant phase noise is a random process sampled at the divider output frequency. Time-domain simulations, which are able to predict the divider's jitter, are, unfortunately, very time consuming and provide little insight into the physical processes at the basis of the jitter generation.

A viable method of estimating the phase noise spectrum of a frequency divider is based on the calculation of the output jitter of the latch. The same approach can be applied to any circuit with switched behaviour.

8.6.1 Link between jitter and phase noise spectrum

The *time jitter* $\sigma_{t_0}^2$ is defined as the variance of the instant t_0 of the output zero crossing. In a digital period signal, the phase is known only at the zero crossing. Therefore, it is intrinsically sampled at frequency $f_{out} = 1/T_{out}$ and it is proportional to the switching instant, according to the expression $\phi = 2\pi f_{out} t_0$. Owing to this sampling operation, any noise component at frequencies higher than $f_{out}/2$ is folded back and the phase spectrum is defined within the Nyquist band $0 - f_{out}/2$. Then, the time jitter can be written in terms of the integral of the *single-sided* power spectral density (PSD) of the phase S_ϕ within the Nyquist band: [22, 25]

$$\sigma_{t_0}^2 = \frac{1}{4\pi^2 f_{out}^2} \int_0^{f_{out}/2} S_\phi(f_m)\, \mathrm{d} f_m. \tag{8.2}$$

For white noise, $S_\phi(f) = W$ is constant and (8.2) gives the link between the jitter and the phase spectrum. The single-sideband-to-carrier ratio (SSCR or $\mathcal{L}$), which is equal to $S_\phi/2$, is given by:

$$\mathcal{L}_W = \frac{W}{2} = 4\pi^2 f_{out} \cdot \sigma_{t_0}^2. \tag{8.3}$$

The linear dependence of $\mathcal{L}_W$ on the output frequency is in accordance with the model proposed by Kroupa. [23] In practice, even if the voltage noise spectrum is not white, the spectrum S_ϕ will be flat in the band $0 - f_{out}/2$, since the sampling frequency f_{out} is usually much lower than the noise bandwidth. In the special case of $1/f$ noise, (8.2) leads to a diverging variance and an alternative definition of variance has to be adopted; see, for instance, [26]. However, flicker can be more conveniently treated in the frequency domain, as discussed further.

8.6.2 Jitter of a synchronous counter

Referring to the master–slave configuration of modulo-2 dividers, only the second latch is responsible for the switching of the output signal, so it is the only stage contributing to the jitter. Analogously, the jitter of a synchronous counter is only caused by noise sources of the second latch of the last flip-flop and to the input clock jitter.

This latch at the switching instant can be approximated by a simple differential pair, as shown in Figure 8.23. In the operation of the counter, the D inputs have already switched completely, before CK starts to switch. Therefore, one of the transistors of the corresponding pair is off and the other one is in the triode region. The same happens to the pair in positive feedback. Neglecting the resistance of those transistors in the triode region, the circuit can be regarded as a differential pair in the balanced condition. The p-MOSFETs have been replaced by two resistors R_L. The capacitor C_L represents the total output capacitance, given by the transistors connected to the output nodes and by the interconnect capacitance. The noise sources affecting the output zero crossings are the thermal noise of the pMOS loads, the noise of the tail current I_T and the noise of the nMOS pair.

The time jitter can be derived from the noise sources of the logic cell. In fact, it can be expressed in terms of variance of the output voltage σ_V^2 as:

$$\sigma_{t_0}^2 = \frac{\sigma_V^2}{(\mathrm{SL})^2}, \tag{8.4}$$

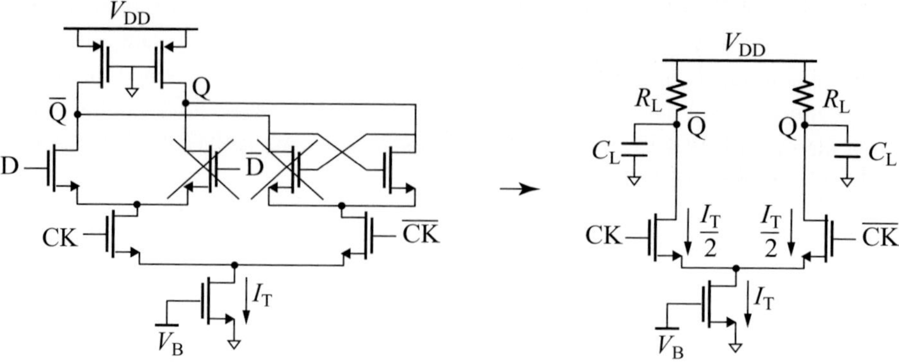

Figure 8.23 Simplified circuit for calculation of latch phase noise

where SL is the output voltage slope at the zero crossings. The *cycle jitter* defined as the variance of the period is twice the time jitter, since the period is the difference between two successive switching instants.

Since both the transients of the output voltage are exponential waveforms with the same time constant $R_L C_L$, the time delay t_0 between the input clock switching and the zero-crossing instant of the output differential signal is $t_0 = R_L C_L \ln 2$. It is interesting to note that any noise affecting the output nodes is band-limited by the $R_L C_L$ parallel network. Thus, the noise fluctuations are slower than the output rise or fall time and cannot trigger multiple commutations of the following stage. The slope of the differential waveform at the zero crossings is $\text{SL} = I_T / C_L$, where I_T is the tail current.

The thermal noise of the load resistors causes voltage noise at the differential output, whose variance is $\sigma_V^2 = 2kT/C_L$. The resulting jitter from (8.4) is, therefore

$$\sigma_{t_0,\,\text{load}}^2 = \frac{2kTC_L}{I_T^2}, \tag{8.5}$$

which can be regarded as the ratio between the power of the capacitors' charge noise and the square of the bias current.

The total phase noise following from (8.3) is:

$$\mathcal{L}_W = 8\pi^2 \frac{kTC_L}{I_T^2} f_{\text{out}} \left(1 + F\right), \tag{8.6}$$

where F is the noise factor accounting for the noise from the tail transistor and the switching pair. Note that, as in oscillators, proper design makes F tend to γ/α, where γ is the thermal noise factor of the transistors of the pair, and α is the ratio between their transconductance and their drain–source conductance at zero V_{DS}. A complete derivation of F can be found in [24].

Since the phase noise is dependent on the tail current, it is appropriate to normalize it to the power consumption of the divider. Equation (8.6) can be simplified by noting that the time constant $R_L C_L$ depends on the output frequency. A higher frequency requires short transients. In particular, $R_L C_L$ should be sufficiently lower than $1/f_{\text{out}}$ to guarantee a correct operation. Therefore, (8.6) can be written as

$$\mathcal{L}_{\mathrm{W}} = 8\pi^2 \frac{kT}{P} \left(\frac{V_{\mathrm{DD}}}{V_{\mathrm{peak}}} \right) K(1+F), \tag{8.7}$$

where P is the dissipated power from the voltage supply V_{DD}, V_{peak} is the peak voltage of the differential output and $K = f_{\mathrm{out}} R_{\mathrm{L}} C_{\mathrm{L}} < 1$.

Thus, phase noise can be traded with the dissipated power. Better noise performance can also be achieved by maximizing the output swing with respect to the voltage supply and increasing the output bandwidth with respect to the output frequency.

An appropriate figure of merit (FoM) can be defined as the inverse of the phase noise–power product: $\mathrm{FoM} = 1/(\mathcal{L}_{\mathrm{W}} P)$. To compare different dividers, the power consumption must be normalized to the number of stages. A ripple divider-by-N typically has $\log_2(N)$ stages.

8.6.3 Flicker noise

The analysis based on jitter cannot easily be applied to flicker noise sources. In this case, it is convenient to operate in the frequency domain. The output voltage noise at frequency $(kf_{\mathrm{out}} \pm f_{\mathrm{m}})$ (with $k = 0, 1, \ldots$ and $f_{\mathrm{m}} < f_{\mathrm{out}}/2$) can be represented as a tone with amplitude V_{m} and random phase. Because of sampling, this voltage tone causes a phase tone at f_{m}, whose amplitude can be calculated as $\phi_{\mathrm{m}} = 2\pi f_{\mathrm{out}} V_{\mathrm{m}}/(\mathrm{SL})$ with SL the slope at the zero crossing. Hence, the phase spectrum can be written as:

$$\mathcal{L}(f_{\mathrm{m}}) = \frac{2\pi^2 f_{\mathrm{out}}^2}{(\mathrm{SL})^2} S_V^{\mathrm{folded}}(f_{\mathrm{m}}), \tag{8.8}$$

where $S_V^{\mathrm{folded}}(f_{\mathrm{m}})$ is the PSD of the output voltage noise folded in the Nyquist band from 0 to $f_{\mathrm{out}}/2$.

If the corner frequency of the voltage spectrum is lower than $f_{\mathrm{out}}/2$, the flicker component is not subject to folding. In this case, $S_V^{\mathrm{folded}}(f_{\mathrm{m}})$ in (8.8) can be replaced by the unfolded spectrum $S_V(f_{\mathrm{m}})$. As a result, the phase noise in the flicker region is proportional to the square of the output frequency f_{out}. This dependence has been experimentally verified. [23]

The flicker of each switching transistor of the latch can be represented as a voltage generator in series with the MOS gate, whose spectrum is $S_{V_{\mathrm{in}}}(f) = K_{\mathrm{F}}/f$. We are assuming that this noise is unaffected by the transistor Switching. This low-frequency input noise can be regarded as a static perturbation ΔV_{in}, which alters the threshold level of the stage. Consequently, the zero crossings of the input and the output shift by the same quantity: $\Delta V_{\mathrm{in}}/\mathrm{SL}$. The corresponding flicker phase noise is given by (8.8):

$$\mathcal{L}_{\mathrm{F}}(f) = 2\pi^2 f_{\mathrm{out}}^2 \left(\frac{C_{\mathrm{L}}}{I_{\mathrm{T}}} \right)^2 \frac{K_{\mathrm{F}}}{f}. \tag{8.9}$$

Instead, the flicker coming from the tail usually causes negligible jitter at the output. A low-frequency tail noise can be viewed as a static variation ΔI_{T} of the bias current. The effect of this perturbation is sketched in Figure 8.24. The solid line represents the unperturbed case, while the dashed line refers to the perturbed case. The lower rail of the output shifts

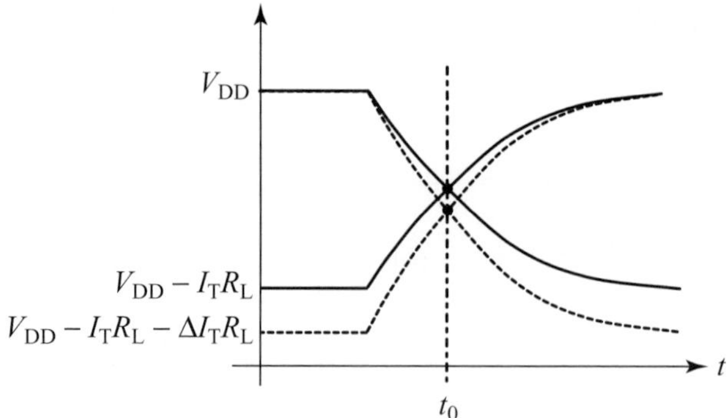

Figure 8.24 Effect of low-frequency tail current perturbation in switching differential pair

by $\Delta I_T R_L$, but the crossing point of the two waveforms occurs at the same instant t_0 and no jitter is produced. It may seem obvious since a differential stage is known to be insensitive to common-mode noise. However, if the tone frequency becomes comparable to the inverse of the transient time (i.e., $R_L C_L \cdot \ln 2 \approx 0.7 \cdot R_L C_L$), this noise cancellation does not occur. The transfer function of this noise source has been derived. [24]

Residual up-conversion of flicker noise of the tail-current generator can still exist. The amplitude noise caused by low-frequency tail noise can be converted into phase noise by non-linear capacitances connected to the output nodes.

8.6.4 Noise reduction through synchronization

In asynchronous counters, the output of one divider is connected to the clock input of the following one. Thus, the time jitter of any stage is transferred to the following stage. Assuming uncorrelated noise for each stage, the jitter at the output of the chain is the quadratic sum of the jitters of each stage plus the jitter of the input clock. [22] Therefore, the output time jitter follows from (8.5):

$$\sigma_{out}^2 = \sigma_{in}^2 + \frac{2kTC_{L1}}{I_{T1}^2}(1 + F_1) + \frac{2kTC_{L2}}{I_{T2}^2}(1 + F_2) + \cdots + \frac{2kTC_{Ln}}{I_{Tn}^2}(1 + F_n). \tag{8.10}$$

A long chain can cause significant jitter accumulation and high output-phase noise. Moreover, in long asynchronous counters the bias current is very often scaled down along the chain. This is done because the speed requirement of each modulo-2 divider is halved. However, the consequence of this power scaling is that the jitter introduced by the last stages is dominant over the other ones (see (8.5)).

For these reasons, it is always better to re-synchronize the output of the asynchronous counter to the input clock. This operation can be performed using a flip-flop whose clock input is given by the high-speed clock and whose data input is the output of the asynchronous counter. The output phase noise using a synchronizer will only be given by the input clock jitter and by the jitter introduced by the synchronizer.

8.7 References

[1] B. Razavi, *RF Microelectronics*, New York, NY: Prentice-Hall, 1997.

[2] W. F. Egan, *Frequency Synthesis by Phase Lock*, New York, NY: Wiley, 2nd edn, 2000, 142–6.

[3] R. Adler, A study of locking phenomena in oscillators, *P. IEEE*, **61**, Oct. 1973, 1380–5.

[4] N. Foroudi and T. A. Kwasniewski, CMOS high-speed dual-modulus frequency divider for RF frequency synthesis, *IEEE J. Solid-St. Circ.*, **30**, Feb. 1995, 93–100.

[5] T.-C. Lee and B. Razavi, A stabilization technique for phase-locked frequency synthesizers, *IEEE J. Solid-St. Circ.*, **38**, Jun. 2003, 888–94.

[6] N.-H. Sheng *et al.*, A high-speed multimodulus HBT prescaler for frequency synthesizer applications, *IEEE J. Solid-St. Circ.*, **26**, Oct. 1991, 1362–7.

[7] P. Larsson, High-speed architecture for a programmable frequency divider and a dual-modulus prescaler, *IEEE J. Solid-St. Circ.*, **31**, May 1996, 744–8.

[8] C. S. Vaucher, I. Ferencic, M. Locher *et al.*, A family of low-power truly modular programmable dividers in standard 0.35-μm CMOS technology, *IEEE J. Solid-St. Circ.*, **35**, Jul. 2000, 1039–45.

[9] J. Craninckx and M. S. J. Steyaert, A 1.75-GHz/3-V dual-modulus divide-by-128/129 prescaler in 0.7-μm CMOS, *IEEE J. Solid-St. Circ.*, **31**, Jul. 1996, 890–7.

[10] N. Krishnapura and P. R. Kinget, A 5.3-GHz programmable divider for HiPerLAN in 0.25-μm CMOS, *IEEE J. Solid-St. Circ.*, **35**, Jul. 2000, 1019–24.

[11] M. H. Perrot, *Techniques for High Data Rate Modulation and Low Power Operation of Fractional-N Frequency Synthesizer*, Ph.D. thesis, Massachusetts Institute of Technology, Cambridge, MA, 1997.

[12] A. Benachour, S. H. K. Embabi and A. Ali, A 1.5 GHz, sub-2 mW CMOS dual modulus prescaler, in *Proc. IEEE Custom Integrated Circuits Conf.*, San Diego, CA, 1999, 613–16.

[13] K. Shu, E. Sánchez-Sinencio, J. Silva-Martínez and S. H. K. Embabi, A 2.4-GHz monolithic fractional-*N* frequency synthesizer with robust phase-switching prescaler and loop capacitance multiplier, *IEEE J. Solid-St. Circ.*, **38**, Jun. 2003, 866–74.

[14] G. C. T. Leung and H. C. Luong, A 1-V 5.2-GHz CMOS synthesizer for WLAN applications, *IEEE J. Solid-St. Circ.*, **39**, Nov. 2004, 1873–82.

[15] M. Zargari *et al.*, A single-chip dual-band tri-mode CMOS transceiver for IEEE 802.11 a/b/g wireless LAN, *IEEE J. Solid-St. Circ.*, **39**, Dec. 2004, 2239–49.

[16] H. Yan, M. Biyani and K. K. O, A high-speed CMOS dual-phase dynamic-pseudo NMOS ($(DP)^2$) latch and its application in a dual-modulus prescaler, *IEEE J. Solid-St. Circ.*, **34**, Oct. 1999, 1400–4.

[17] J. Yuan and C. Svensson, High-speed CMOS circuit technique, *IEEE J. Solid-St. Circ.*, **24**, Feb. 1989, 62–70.

[18] Q. Huang and R. Rogenmoser, Speed optimization of edge-triggered CMOS circuits for gigahertz single-phase clocks, *IEEE J. Solid-St. Circ.*, **31**, Mar. 1996, 456–65.

[19] B. De Muer and M. Steyaert, A single ended 1.5 GHz 8/9 dual modulus prescaler in 0.7-μm CMOS technology with low phase noise and high input sensitivity, in *Proc. 1998 Eur. Solid-State Circuits Conf.*, 1998, 256–9.

[20] J. Navarro Soares and W. A. M. Van Noije, A 1.6-GHz dual modulus prescaler using the extended true-single-phase-clock CMOS circuit technique (E-TSPC), *IEEE J. Solid-St. Circ.*, **34**, Jan. 1999, 97–102.

[21] M. Shoji, FET scaling in domino CMOS gates, *IEEE J. Solid-St. Circ.*, **20**, Oct. 1985, 1067–71.

[22] W. F. Egan, Modeling phase noise in frequency dividers, *IEEE T. Ultrason. Ferr.*, **37**, Jul. 1990, 307–15.

[23] V. F. Kroupa, Jitter and phase noise in frequency dividers, *IEEE T. Instrum. Meas.*, **50**, Oct. 2001, 1241–3.

[24] S. Levantino, L. Romanò, S. Pellerano, C. Samori and A. L. Lacaita, Phase noise in digital frequency dividers, *IEEE J. Solid-St. Circ.*, **39**, May 2004, 775–84.

[25] J. G. Sneep and C. J. M. Verhoeven, A new low-noise 100-MHz balanced relaxation oscillator, *IEEE J. Solid-St. Circ.*, **25**, Jun. 1990, 692–8.

[26] J. Rutman and F. L. Walls, Characterization of frequency stability in precision frequency sources, *P. IEEE*, **79**, Jun. 1991, 952–60.

9 Phase comparison

9.1 Introduction

The phase comparison path in a phase-locked loop detects the phase delay between the reference and the feedback oscillation. Since the time delay between the edge of the reference signal and the subsequent edge of the divider signal is measured, the phase delay is a signal sampled at the reference frequency.

The phase comparison can be performed either in the analogue or in the digital domain. The analogue implementations, although they use logic circuits such as the XOR gate or the tri-state PFD, provide an analogue value of the phase error. Instead, digital implementations derive the phase error by subtracting the instantaneous count of the reference counter from that of the feedback counter. After the phase error is detected, it is filtered and applied to the VCO tuning port.

The type of phase detection affects the in-band noise, the level of spurious tones and the lock acquisition behaviour. The non-linearity of the phase comparison path may seriously degrade both noise and spurs. Thus, it has to be accurately examined and attenuated.

9.2 Phase comparison path

An analogue phase detector is, typically, a logic circuit, which generates pulses of duration equal to the time difference between the REF and the DIV edges. The phase comparison path is sketched in Figure 9.1(a). The phase detector output is low-pass filtered, to extract its average value, or it is accumulated into an integrator in a type-II loop.

In the design of a type-II PLL, the adoption of the charge pump driven by the phase detector simplifies the implementation of the integrator within the loop, as shown in Figure 9.1(a). The charge pump is, essentially, a transconductor operated in hard switching. It converts the phase-detector voltage pulses into current impulses. If the finite output impedance of the charge pump is too low, the PLL open-loop gain may be insufficient at d.c. An operation amplifier can still be used to maintain constant voltage at the charge pump output. Alternatively, the charge pump output impedance can be improved by employing cascoded topologies.

(a)

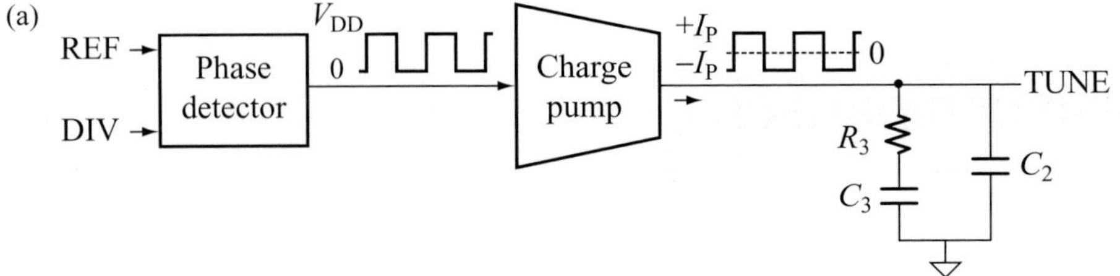

(b)

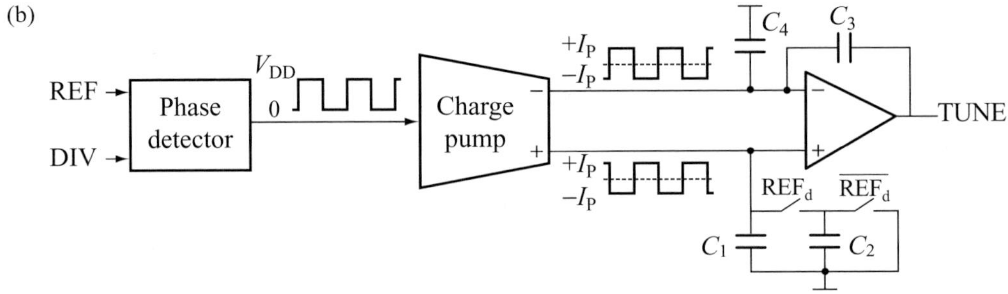

Figure 9.1 Analogue phase comparison: (a) RC filter and (b) active switched-capacitor filter

Narrowband PLLs may require low-frequency singularities and large values of resistances and capacitances. For this reason, the loop filter is often placed off chip. This choice has the additional advantage of allowing more accuracy in placing the loop singularities and more flexibility in the choice of bandwidth. However, when a full integration of the synthesizer is desired, methods exist which avoid the use of large-value passive components. For instance, the dual-charge-pump topology allows synthesis of the low-frequency zero with two parallel signal paths. [1, 2]

Another approach involves synthesizing the resistor with a switched-capacitor network, as shown in Figure 9.1(b). [3] Such a solution has not only the advantage of synthesizing a large resistance by using a moderate capacitance, but also that of making the singularity positions dependent on capacitance ratios and not on their absolute values. The clock signal driving the switches is given by the PLL reference. Having the loop bandwidth much narrower than the reference frequency prevents aliasing problems.

Phase comparison can also be performed in the digital domain. [4] This is common in PLL generating the clock signal of digital systems. As shown in Figure 9.2(a), the phases of the reference and the feedback signal can be detected by reading the running digital word of the counters. After resampling the feedback count at the reference frequency, the phase delay is obtained by digitally subtracting the two numbers. Alternatively, the arrangement in Figure 9.2(b) uses accumulators instead of counters to measure the signal's phase.

Since the phase delay is quantized, in-band quantization noise appears when digital phase comparators are used in a PLL. Techniques for interpolating the phase error are known and have been implemented successfully in all-digital PLLs (ADPLLs). [5, 6]

The phase error may be either converted from digital to analogue domain, filtered and fed to the VCO tuning voltage, as shown in Figure 9.2(a), or directly applied to a digital filter

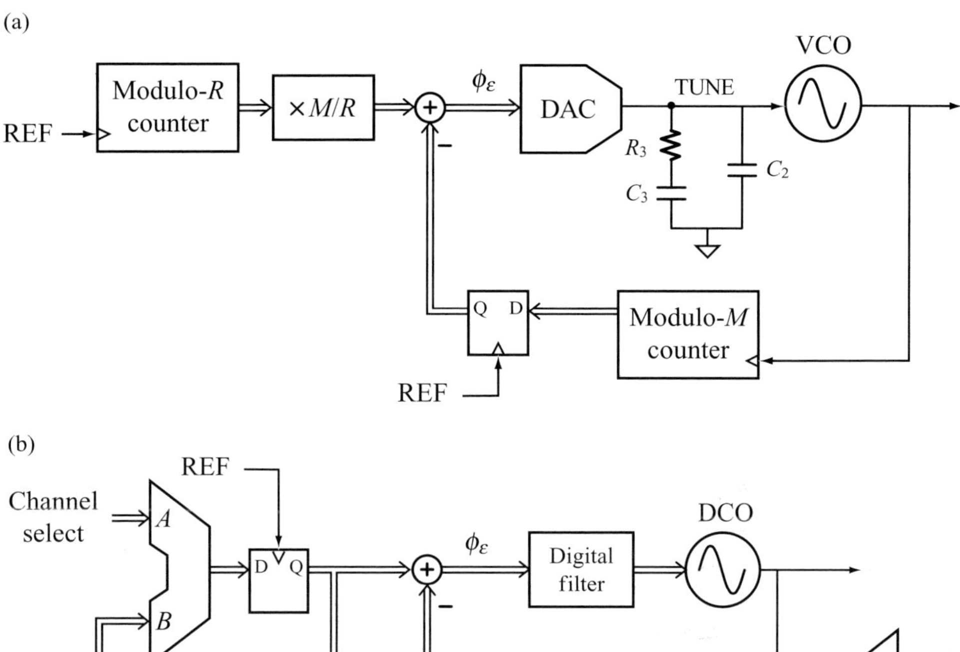

Figure 9.2 PLL with digital phase comparison: (a) phase measurement based on counters and digital-to-analogue conversion of phase error for VCO tuning; (b) phase measurement based on accumulators and direct use of digital phase error for DCO tuning

and a digitally controlled oscillator (DCO), as in Figure 9.2(b). The latter allows one to avoid the DAC circuit and the off-chip filter. The DCO can be implemented as an LC oscillator employing the switched tuning described in Section 7.4.1, with no analogue tuning control. [6]

Example 9.1 Switched-capacitor loop filter

The transfer function of the switched-capacitor filter in Figure 9.1(b) can be approximated after replacing the switched capacitor C_2 with an equivalent resistor $R_2 = T_{\text{ref}}/C_2$. Applying the principle of superposition, the tuning voltage V_{tune} is given by:

$$V_{\text{tune}} = -\frac{1}{sC_3}I_- + \frac{R_2}{1+sR_2C_1}\left(1+\frac{C_4}{C_3}\right)I_+, \tag{9.1}$$

where I_- and I_+ are the currents injected by the CP into the minus and the plus inputs of the opamp, respectively. The open-loop gain of this filter has been implicitly assumed to be much greater than one. Since the input currents are differential, $I_- = -I_P$ and $I_+ = I_P$.

Substituting the last derivations in (9.1), it possible to write the tuning voltage as a function of the charge-pump peak current. Then, the filter transfer function is given by:

$$Z(s) = \frac{V_{\text{tune}}}{I_{\text{P}}} = \frac{1 + sT_{\text{ref}} \cdot (C_1 + C_3 + C_4)/C_2}{sC_3(1 + sT_{\text{ref}} \cdot C_1/C_2)}.$$

As previously predicted, both the pole and the zero frequencies are accurate, since they depend on capacitance ratios and on the reference period. Both parameters are well controlled: the first one relies on capacitance matching in integrated processes and the second one on the accuracy of a crystal oscillator. The filter can be designed by following the procedure described in Example 2.4.

9.3 Phase/frequency detectors

As shown in Chapter 2, phase-locked loops employing pure phase detectors have a narrow acquisition range and a slow locking behaviour. To aid and speed up the acquisition, several techniques can be employed. A frequency loop based on a frequency detector can be combined with the phase loop. For instance, adding the output of a digital quadricorrelator to that of the phase detector produces a proper correction voltage when the frequency difference is large and the phase detector skips cycles (see, for instance, [7] or [8]).

Alternatively, the acquisition range can be extended by employing a phase/frequency detector (PFD). It is essentially a phase detector. However, thanks to the non-periodic dependence of its average output voltage on the input phase difference, it is able to detect which input is faster. The PFD average output remains positive (negative), if the phase difference increases (decreases), that is, if the reference frequency is higher (lower) than the divider output frequency.

Hence, PFDs can be derived by modifying pure phase detectors. For instance, the XOR gate shown in Chapter 2 can be extended to become a PFD by adding frequency-detection logic; the single flip-flop used as a phase detector can be modified into a tri-state PFD by adding a second flip-flop. Both the XOR-based PFD and the tri-state will now be reviewed and their performance in terms of linearity and spur performance will be discussed.

9.3.1 XOR-based PFD

The main drawback of an XOR gate is the limited phase range where the output average voltage is proportional to the input phase difference. In fact, out of this range, the slope of the transfer function has the opposite sign. Consequently, the loop cannot operate in those regions. This phenomenon sets a limit to the lock range.

To extend this range, two divide-by-two circuits can be connected to the REF and the DIV inputs, as shown in Figure 9.3. Thus, a phase difference of $\pi/2$ between the XOR inputs is equivalent to π at the divider inputs. Figure 9.4 illustrates the characteristic of the XOR gate as a function of the phase difference between REF and DIV. The stable and unstable regions of the XOR gate can be discerned by the value of its output A on the rising transitions of REF and DIV. In the stable region, the XOR output is low before REF goes

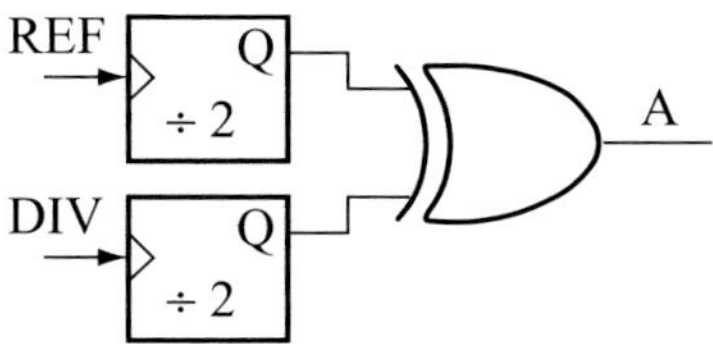

Figure 9.3 Divide-by-two circuits and XOR gate as phase detector

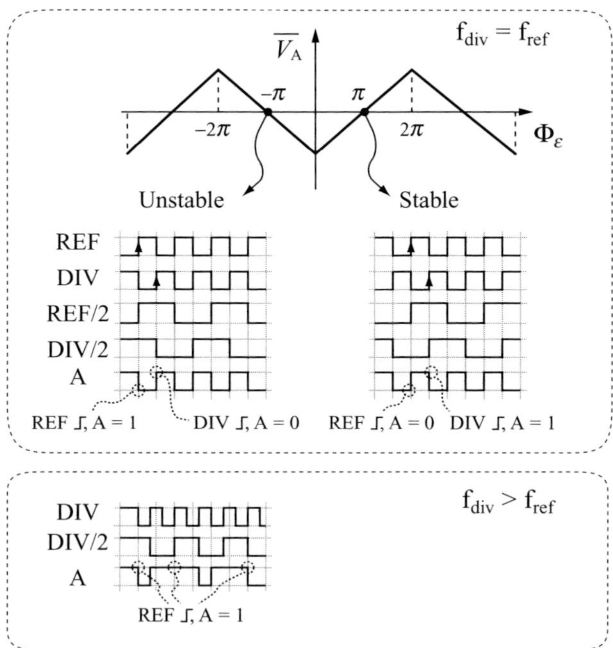

Figure 9.4 Phase characteristic of ensemble of divide-by-two circuits and XOR gate: logic states in unstable and stable regions of operation

high and it is high before DIV goes high. The opposite conditions are true in the unstable region. [9]

The XOR detector can become a PFD if its output is altered for phase differences originally placed in the unstable regions. The desired transfer characteristics are shown in Figure 9.5. In the unstable regions of the XOR gate, the output is forced to one or zero, depending on the relative frequency between the REF and DIV signals. As a result, if, for instance, $f_{div} > f_{ref}$, the average output voltage $\overline{V}_{out}$ is positive and tends to correct this frequency difference.

The desired state diagram can be achieved by looking at the logic signals in Figure 9.4. If the XOR output A at the REF and DIV edges is monitored, it is possible to determine whether the divider frequency equals the reference frequency, or which of them is greater. If $f_{div} > f_{ref}$, the XOR output is the one at the REF positive edges. Therefore, if this case is detected, the finite-state machine transitions to a new state and the PFD output OUT is

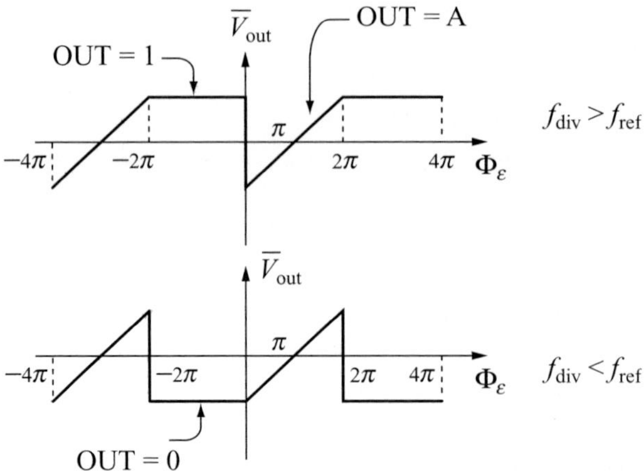

Figure 9.5 Desired phase characteristic of XOR-based PFD for $f_{ref} > f_{div}$ and $f_{ref} < f_{div}$

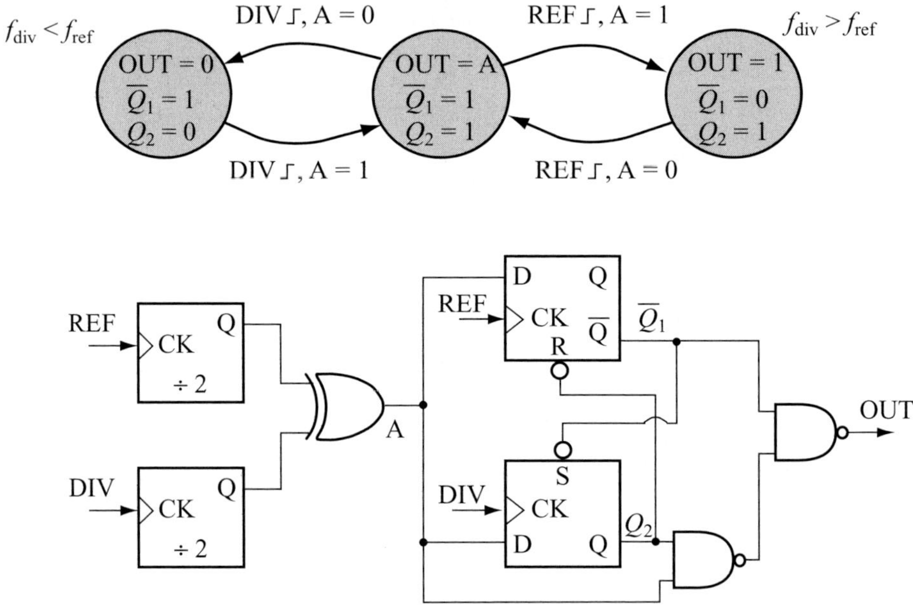

Figure 9.6 State diagram and implementation of XOR-based PFD

forced to one (OUT = 1). If, instead, A is zero at the REF edges, meaning that the two frequencies are equal, the machine returns to the initial state and OUT is connected to the XOR output (OUT = A). On the other hand, if $f_{div} < f_{ref}$, the XOR output tends to be zero at the DIV edges. Therefore, the machine transitions to a third state where the output is forced to zero (OUT = 0).

The state diagram and the logic circuit implementing it are shown in Figure 9.6. Two flip-flops, acting as a frequency detector, have been added to the divide-by-two circuits and the XOR gate. Under locked conditions, the state of the two flip-flops is $\overline{Q_1}Q_2 = 11$ and the XOR output passes directly to the output of the PFD. If the XOR enters its unstable region, the circuit transitions either to state $\overline{Q_1}Q_2 = 01$ or to state 10. The output is set to either one or zero, respectively.

This PFD is also known as a square-wave PFD, [7] because of the shape of its transfer characteristic, shown in Figure 9.5. Like the XOR-gate phase detector, this PFD has a 50% duty-cycle output in the locked condition. A charge-pump circuit after the PFD can inject a current, whose sign depends on the PFD output, into a loop filter.

Example 9.2 Reference spur

In the locked state, the two inputs of the XOR-based PFD are 180 degrees out of phase and the output of the PFD is a 50% duty-cycle square wave with a period equal to the reference period. This result has been obtained by introducing two divide-by-two circuits preceding the two XOR-gate inputs.

The charge-pump current is also a square wave with a 50% duty cycle and its fundamental harmonic component is $(4/\pi)\, I_{\mathrm{P}}$. For the third-order type-II PLL using the charge pump and a passive filter, the reference spur level can be calculated by relying on (2.16):

$$\mathrm{SFDR} = 20\log_{10}\left(\frac{K_{\mathrm{VCO}}\left|I\left(\omega_{\mathrm{ref}}\right)\right|\left|Z\left(\mathrm{j}\omega_{\mathrm{ref}}\right)\right|}{2\omega_{\mathrm{ref}}}\right).$$

The filter impedance Z at the reference frequency is given approximately by the impedance of the capacitance C_2. Therefore, it is

$$\mathrm{SFDR} \approx 20\log_{10}\left(\frac{2}{\pi}\frac{K_{\mathrm{VCO}} I_{\mathrm{P}}}{\omega_{\mathrm{ref}}^2 C_2}\right) = 20\log_{10}\left(4N\frac{\omega_{\mathrm{u}}\omega_{\mathrm{P}}}{\omega_{\mathrm{ref}}^2}\frac{b-1}{b}\right), \tag{9.2}$$

where ω_{u} is the unity-gain frequency, which can be used as an approximation of the close-loop bandwidth; ω_{P} is the filter pole frequency, b is the ratio between pole and zero frequency and N is the frequency division factor.

A further simplification can be obtained if b is at least ten and the unity-gain frequency is set to the geometric mean of the zero and the pole frequency ($\omega_{\mathrm{u}} = \sqrt{\omega_{\mathrm{Z}}\omega_{\mathrm{P}}} = \omega_{\mathrm{P}}/\sqrt{b}$):

$$\mathrm{SFDR} \approx 20\log_{10}\left[4\sqrt{b}N\left(\frac{\omega_{\mathrm{u}}}{\omega_{\mathrm{ref}}}\right)^2\right]. \tag{9.3}$$

The last expression shows that a tolerable spur level can be achieved by narrowing the loop bandwidth with respect to the reference frequency. A higher reference frequency also leads to lower values of the loop frequency division N, which further lowers the spur level.

As an example, a fractional-N synthesizer with a 2.4 GHz output and a 40 MHz reference requires a division factor N of about 60. To maintain the reference spur lower than −50 dBc, the closed-loop bandwidth ω_{u} must be lower than about 46 kHz (with $b = 10$).

Higher attenuation of the reference spur with the same bandwidth can be obtained by adding a fourth pole, as shown in Example 2.4.

9.3.2 Tri-state PFD

The main drawbacks of the XOR-based PFD are the high harmonic content at the output at steady state and the charge-pump current noise injected into the loop filter for 50% of the time. These drawbacks are overcome by the tri-state PFD, whose output is, ideally, in

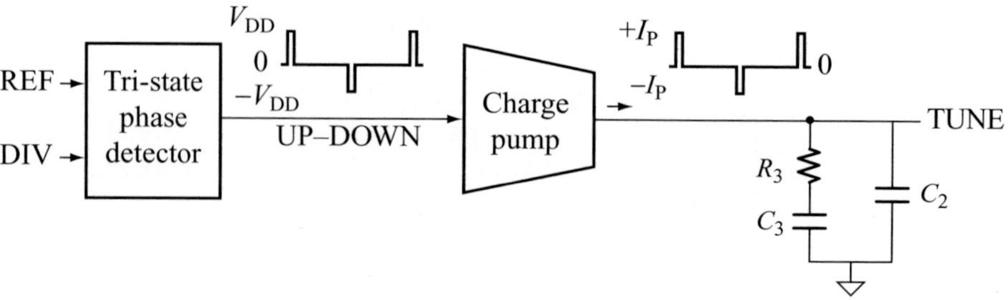

Figure 9.7 Phase comparison path utilizing tri-state PFD

'tri-state', that is in high-impedance mode, at the steady state. A phase comparison path employing a tri-state PFD is shown in Figure 9.7. This is the most typical implementation of PFDs in fully integrated CMOS synthesizers and it has been already discussed in Chapter 2. Some non-idealities, however, which affect this type of phase detector in its plain implementation, will now be reviewed, along with the common modifications to overcome them.

The charge pump, which converts the positive and the negative voltage pulses into current pulses, has a finite switching delay. The capacitive load of the charge pump and its intrinsic switching delay are typically increased by the use of large devices within the charge pump, which allows for large charge-pump currents, good matching, low flicker noise and large output-voltage headroom. In practice, the charge-pump current reaches its steady-state value after a certain time delay τ_c. This transient can be approximated by a piecewise waveform, as shown in Figure 9.8.

In the locked state, the output current averaged over the closed-loop bandwidth is zero. Therefore, this forces the REF and DIV signal edges to be aligned and no charge is injected into the loop filter. When a time error lower than τ_c exists between REF and DIV, the corresponding UP or DOWN signal is produced but the charge-pump current has not enough time to reach its final value. As shown in Figure 9.8, a nearly triangular current pulse is produced, whose area is smaller than the expected value and is equal to $\overline{Q} = I_P t_e^2/\tau_c$. Moreover, the dependence of this charge on the time error t_e is not linear, as in the ideal case.

For time errors t_e greater than τ_c, the injected charge over a reference period $\overline{Q} = I_P t_e$ is linear. The resulting transfer function of the phase detector in the piecewise current approximation is shown in Figure 9.8. Two linear regions for $|t_e| > \tau_c$ and one flat region for $|t_e| < \tau_c$ can be identified in this transfer function. In the flat region, the gain is lower than the ideal one and it tends to zero. For this reason, it is usually called the dead zone.

The dead zone should always be avoided in the design of a PLL. The open-loop gain would be highly reduced at small phase errors. Consequently, the phase margin and the bandwidth would be lowered with potential stability problems and lower VCO noise filtering, respectively. The phase error at the steady state may wander around the region where the gain is almost zero. Therefore, idle states can take place in the loop dynamics, which translate into spurious tones at the VCO output.

Another issue related to the dead zone and, in general, to the presence of any non-linearity in the phase comparison path is the potential folding of out-of-band noise. For instance,

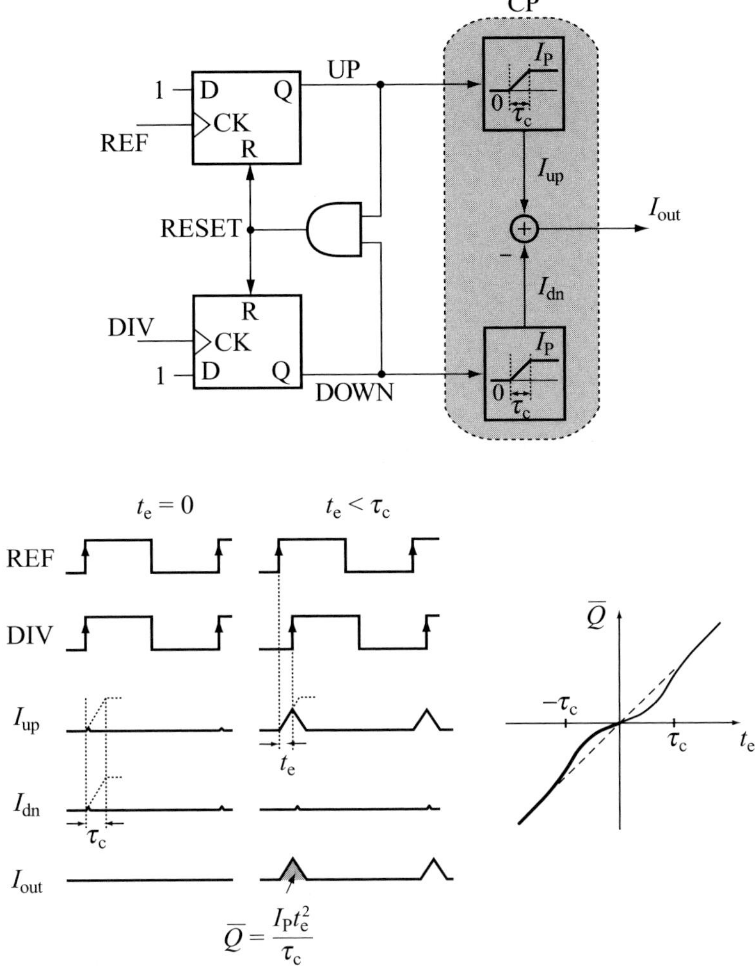

Figure 9.8 (Top) tri-state PFD and charge pump with finite rise-time and fall-time of CP pulses. (Bottom left) logic signals for zero time error and for time error t_e lower than CP rise time τ_c. (Bottom right) average output charge $\overline{Q}$ versus input time error t_e

the high-pass-shaped quantization spurs of the $\Delta\Sigma$ modulator around $f_{REF}/2$ can be down-converted to d.c., because of the phase comparison non-linearity. [10, 11] Similarly, the wideband phase noise generated by the phase detector, charge pump and frequency divider may be folded, thus degrading the PLL in-band noise.

9.3.3 Avoidance of dead zone and crossover distortion

A delay τ higher than τ_C added to the reset delay of the flip-flops in the PFD eliminates the dead zone in the transfer characteristic. This modification of the PFD can be referred to as an *overlapping tri-state* PFD. As shown in Figure 9.9, at the steady state ($t_e = 0$), two

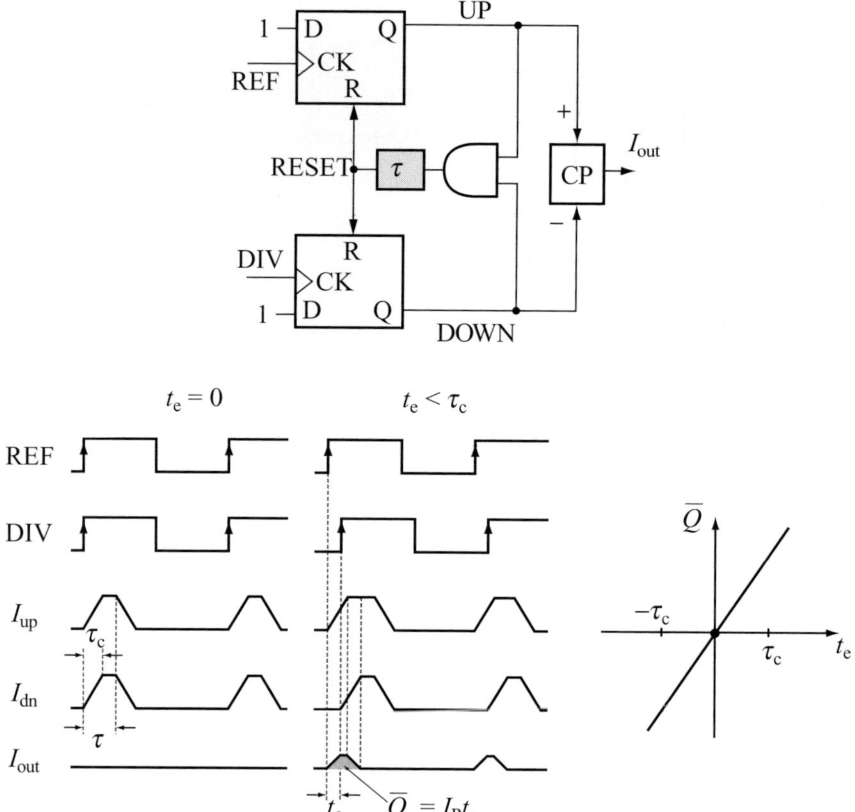

Figure 9.9 Elimination of dead zone in input–output characteristic of tri-state PFD and CP by insertion of PFD reset delay $\tau > \tau_C$

synchronous UP and DOWN pulses lasting τ are generated, which cancel each other out. Therefore, no net charge is injected into the loop.

Thanks to the reset delay, the charge-pump pulses can always reach their final value I_P. Therefore, for time errors lower than the charge-pump rise time τ_c, the two UP and DOWN pulses are delayed by t_e and a net charge equal to

$$\overline{Q} = (I_P/\tau_c)t_e^2 + (I_P/\tau_c)t_e(\tau_c - t_e) = I_P t_e$$

is injected into the loop. The resulting transfer characteristic is linear.

Better linearity has been traded with potentially higher in-band noise. For a zero time error, even if the UP and DOWN current pulses cancel out, their uncorrelated noises add in power. The longer the reset delay τ, the higher the charge noise injected into the loop. A quantitative estimation of the resultant phase noise can be found in Section 9.5, which deals with the noise of the phase comparison path.

Besides the finite switching time, the charge pump also suffers from mismatches between the positive and the negative current peaks. This non-ideality is shown schematically in Figure 9.10. By designating ε as the relative current mismatch $\Delta I_P/I_P$ between positive

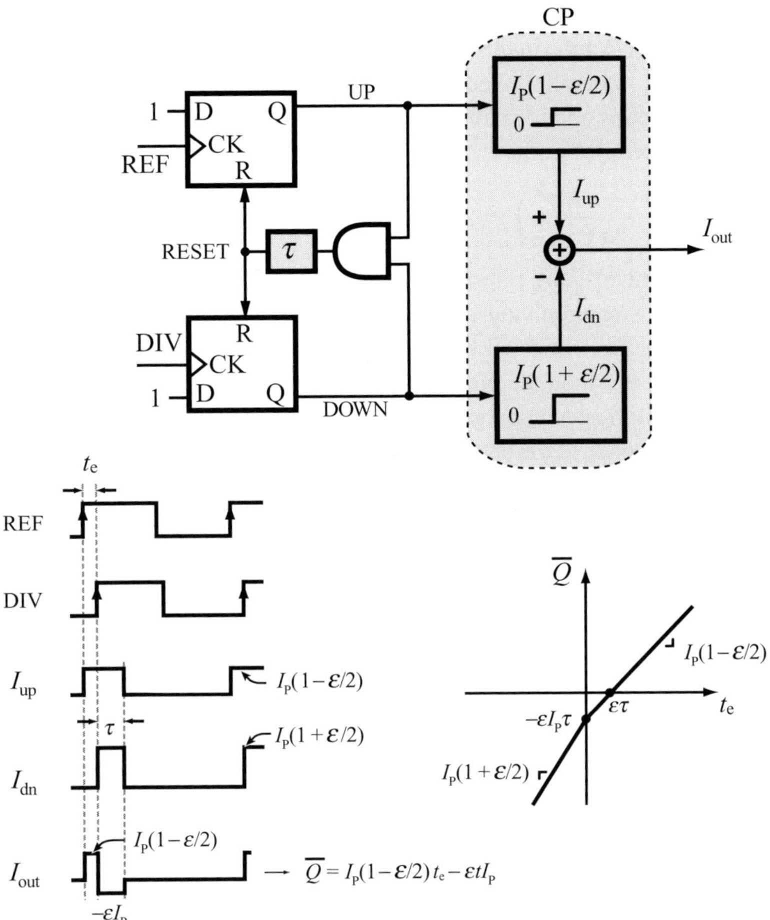

Figure 9.10 Crossover distortion caused by mismatch between UP and DOWN current pulses of charge pump

and negative currents, the UP and DOWN currents can be written as $I_P(1-\varepsilon/2)$ and $I_P(1+\varepsilon/2)$. At the steady state, the DIV and the REF edges cannot be perfectly aligned, since the positive injected charge would not equal the negative charge. Therefore, a time shift $t_e = \Delta t_s$ between DIV and REF takes place, so that the pulse corresponding to the lower charge-pump current (I_{up} in this case) is widened and its area is equalized to that of the other pulse (I_{dn} in this case). By equating the positive and the negative charges

$$I_P(1-\varepsilon/2)(\tau+\Delta t_s) = I_P(1+\varepsilon/2)\tau,$$

the expression for the time offset follows:

$$\Delta t_s = \frac{\varepsilon\tau}{1-\varepsilon/2} \approx \varepsilon\tau.$$

The last simplification holds true for small current offsets ($\varepsilon \ll 1$).

The first consequence of the charge-pump current mismatch is the growth of a reference spur. Even in an integer-N loop in the locked state, the injected charge is not instantaneously zero, but it is zero if integrated over a reference period. As shown in Section 2.4.3, the resultant reference spur is obtained after calculating the harmonic component of the current at the reference frequency. For the type-II third-order PLL, the spur level is given by:

$$\mathrm{SFDR} = 20\log_{10}\left(\frac{K_{\mathrm{VCO}}\varepsilon I_{\mathrm{P}}\tau^2}{4\pi C_2}\right) \approx 20\log_{10}\left(N\frac{\omega_{\mathrm{u}}\omega_{\mathrm{P}}\varepsilon\tau^2}{2}\right). \tag{9.4}$$

As can be seen intuitively, the reference spur is higher for larger delays τ and current mismatches ε. Moreover, large values of the reset delay τ may degrade the in-band noise, since some charge noise is injected by the charge pump at the steady state. As the reset delay τ has to be greater than the charge-pump rise time τ_{c}, it is important to reduce the latter by employing high-speed charge-pump topologies.

The second consequence of the charge-pump current mismatch is the crossover distortion of the charge-time transfer function. By applying a time error t_{e} onto the DIV positive edge, the positive edge of the DOWN pulse is shifted by t_{e}, so the negative edges of the DOWN and the UP pulses follow after τ. Therefore, the net charge injected into the filter is $\overline{Q} = I_{\mathrm{P}}(1-\varepsilon/z)t_{\mathrm{e}} - \varepsilon\tau I_{\mathrm{P}}$. However, if the time error is negative (DIV leads REF), the net current is always negative and the net charge is $\overline{Q} = -I_{\mathrm{P}}(1+\varepsilon/z)t_{\mathrm{e}} - \varepsilon\tau I_{\mathrm{P}}$.

The charge-time transfer function is sketched in Figure 9.10. Although this distortion is actually 'near crossover', the time error in the locked state may be larger than $\varepsilon\tau$. Both factors are kept very low, to reduce noise and reference spurs. In fractional-N loops, the phase error is zero only if integrated over several reference periods. However, the phase deviation at a certain reference edge may be larger than the offset time. Therefore, it may experience the non-linearity of the phase comparison path.

9.3.4 Offset tri-state PFD

As shown in the preceding analysis, the presence of a time offset between the UP and DOWN currents at the steady state guarantees that small deviations of the phase error are compensated for by either the DOWN or the UP current only. Therefore, even in the presence of current mismatch, only a linear region of the transfer characteristic is used at the steady state.

One of the most obvious methods of guaranteeing a linear characteristic would be to have a time offset larger than the maximum expected time error in the locked state. As already highlighted, enlarging the reset delay τ degrades in-band noise and spurs. Similarly, adding a constant current I_{s} to the charge-pump output creates a steady-state time offset of $\Delta t_{\mathrm{s}} = (I_{\mathrm{s}}/I_{\mathrm{P}})\,T_{\mathrm{ref}}$, but it would increase noise and spurs. An estimation of the spur due to a constant current has already been provided in Section 2.4.4 for the type-II third-order PLL. The leakage current considered in that case is a constant current added deliberately to the charge-pump output in this case. The resulting reference spur scales as the time offset:

$$\mathrm{SFDR} = 20\log_{10}\left(\frac{K_{\mathrm{VCO}}I_{\mathrm{s}}}{\omega_{\mathrm{ref}}^2 C_2}\right) \approx 20\log_{10}\left(N\frac{\omega_{\mathrm{u}}\omega_{\mathrm{P}}}{\omega_{\mathrm{ref}}}\Delta t_{\mathrm{s}}\right). \tag{9.5}$$

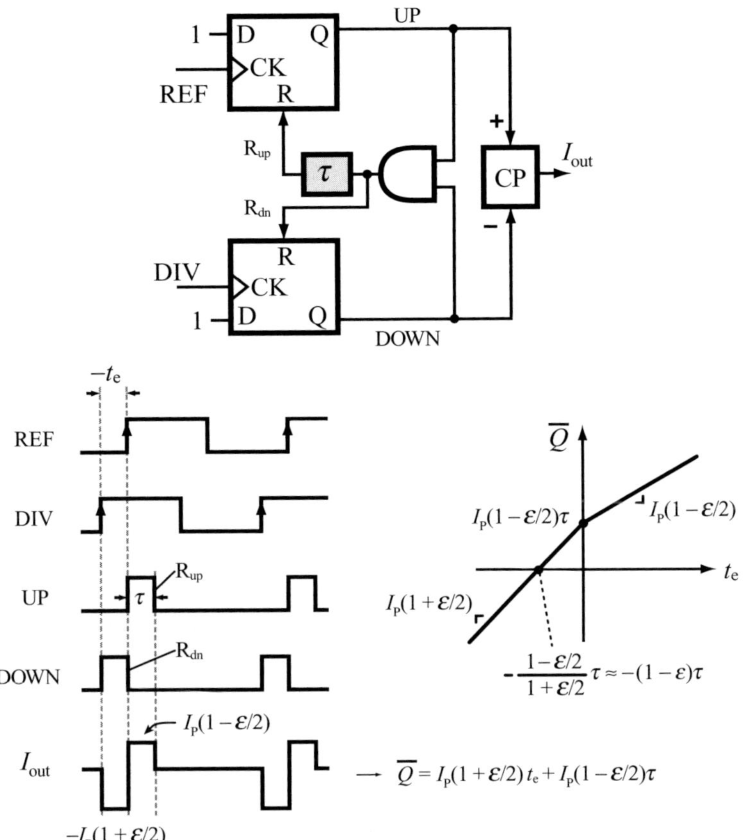

Figure 9.11 Offset tri-state PFD: elimination of crossover distortion by offsetting of REF and DIV at steady state

Some noise is associated with the constant current added to the charge-pump output, which contributes to the in-band noise of the PLL. A better way of offsetting REF and DIV signals consists of adding a gated current to the charge-pump output. By doing so, the noise associated with this current is injected for only a short time.

An easy way to enforce, by design, a time offset is shown in Figure 9.11. [12, 13] The overlapping tri-state PFD is slightly modified, since only the reset signal of one of the flip-flops is delayed by τ. This PFD can be referred to as an *offset tri-state* PFD. The consequence of this topological modification is that the steady-state time offset between REF and DIV is equal to $-(1-\varepsilon)\tau$, as shown in the signal diagrams in Figure 9.11. As long as the time error is negative the charge injected into the loop is proportional to the DOWN current. The change of slope of the transfer characteristic takes place when REF leads DIV (i.e., for positive time errors). In order to span the PFD characteristic in its linear region only, the time offset $\Delta t_s = -(1-\varepsilon)\tau \approx -\tau$ can be set, by design, to be larger than the maximum time-error deviation.

The reference spur for the type-II third-order PLL can be derived from (2.17):

$$\mathrm{SFDR} = 20\log_{10}\left(\frac{K_{\mathrm{VCO}} I_{\mathrm{P}} \tau^2}{2\pi C_2}\right) \approx 20\log_{10}\left(N\omega_{\mathrm{u}}\omega_{\mathrm{P}}\Delta t_{\mathrm{s}}^2\right). \tag{9.6}$$

Comparing (9.6) and (9.5), the offsetting PFD produces a spur lower than the overlapping PFD with an added constant current, if:

$$\Delta t_{\mathrm{s}} < \frac{1}{\omega_{\mathrm{ref}}} = \frac{T_{\mathrm{ref}}}{2\pi}. \tag{9.7}$$

The last expression holds true in any practical case, since the enforced time offset is typically much shorter than the reference period.

9.4 Charge pump

9.4.1 Single-ended input

Charge-pump single-ended-input topologies are typically implemented by current mirrors. The UP and DOWN inputs enable or disable the mirror output branch. Figure 9.12(a) shows the layout of a single-ended charge pump employing source switching. To avoid saturation of the output stage due to a slight difference between the UP and DOWN currents, an operational amplifier senses the output voltage and regulates the UP current. The currents are matched even when $\mathrm{M_U}$ is in the triode region.

Modified versions of this topology employing cascoded current mirrors are conventionally used to increase the output impedance, so that the current is less sensitive to the output voltage. An example of a cascoded source-switching topology taken from [14] is shown in Figure 9.12(b). The feedback is realized not by an operational amplifier, but simply by the pFET $\mathrm{M_{FB}}$.

Alternatively, gate switching, as shown in Figure 9.12(c), can be employed. [15] A fast switching time can be achieved by using the capacitors C_{U} and C_{D}. They reduce the charge coupling to the gates and C_{U} compensates the opamp feedback loop.

However, drain switching is scarcely used, due to its poor linearity. When DOWN goes high and the switch is turned on, the drain of $\mathrm{M_D}$ increases from 0 V to the loop filter voltage. Therefore, $\mathrm{M_D}$ is in the triode region until the voltage at its drain is higher than the minimum saturation voltage. During this time, a high peak current is generated at the output. [16] These current overshoots cannot be matched between UP and DOWN currents. Thus, they cause distortion in the transfer characteristic.

9.4.2 Differential input

A differential topology of CP is shown in Figure 9.13. [17] The UP/$\overline{\mathrm{UP}}$ and the DOWN/$\overline{\mathrm{DOWN}}$ signals deliver the currents of two generators either to the CP output or to

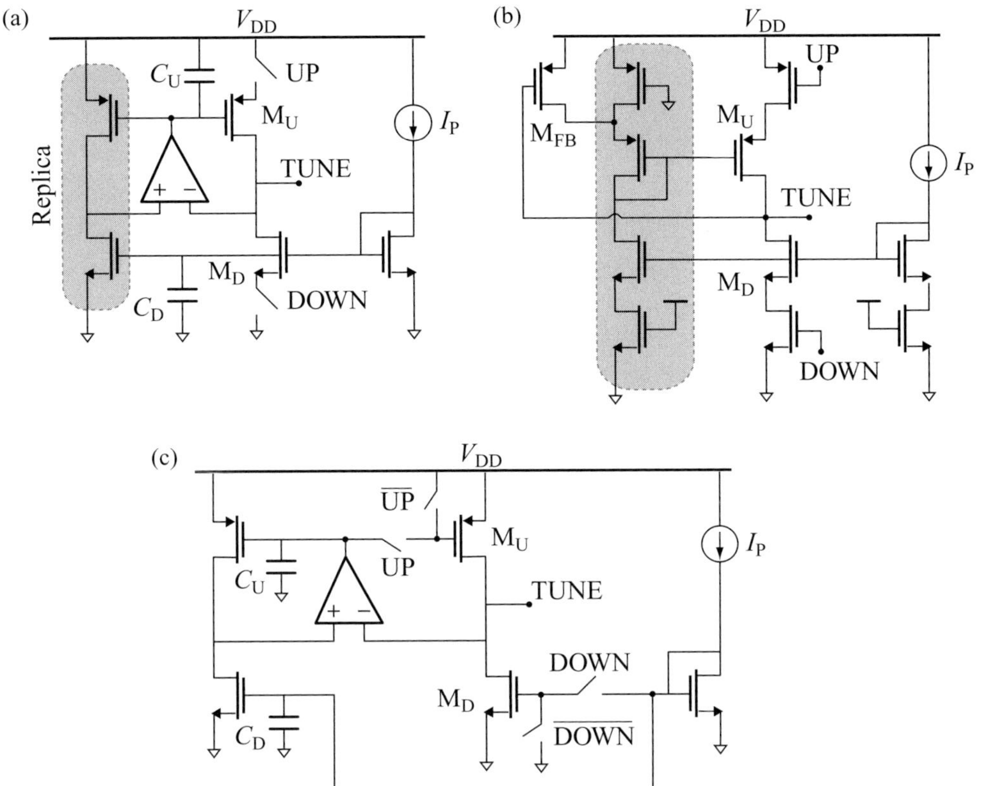

Figure 9.12 Charge pump with (a) source switching, (b) cascoded source switching and detailed feedback network and (c) gate switching

a d.c. bias. Matching between the UP and DOWN currents can be obtained by employing replica bias and feedback, similar to that shown in the schematic in Figure 9.12(a).

A plain implementation of the circuit in Figure 9.13 suffers from the charge-sharing effects depicted in Figure 9.14. In this scheme, the charge pump is assumed to be driven by the overlapping PFD. When UP is high and DOWN is low (phase 1), the loop filter is at a certain voltage V_0, while the parasitic capacitance C_P is at $V_{DD}/2$. In phase 2, when DOWN goes high, the current from one generator is carried by the other one. However, because of the initial difference between the output voltage and the voltage across C_P, the filter capacitance C and the parasitic capacitance C_P share the total charge. As a result, a charge approximately equal to $C_P\,(V_0 - V_{DD}/2)$ is spilled out of the capacitor C. This error depends on the voltage stored in the loop filter and, in turn, on the PLL output frequency.

To solve this issue, a unity-gain buffer is very often connected from the CP output to the symmetric node, as shown in Figure 9.15. By doing so, the symmetric-node voltage is always kept equal to the CP output voltage. In principle, no charge sharing should take place. The operational amplifier implementing this buffer needs rail-to-rail headroom at both input and output ports.

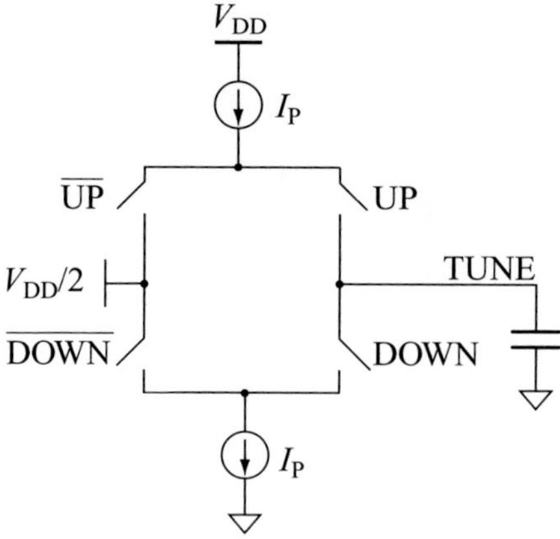

Figure 9.13 Charge pump with differential input

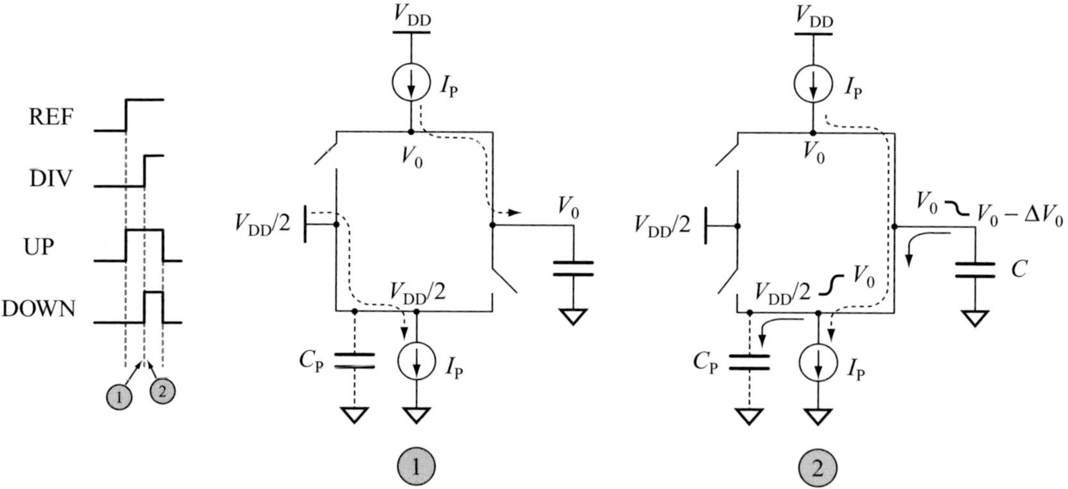

Figure 9.14 Charge-sharing phenomenon

9.4.3 Dynamic effects affecting linearity

Although strongly mitigated by the adoption of the buffer, the phenomenon of spurious charge injection is not eliminated at small time errors. The limited gain–bandwidth product of the operation amplifier is responsible for charge sharing, as shown in Figure 9.16. Initially (phase 0), both UP and DOWN are low and the current I_P flows into the left branch of the charge pump. If the REF signal edge arrives t_e seconds before the DIV signal edge, the UP current is switched to the filter and the DOWN current is drained from the buffer (phase 1). However, because of its limited bandwidth, the buffer output voltage V_d suddenly drops by $R_0 I_P$, where R_0 is the open-loop output impedance of buffer. After a time t_0

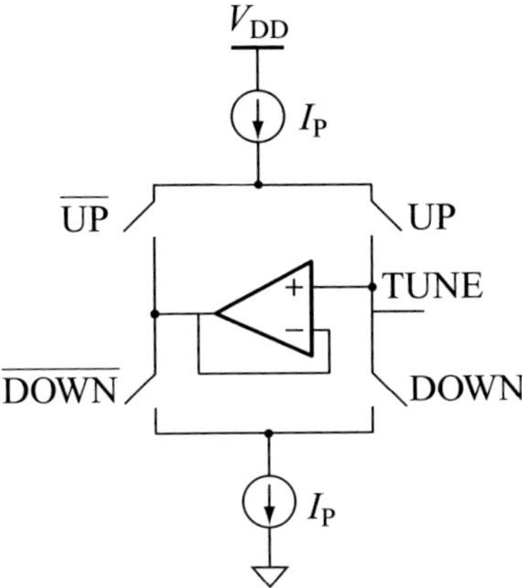

Figure 9.15 Charge pump with buffer to limit charge sharing

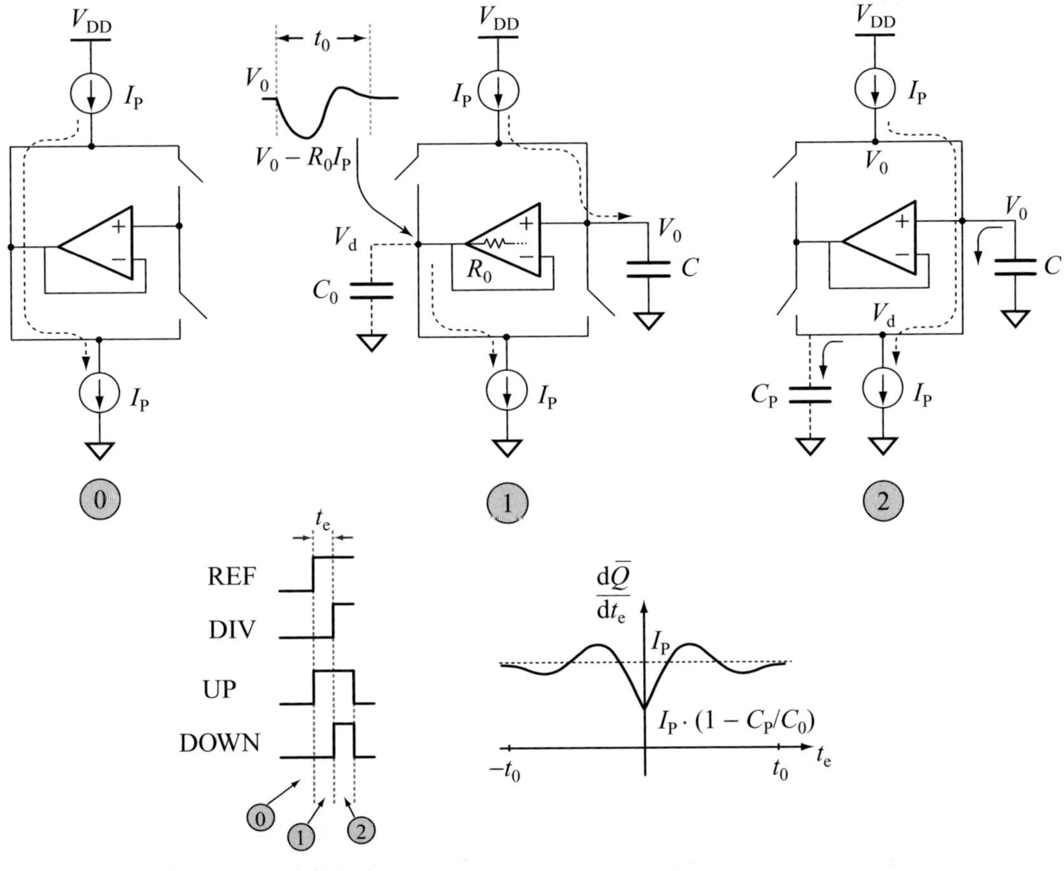

Figure 9.16 Charge sharing for small phase errors owing to finite opamp gain-bandwidth product

dependent on the buffer closed-loop bandwidth, the feedback enforces V_0 at the output of the buffer.

After t_e, the DOWN signal goes high (phase 2) and the voltage $V_d(t_e)$ is sampled by the parasitic capacitance C_P. If the time error t_e is shorter than t_0, the voltage $V_d(t_e)$ differs from the filter voltage V_0 and charge sharing occurs between C_P and C. The charge coming from the capacitance C is $C_P\,[V_0 - V_d(t_e)]$ and the gain of the PFD/CP transfer characteristic is given by differentiating the total charge injected into the filter during t_e:

$$\frac{dQ}{dt_e} = \frac{d}{dt_e}\{I_P t_e - C_P\,[V_0 - V_d(t_e)]\} = I_P + C_P \frac{dV_d(t_e)}{dt_e}.$$

This PFD/CP gain, which is plotted schematically in Figure 9.16, differs from the ideal constant gain, if $t_e < t_0$. This dynamical effect is, therefore, responsible for distortion in the PFD/CP characteristic.

The first part of the transient of V_d is roughly an exponential decrease with constant time R_0C_0. The second part results from the two slower poles of the unity-gain buffer. Therefore, for low values of t_e, the first derivative of V_d has, approximately, a step equal to $-I_P/C_0$ and the CP gain is:

$$\left.\frac{dQ}{dt_e}\right|_{t_e \approx 0} = I_P - C_P\frac{I_P}{C_0} = I_P\left(1 - \frac{C_P}{C_0}\right).$$

The last expression shows an intuitive result. Lower distortion can be achieved by increasing the capacitance C_0 of the dummy node with respect to the parasitic capacitance C_P. In other words, this non-linear charge injection can be avoided by preventing the glitch on the output node of the buffer. This can be accomplished by placing a capacitance on the output of the buffer, which instantaneously provides the charge required by the current generator.

Example 9.3 PFD/CP transfer function derivation

The transfer characteristic of the PFD/CP block can be derived by circuit simulations. Since the loop bandwidth is much narrower than the reference frequency, the output voltage of the charge pump can be considered to be constant. In simulations, it can be set to $V_{DD}/2$ by an ideal voltage source. The current drained or injected into this voltage source by the charge pump is sensed and injected into a capacitance C, which is initially discharged. For each value of the input delay t_e between REF and DIV, the charge dumped from the charge pump is CV_C, where V_C is the voltage across C, and the PFD/CP gain can be evaluated as (CV_C/t_e).

Figure 9.17 shows the gain from circuit simulations of the charge pump in Figure 9.15 without and with the introduction of a capacitance $C_0 = 10\,\text{pF}$ at the output of the unity-gain buffer (circles and squares, respectively). The charge pump has been designed in a 0.25 μm CMOS to provide a pump current of 1 mA. The reset delay of the overlapping PFD has been set to about 0.4 ns. For input time errors between 0.1 and 0.3 ns, the gain of the original charge pump is lower than the ideal value. The higher gain for time errors lower than 0.1 ns, which was not predicted in the intuitive sketch in Figure 9.16, is caused

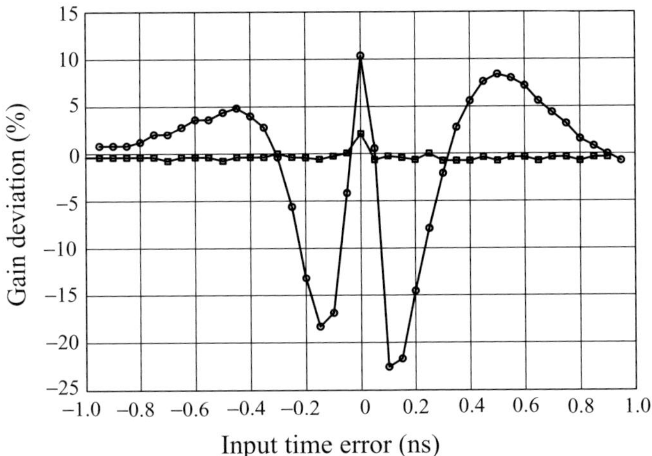

Figure 9.17 Circuit simulations of charge pump without (circles) and with (squares) introduction of capacitance C_0 at the output of the buffer

by an initial positive glitch in the simulated voltage waveform, V_d. The non-linearity of the charge pump is strongly mitigated by the introduction of the capacitance C_0.

9.4.4 Differential topology with improved linearity

The introduction of a capacitance C_0 limits the charge sharing but it does not eliminate it. Moreover, it may require a large silicon area. Figure 9.18 shows a smarter solution, which avoids the dynamic charge sharing and requires one additional unity-gain buffer. The three phases are the same as those described for the conventional differential-input CP. Unlike the conventional charge pump, from phase 0 to phase 1, the DOWN current is no longer diverted from the current generator to the buffer output. Therefore, the buffer output voltage does not suffer from glitches and the previously described charge sharing is avoided. A similar topology employing the same concept can be found in [18].

9.5 Phase-detection noise

An ideal tri-state PFD would not inject noise into the filter at the steady state. However, it is never used, because of its non-linear transfer characteristic. As shown in the previous section, the elimination of the dead zone and the crossover distortion requires an additional delay τ on one flip-flop reset. When the PLL is in the locked state, the PFD produces two synchronous UP and DOWN pulses lasting τ. The noise associated with these pulses is injected into the filter at each reference cycle. The same situation occurs using the overlapping PFD, since the noise associated with the two pulses is uncorrelated.

Other contributions to the PLL in-band noise may arise from the PFD and the reference buffer noise.

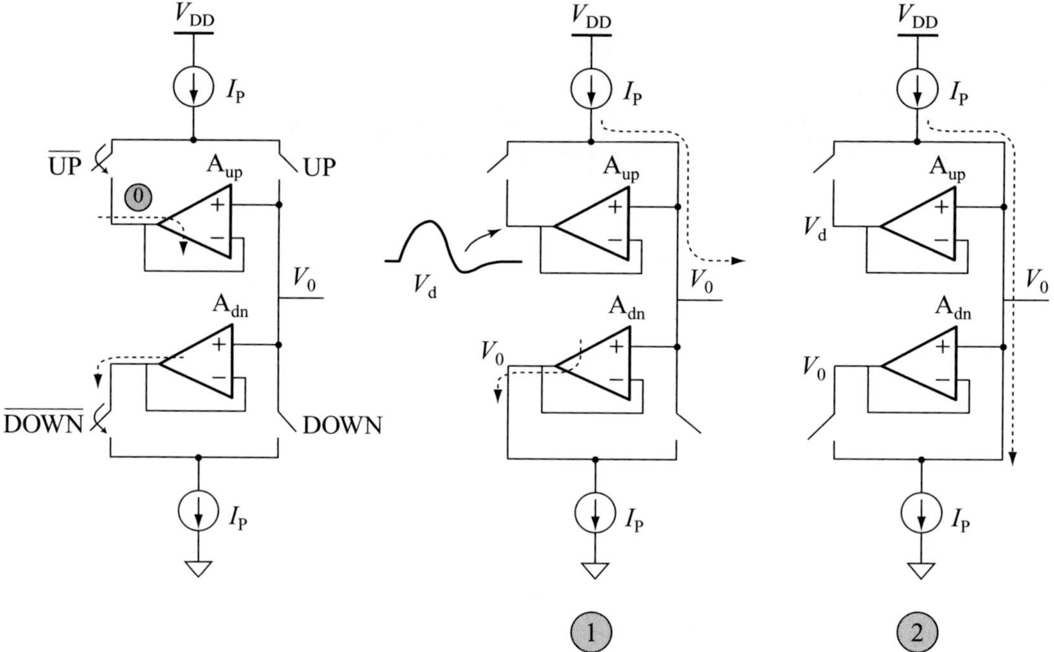

Figure 9.18 Charge-pump topology with improved linearity

9.5.1 Charge-pump noise

Figure 9.19 shows, schematically, the injection of charge-pump current noise in the locked state. If an overlapping tri-state is employed, noise associated with both the UP and the DOWN currents is injected into the loop filter. Therefore, the circuit can be simplified, as shown in Figure 9.19. The current noise generator i_{CP} represents the combination of the UP and DOWN current noises ($i_{n,u}$ and $i_{n,d}$), which are mainly caused by the current generators and the replica bias network. In general, the total noise power has to be calculated by taking noise correlation into account.

The continuous-time current noise i_{CP} is windowed by the switch action. Therefore, the output-switched noise i_{out} can be written as the input noise multiplied by the weighting function $w(t)$. The average noise spectrum[1] is calculated by convolving the spectrum of i_{CP} with the squared magnitude of the Fourier transform of the weighting function, as shown in Figure 9.19:

$$\frac{\langle i_{out}^2 \rangle}{\Delta f} = \frac{\langle i_{CP}^2 \rangle}{\Delta f} * \frac{\tau^2}{T_{ref}^2} \sum_{k=-\infty}^{\infty} \left[\text{sinc}^2\left(\frac{k\tau}{T_{ref}}\right) \cdot \delta\left(f - \frac{k}{T_{ref}}\right) \right]. \tag{9.8}$$

[1] The output noise autocorrelation is obtained by multiplying the input noise autocorrelation by the autocorrelation of the weighting function. The output current noise is a cyclostationary process and its spectrum is time variant. In practice, we are interested in the average noise spectrum. In fact, the current from the charge pump is converted into voltage by a low-pass filter.

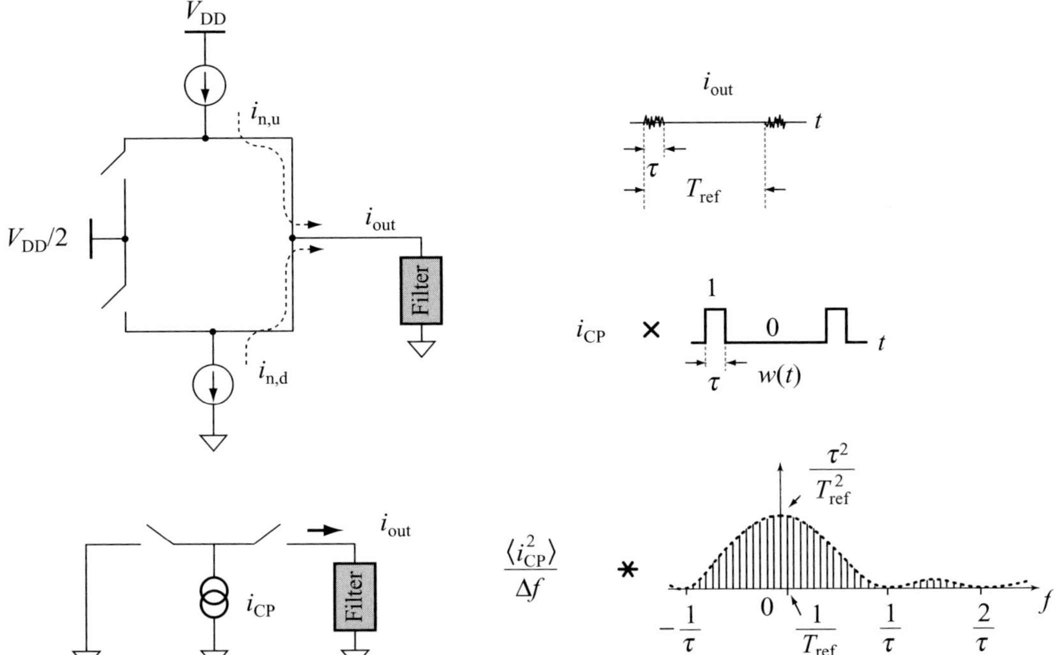

Figure 9.19 Charge-pump current noise: weighted noise injection and weighting function

The noise sampling from the periodic switching corresponds to folding of the noise spectrum. The white component of the charge-pump noise can be obtained by observing that each Dirac delta function in (9.8) folds the noise spectrum. Therefore, the noise-folding factor can be obtained by summing the series in (9.8). The resultant white noise is:

$$\frac{\langle i_{\mathrm{out,W}}^2 \rangle}{\Delta f} = \frac{\langle i_{\mathrm{CP,W}}^2 \rangle}{\Delta f} \frac{\tau^2}{T_{\mathrm{ref}}^2} \frac{T_{\mathrm{ref}}}{\tau} = \frac{\langle i_{\mathrm{CP,W}}^2 \rangle}{\Delta f} \cdot \frac{\tau}{T_{\mathrm{ref}}}. \tag{9.9}$$

The switched-noise power is obtained from the continuous-time noise by scaling it by the switching duty cycle (τ/T_{ref}).

The flicker corner frequency of the charge-pump current noise may be lower than $f_{\mathrm{ref}}/2$. In such a case, the charge-pump flicker noise is not subject to folding. Therefore, the output flicker spectrum is given by:

$$\frac{\langle i_{\mathrm{out,F}}^2 \rangle}{\Delta f} = \frac{\langle i_{\mathrm{CP,F}}^2 \rangle}{\Delta f} \cdot \frac{\tau^2}{T_{\mathrm{ref}}^2}. \tag{9.10}$$

Since flicker and white noise components have different gain factors, the output flicker corner differs from the CP flicker corner, and it is $f_{\mathrm{c,out}} = f_{\mathrm{c,CP}}\,(\tau/T_{\mathrm{ref}})$. Since $(\tau/T_{\mathrm{ref}}) < 1$, the output corner frequency is lower than that of the charge-pump noise sources.

Equations (9.9) and (9.10) may be useful for estimating the output noise of the charge pump, after the continuous-time charge-pump noise spectrum is obtained by a.c. noise

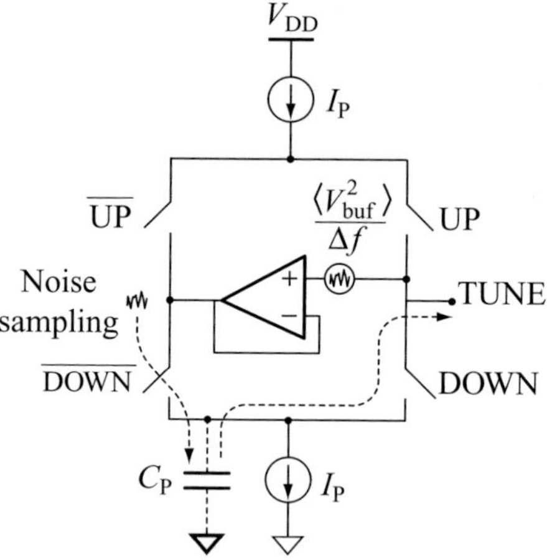

Figure 9.20 Charge-pump noise from the unity-gain buffer

analyses. Direct simulation of the switched charge-pump noise would, instead, require a periodic noise analysis.

The PLL input-referred phase noise due to the charge-pump current noise can be obtained by dividing the current noise spectrum by the squared CP gain ($I_P/2\pi$):

$$\mathcal{L}_{CP,W} = \frac{S_\phi}{2} = 2\pi^2 \left(\frac{\tau}{T_{ref}}\right) \left(\frac{\langle i_{CP,W}^2 \rangle / \Delta f}{I_P^2}\right). \tag{9.11}$$

The buffer, which has been introduced in the charge pump in Figure 9.15 to reduce charge sharing, contributes to the output noise as shown in Figure 9.20. Its noise is sampled onto the capacitance C_P and then the stored noise charge is transferred to the output during the DOWN pulse.

9.5.2 PFD noise

Not only the charge pump, but also the PFD, injects noise into the loop filter. In the locked state, the charge pump generates two synchronous pulses lasting τ, if the overlapping PFD is employed. However, the pulse duration τ is set by the reset added delay and by the AND gate controlling the reset. Such logic stages add timing noise to τ. An estimation of such noise can be obtained by following the same arguments used for the frequency divider noise in Chapter 8. The input-referred white phase noise from the jitter $\sigma_{\tau,W}^2$ of the reset delay is given by:

$$\mathcal{L}_{delay,W} = 4\pi^2 f_{ref} \cdot \sigma_{\tau,W}^2. \tag{9.12}$$

Moreover, the UP and DOWN pulses are generated by the D-type flip-flops of the PFD. Therefore, the flip-flops add noise to both the positive and the negative pulse edges. The contribution to the input-referred phase noise is

$$\mathcal{L}_{PFD,W} = 4\pi^2 f_{ref} \cdot 4\sigma_{t_{latch},W}^2. \tag{9.13}$$

The jitter of the single latch $\sigma^2_{t_{\text{latch}},\text{W}}$ is multiplied by four, since both the UP and DOWN pulses contribute to phase noise and each pulse has two edges.

The PLL input-referred white noise from the PFD/CP can be obtained by combining (9.11), (9.12) and (9.13):

$$\mathcal{L}_{\text{PFD/CP,W}} = 4\pi^2 f_{\text{ref}} \cdot \left(\frac{\tau}{2} \cdot \frac{\langle i^2_{\text{CP,W}} \rangle / \Delta f}{I_{\text{P}}^2} + \sigma^2_{\tau,\text{W}} + 4\sigma^2_{t_{\text{latch}},\text{W}} \right). \tag{9.14}$$

9.5.3 Reference buffer noise

The reference buffer typically interfaces the off-chip crystal oscillator with the REF input of the PFD. The crystal oscillator provides a sinusoidal signal, which is coupled to the reference buffer input. The purpose of the reference buffer is to convert the sinusoid to a near rail-to-rail signal to provide to the PFD. The noise contribution of the reference buffer can be calculated as shown for the frequency dividers, if it is implemented as a hard-switching differential stage.

As shown in Chapter 8, flicker noise of the differential pair is converted into phase noise because of the limited slope of the input signal (8.8):

$$\mathcal{L}_{\text{buf}}(f) = \frac{2\pi^2 f^2_{\text{ref}}}{\text{SL}^2_{\text{ref}}} \left[S_{V,\text{F}}(f) + \frac{\sigma^2_{V,\text{W}}}{f_{\text{ref}}/2} \right], \tag{9.15}$$

where SL_{ref} is the crossover slope of the reference input voltage, $S_{V,\text{F}}(f)$ is the input-referred flicker noise spectrum of the buffer and $\sigma^2_{V,\text{W}}$ is the variance of the input-referred white noise. Since the reference is a sinusoid, its slope depends on the oscillation amplitude: $\text{SL}_{\text{ref}} = \omega_{\text{ref}} A_{\text{ref}}$, where A_{ref} is the reference peak amplitude. Substituting in (9.15),

$$\mathcal{L}_{\text{buf}}(f) = \underbrace{\frac{1}{2} \cdot \frac{S_{V,\text{F}}(f)}{A^2_{\text{ref}}}}_{\text{flicker}} + \underbrace{\frac{1}{f_{\text{ref}}} \cdot \frac{\sigma^2_{V,\text{W}}}{A^2_{\text{ref}}}}_{\text{white}}. \tag{9.16}$$

As is intuitive, the input-referred phase noise depends on the ratio between the reference buffer voltage noise and the input reference power.

9.5.4 Other contributions to in-band noise

The noise sources so far considered in the phase comparison path are responsible for the in-band phase noise of the PLL. If the PLL bandwidth is properly designed, these noise sources are negligible out of the PLL bandwidth. However, this result is based on the assumption of linear PLL blocks. As mentioned already, if the phase comparison is non-linear, some out-of-band noise may be folded back to the PLL bandwidth. The consequence is the presence of another source of in-band PLL noise.

The estimation of the noise spectral regrowth is not trivial. [19–21] Those methods of analysis or simulations based on calculating the effect of non-linearity on the noise source at open loop are questionable.

The $\Delta\Sigma$ modulator may also be responsible for additional noise from cross-talk. For this reason, the VCO should be laid out far away and should be isolated by guard rings of substrate contacts or oxide trenches when available. Cross-talk between the $\Delta\Sigma$ modulator and the PFD/CP can also produce spurs and degrade phase noise. Since the duty cycle of the divider output is typically less than 10%, it is useful to trigger the $\Delta\Sigma$ modulator and the flip-flops presetting the division factor by the negative edges. Doing so, both circuits have enough time to settle before the PFD/CP becomes active and cross-talk is reduced. [22]

9.6 References

[1] B. De Muer and M. S. J. Steyaert, A CMOS monolithic $\Delta\Sigma$-controlled fractional-*N* frequency synthesizer for DCS-1800, *IEEE J. Solid-St. Circ.*, **37**, Jul. 2002, 835–44.

[2] T.-C. Lee and B. Razavi, A stabilization technique for phase-locked frequency synthesizers, *IEEE J. Solid-St. Circ.*, **38**, Jun. 2003, 888–94.

[3] M. H. Perrot, M. D. Trott and C. G. Sodini, A modeling approach for $\Delta\Sigma$ fractional-*N* frequency synthesizers allowing straightforward noise analysis, *IEEE J. Solid-St. Circ.*, **37**, Aug. 2002, 1028–38.

[4] M. Zuta, A new PLL with fast settling time and low phase noise, *Microw. J.*, Jun. 1997, 94–108.

[5] T.-Y. Hsu, T.-R. Hsu, C.-C. Wang, Y.-C. Liu and C.-Y. Lee, Design of a wide-band frequency synthesizer based on TDC and DVC techniques, *IEEE J. Solid-St. Circ.*, **37**, Oct. 2002, 1244–55.

[6] R. B. Staztweki, K. Muhammad, D. Leipold *et al.*, All-digital TX frequency synthesizer and discrete-time receiver for Bluetooth radio in 130-nm CMOS, *IEEE J. Solid-St. Circ.*, **39**, Dec. 2004, 2278–91.

[7] W. F. Egan, *Frequency Synthesis by Phase Lock*, New York, NY: J. Wiley & Sons, Inc., 1990, 2nd edn, 2000.

[8] B. Razavi, *Design of Analog CMOS Integrated Circuits*, New York, NY: McGraw-Hill, 2000.

[9] M. H. Perrot, *Techniques for High Data Rate Modulation and Low Power Operation of Fractional-N Frequency Synthesizers*, Ph.D. thesis, Massachusetts Institute of Technology, Cambridge, MA, Sep. 1997.

[10] E. Temporiti, G. Albasini, I. Bietti, R. Castello and M. Colombo, A 700-kHz bandwidth $\Delta\Sigma$ fractional synthesizer with spurs compensation and linearization techniques for WCDMA applications, *IEEE J. Solid-St. Circ.*, **39**, Sep. 2004, 1446–54.

[11] H. Huh, Y. Koo, K.-Y. Lee *et al.*, Comparison frequency doubling and charge pump matching techniques for dual-band $\Delta\Sigma$ fractional-*N* frequency synthesizer, *IEEE J. Solid-St. Circ.*, **40**, Nov. 2005, 2228–36.

[12] S. E. Meninger and M. H. Perrot, A 1 MHz bandwidth 3.6-GHz 0.18-μm CMOS fractional-*N* synthesizer utilizing a hybrid PFD/DAC structure for reduced broadband phase noise, *IEEE J. Solid-St. Circ.*, **41**, Apr. 2006, 966–80.

[13] S. F. Gillig, *Linearized Three State Phase Detector*, U.S. Patent 4970475, Nov. 1990.

[14] E. Hegazi and A. A. Abidi, A 17-mW transmitter and frequency synthesizer for 900-MHz GSM fully integrated in 0.35-μm CMOS, *IEEE J. Solid-St. Circ.*, **38**, May 2003, 782–92.

[15] M. Zargari *et al.*, A single-chip dual-band tri-mode CMOS transceiver for IEEE 802.11a/b/g wireless LAN, *IEEE J. Solid-St. Circ.*, **39**, Dec. 2004, 2239–49.

[16] W. Rhee, Design of high-performance CMOS charge pumps in phase-locked loops, *Proc. 1999 IEEE International Symposium on Circuits and Systems*, **2**, Jun. 1999, 545–8.

[17] H. R. Rategh, H. Samavati and T. H. Lee, A CMOS frequency synthesizer with an injection-locked frequency divider for a 5-GHz wireless LAN receiver, *IEEE J. Solid-St. Circ.*, **35**, May 2000, 780–7.

[18] I. Vassiliou, K. Vavelidis, T. Georgantas *et al.*, A single-chip digitally calibrated 5.15–5.825-GHz 0.18-μm CMOS transceiver for 802.11a wireless LAN, *IEEE J. Solid-St. Circ.*, **38**, Dec. 2003, 2221–31.

[19] T. A. D. Riley, N. M. Filiol, Q. Du and J. Kostamovaara, Techniques for in-band phase noise reduction in $\Delta\Sigma$ synthesizers, *IEEE T. Circuits-II*, **50**, Nov. 2003, 794–803.

[20] B. De Muer and M. S. J. Steyaert, On the analysis of $\Delta\Sigma$ fractional-N frequency synthesizers for high-spectral purity, *IEEE T. Circuits-II*, **50**, Nov. 2003, 784–93.

[21] H. Arora, N. Klemmer, J. C. Morizio and P. D. Wolf, Enhanced phase noise modeling of fractional-N frequency synthesizers, *IEEE T. Circuits-I*, **52**, Feb. 2005, 379–95.

[22] P. Zhang, L. Der, D. Guo *et al.*, A single-chip dual-band direct-conversion IEEE 802.11a/b/g WLAN transceiver in 0.18-μm CMOS, *IEEE J. Solid-St. Circ.*, **40**, Sep. 2005, 1932–9.

Index